Design theory

Design theory

D. R. Hughes and F. C. Piper

School of Mathematical Sciences
Queen Mary College, London

Department of Mathematics
Royal Holloway College, London

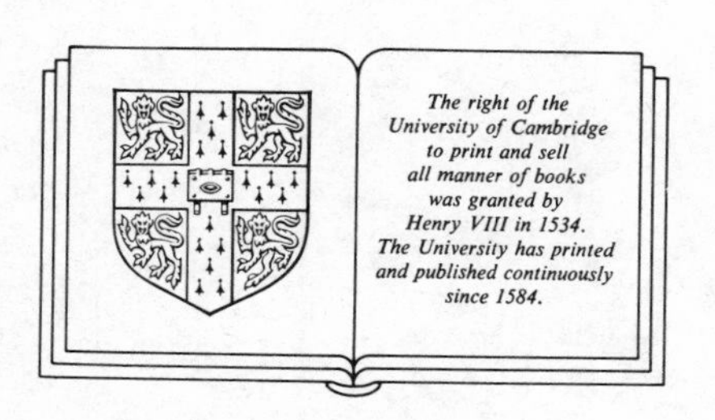

Cambridge University Press

Cambridge
London New York New Rochelle
Melbourne Sydney

Published by the Press Syndicate of the University of Cambridge
The Pitt Building, Trumpington Street, Cambridge CB2 1RP
32 East 57th Street, New York, NY 10022, USA
10 Stamford Road, Oakleigh, Melbourne 3166, Australia

First published 1985

Printed in Great Britain at
the University Press, Cambridge

Library of Congress catalogue card number: 84-14309

British Library cataloguing in publication data

Hughes, D. R.
Design theory.
1. Combinatorial designs and configurations
I. Title II. Piper, F. C.
511′.6 QA166.25

ISBN 0 521 25754 9

MP

Contents

Introduction

The subject of design theory has grown out of several branches of mathematics, and has been increasingly influenced in recent years by developments in other areas. Its statistical origins are still evident in some of its standard terminology (thus 'v' for the number of points in a structure comes from 'varieties'). Today it has very fruitful connections with group theory, graph theory, coding theory and geometry; these ties have been two-way, by and large.

We have attempted in this book to lay the groundwork for an understanding of designs, with advanced undergraduate or postgraduate students in mind. Our aim is to prepare the reader to use designs in other fields or to enter the active field of designs themselves. Finite projective and affine geometries are central to design theory, and are introduced early in the book. Since classical geometry is a very large field, the student with a background in this subject will be at an advantage, but we have tried to present a treatment sufficiently self-contained to answer the needs of a reader with a reasonable knowledge of linear algebra. The subject of symmetric designs is also introduced early, and its important aspects (the Bruck–Ryser–Chowla Theorem, Singer groups and difference sets, Hadamard 2-designs, etc.) are developed. The first four chapters, covering basic definitions, geometry and symmetric designs, are designed to be part of any course based on the book.

The other four chapters can be studied more or less independently of one another. Chapter 5 covers resolvable and affine designs; Chapter 6 introduces 2-designs other than those met already; Chapter 7 deals with 1-designs and an introduction to generalised quadrangles; finally Chapter 8 studies the large Mathieu designs and groups.

Certain families of designs (e.g., projective planes or Hadamard designs) occur so often in design theory that they merit special treatment. Sometimes even particular single designs play such an important role that we have dealt with them in considerable detail (e.g. the 2-design for (11, 5, 2), the biplanes on 16 points, the projective plane of order 4). The little Mathieu designs (in Chapter 4) and the large Mathieu designs (in Chapter 8) take up a large amount of the book. We have chosen also to emphasise the connections with group theory, where some of the most beautiful and important theorems of design theory can be demonstrated. But we have tried to isolate the less elementary group theory so that a reader without the necessary background (which is not very great, in any case) can cope with it satisfactorily. Another important connection is between graph theory and designs: we include a self-contained introduction to the elementary theory of strongly regular graphs in Chapter 3.

The authors have found that an undergraduate one-term course can be based on the first four chapters. A slightly more advanced (or longer) course can add one of the later chapters. At a substantially higher level much, or even all, of the last four chapters can be covered.

Many different notational and terminological usages have grown up in the theory of designs. We have attempted to lay out a consistent scheme for both of these and, excepting in comments, we have ignored the notations etc. not used in the book. 'Structures' are very primitive objects, perhaps having the relationship to designs that groupoids do to groups. We feel that repeated blocks are not of central interest in the theory (and are only of marginal interest to the 'users' of design theory) and, except in certain very special circumstances, the book reflects this attitude. Some terminologies from the past might best be altered (e.g. 'group-divisible designs', which have nothing to do with groups, and can merely be called 'divisible designs'). We have almost universally used upper case Latin letters for points and lower case for blocks (an exception is in geometries); structures, designs and sets in general, are normally upper case script letters, groups are upper case Latin but their elements either lower case Latin or lower case Greek, while graphs are usually indicated by upper case Greek letters (an exception being the Hussain graphs of Chapter 3).

We have indicated with a '*' those exercises whose solutions are more difficult than the ordinary, or perhaps just considerably longer. There are also a few 'Problems': this means that we do not know the answer but that there might be some profit in investigating the situation.

1 Basic concepts

1.1 Introduction

This chapter is the introduction to structures and designs and, while it is completely elementary, it is essential to the rest of the book. Section 1.2 contains the basic definitions. In Section 1.3 we then give a number of examples. We begin by listing some small carefully chosen ones to illustrate the meanings of the earlier definitions but then go on to examples based on projective and affine geometry. Obviously knowledge of classical geometry will help the reader to follow and understand these examples, but we have tried to make our explanations as full as possible and to make the entire chapter self-contained. Nevertheless the importance of finite projective and affine geometry to design theory cannot be overemphasised and we include some excellent references for further reading. In Section 1.4 we return to definitions and results about arbitrary structures, in particular relating a structure to others which can be constructed from it or from which it can be constructed. Section 1.5 studies the incidence matrix of a structure, already introduced in Section 1.2, and uses it to prove a number of basic theorems: Fisher's Inequality and properties of square structures and symmetric designs in particular. Polarities are introduced and the incidence matrix is exploited to prove a number of their basic and important properties. In Section 1.6 the notion of tactical decomposition of a structure is introduced, Block's Lemma is proved, and applications to automorphism groups (in particular the Orbit Theorems) are deduced. Resolutions and parallelisms are briefly introduced as well. Section 1.7 contains a brief discussion of graph theory and some of its connections with the theory of structures and designs.

The reader should bear in mind that most of the concepts in this chapter are extremely important, not only in the rest of the book, but also in the literature. The 'small' examples in Section 1.3 are mainly pedagogical, but the projective and affine geometries also introduced in that section occur again and again in design theory. The notion of a restriction or residue in Section 1.4 is also very important. The incidence matrix is central to some of the most important theorems in the subject (some of which, like Fisher's Inequality and the results about symmetric designs, are proved in Section 1.5). Automorphism groups, while they are only introduced in a very general and elementary way in this chapter, are of the greatest importance in the entire subject, and arise repeatedly throughout the book.

1.2 Basic definitions and properties

A *structure* (or sometimes an *incidence structure*) is two finite sets of objects, called *points* and *blocks*, with an *incidence relation* $\mathscr{I}$ between them. Usually we shall use upper case Latin letters to denote points and lower case Latin letters to denote blocks. Given any block there is a set of points incident with it, and we shall frequently find it convenient to identify the block with this point set. Thus if a point P is incident with a block y, we can write either $P\mathscr{I}y$ or $P \in y$, and we will use any convenient expression such as 'P is on y', 'y contains P', 'y is on P', etc. An equivalent approach is to regard a structure as a finite point set $\mathscr{P}$, a finite block set $\mathscr{B}$, and the incidence given by a subset $\mathscr{I} \subseteq \mathscr{P} \times \mathscr{B}$; then a point P is on the block y if and only if $(P, y) \in \mathscr{I}$. In this case the structure is denoted by $(\mathscr{P}, \mathscr{B}, \mathscr{I})$.

If P is a point and y is a block of a structure $\mathscr{S}$, then $\langle P \rangle$ and $\langle y \rangle$ are often used to denote the set of blocks on P and the set of points on y, respectively. As usual, we write $|\mathscr{U}|$ for the number of elements in a set $\mathscr{U}$, but we will write $|P|$ for $|\langle P \rangle|$ and $|y|$ for $|\langle y \rangle|$.

Clearly these definitions could be extended to 'infinite structures', but much of the interesting theory and applications is different for the finite and infinite cases. We restrict our study to the finite case, though we may from time to time point out possible extensions to the infinite case.

If $\mathscr{S}$ is a structure then we usually denote the number of points and blocks of $\mathscr{S}$ by v and b respectively. Clearly, for a structure to be interesting it should have non-empty sets of both points and blocks, and we therefore assume that any structure has $v \neq 0$, $b \neq 0$, unless the contrary is explicitly stated or allowed.

When we defined a structure above we said that it would often be convenient to identify a block with the set of points on it. If we do this, however, there is no reason why two distinct blocks might not be identified with the same set of points. Indeed, this will often occur, and when it does we say that the structure has *repeated blocks*. However, if $\mathscr{S}$ has repeated blocks, we can construct from $\mathscr{S}$ a new structure which does not, simply by 'deleting' all but one of a set of blocks which are incident with the same set of points. More formally we can define an equivalence relation R on the set of blocks by: xRy if $\langle x\rangle=\langle y\rangle$. The *multiplicity* of the block x is the size of the R-equivalence class of x, and if x has multiplicity greater than 1, then we say that x is a *repeated* block. Then we define a new structure, the *reduction* of $\mathscr{S}$ (or '$\mathscr{S}$ reduced'), written $\mathscr{S}/R$, whose points are the points of $\mathscr{S}$ and whose blocks are the R-equivalence classes of blocks of $\mathscr{S}$, with P on the equivalence class of x if P is on x in $\mathscr{S}$. Most of the time we shall be interested only in reduced structures, although frequently an important part of a proof, say, will consist of showing that a structure has no repeated blocks.

If an element of a structure $\mathscr{S}$ is incident with 0 or 1 other elements of $\mathscr{S}$, then we call this element *isolated*; on the other hand, a *full* element is one which is incident with every other possible element (i.e. point with blocks or block with points). We can successively remove all full elements, then all isolated elements, then all full elements, etc., to reach a new structure $\bar{\mathscr{S}}$ (which might be empty!), called '*$\mathscr{S}$ standardised*'. Finally, a structure is *totally reduced* if it is both standardised and reduced: that is, it has no full or isolated elements, and no repeated blocks. These concepts, 'isolated', 'full', 'standardised' and 'totally reduced' are usually of very little relevance, and in this book will only arise in dealing with some examples in early chapters.

Suppose that $\mathscr{S}$ is a structure with v points, b blocks, where $v>0$ and $b>0$, and that the points of $\mathscr{S}$ are indexed $P_1, P_2, \ldots, P_v$, while the blocks are $x_1, x_2, \ldots, x_b$. Then an *incidence matrix* $A=(a_{ij})$ for $\mathscr{S}$ is a v by b matrix where $a_{ij}=1$ if P_i is on x_j and $a_{ij}=0$ otherwise. Obviously all the information about $\mathscr{S}$ (up to 'isomorphism', a term to be defined later) is contained in A. (A matrix is called a *(0, 1)-matrix* if all its entries are 0 or 1; clearly any incidence matrix is a (0, 1)-matrix.) It is important to note that A is certainly not uniquely determined by $\mathscr{S}$ since it depends on the indexing of the points and blocks. Different indexings give rise to different incidence matrices. However, as one might expect, there is a close relation between the incidence matrices for a given structure $\mathscr{S}$. Suppose that A is the matrix of $\mathscr{S}$ given by an indexing $P_1, \ldots, P_v$ and $x_1, \ldots, x_b$ and that B is the matrix given by $Q_1, \ldots, Q_v$ and $x_1, \ldots, x_b$ (i.e. the points are labelled differently but the blocks are the same). Then each Q_i is one of the P_j so we get a permutation θ on $\{1, \ldots, v\}$ given by $i^\theta=j$ if and only if $Q_i=P_j$. But this means that, for any i, row i of B is row i^θ of A. In other words, B is obtained from A by permuting its rows. This argument can be extended to solve the following:

Exercise 1.1. If A and B are two incidence matrices for a structure $\mathscr{S}$, show that there exist permutation matrices P and Q such that $PAQ=B$.

Thus if $\mathscr{S}$ is a structure then any two incidence matrices for $\mathscr{S}$ are equivalent and, consequently, have the same rank. This enables us to define the *rank* of a structure $\mathscr{S}$ to be the rank of one (and hence all) of its incidence matrices (over the field of rationals).

If $\mathscr{S}$ and $\mathscr{T}$ are two structures we define an *isomorphism* α from $\mathscr{S}$ onto $\mathscr{T}$ to be a one-to-one mapping from the points of $\mathscr{S}$ onto the points of $\mathscr{T}$ and the blocks of $\mathscr{S}$ onto the blocks of $\mathscr{T}$ such that P is on x if and only if P^α is on x^α. If there is an isomorphism from $\mathscr{S}$ onto $\mathscr{T}$ then we say that $\mathscr{S}$ and $\mathscr{T}$ are isomorphic and write $\mathscr{S} \cong \mathscr{T}$. Clearly, as always, isomorphism is an equivalence relation.

Lemma 1.1. Any (0, 1)-matrix is an incidence matrix of a structure. If A and B are (0, 1)-matrices then they are incidence matrices for isomorphic structures if and only if there exist permutation matrices P, Q with $PAQ = B$.

Proof. If A is a v by b (0, 1)-matrix then we define a set of points $P_1, \ldots, P_v$ and a set of blocks $x_1, \ldots, x_b$, where P_i is incident with x_j if and only if the (i, j) entry of A is 1. This is clearly a structure.

Suppose that A is the incidence matrix of a structure $\mathscr{S}$ with points $P_1, \ldots, P_v$ and blocks $x_1, \ldots, x_b$ such that $a_{ij} = 1$ if and only if P_i is on x_j. If B is an incidence matrix for a structure $\mathscr{T}$ which is isomorphic to $\mathscr{S}$ than, clearly, B is also a v by b matrix. Let $Q_1, \ldots, Q_v$ and $y_1, \ldots, y_b$ be the labelling such that $b_{ij} = 1$ if and only if Q_i is on y_j. If we label the elements of $\mathscr{T}$ as $R_1, \ldots, R_v, z_1, \ldots, z_b$, where $R_i = P_i^\alpha$, $z_j = x_j^\alpha$ $(1 \leqslant i \leqslant v,\ 1 \leqslant j \leqslant b)$, then, since α is an isomorphism, $a_{ij} = 1$ if and only if R_i is on z_j. Thus A is an incidence matrix for $\mathscr{T}$ and, by Exercise 1.1, there exist permutation matrices P, Q with $PAQ = B$.

Now suppose that A, B are v by b (0, 1)-matrices such that there exist permutation matrices P, Q with $PAQ = B$. Let θ be the permutation on $\{1, \ldots, v\}$ and let ϕ be the permutation on $\{1, \ldots, b\}$ given by $a_{i^\theta j^\phi} = b_{ij}$ (i.e. θ is determined by P and ϕ by Q). Let $\mathscr{S}$ be the incidence structure such that it has A as an incidence matrix when its elements are labelled $P_1, \ldots, P_v, x_1, \ldots, x_b$ and let $\mathscr{T}$ have B as an incidence matrix when its elements are labelled $Q_1, \ldots, Q_v, y_1, \ldots, y_b$. Now define α from $\mathscr{S}$ onto $\mathscr{T}$ by $P_i^\alpha = Q_j$, where $j^\theta = i$ and $x_l^\alpha = y_m$, where $m^\phi = l$. Thus $\alpha: P_{j^\theta} \rightarrow Q_j$ and $x_{m^\phi} \rightarrow y_m$. Since α is clearly a bijection (one-to-one and onto mapping) from the points and blocks of $\mathscr{S}$ onto the points and blocks of $\mathscr{T}$, in order to show that α is an automorphism we have only to show the P_i^α is on x_l^α if and only if P_i is on x_l, i.e. that P_{j^θ} is on x_{m^ϕ} if and only if Q_j is on y_m. But P_{j^θ} is on x_{m^ϕ} whenever $a_{j^\theta m^\phi} = 1$ and Q_j is on y_m if $b_{jm} = 1$. But, from the definitions on θ and ϕ, $a_{j^\theta m^\phi} = b_{jm}$ and the lemma is proved. □

This lemma illustrates what an exceedingly simple object a structure is. It is, in fact, too simple to be of any interest by itself, but as soon as we begin to impose extra conditions then it becomes both interesting and useful.

Note, by the way, that if A is the incidence matrix of a structure $\mathscr{S}$ given by the indexing $P_1, \ldots, P_v, x_1, \ldots, x_b$ then the zero entries in row i are in those columns which represent the blocks which are not incident with P_i. Thus an isolated element will lead to a row (or column) of A with at most one non-zero entry. Similarly, a full element is equivalent to a row (or column) where every entry is 1. Finally we observe again that two identical columns in A means that $\mathscr{S}$ has a repeated block. Thus a (0, 1)-matrix A is the incidence matrix of a totally reduced structure if

(i) each row and column has at least one zero entry and at least two ones, and

(ii) A has no two identical columns.

We now introduce the first, and most common, condition which we impose on our structures.

A structure $\mathscr{S}$ will be called *uniform* if its block set is non-empty and if each of its blocks contains exactly $k>0$ points. The uniformity of a structure can be recognised from its incidence matrix in a very simple way.

For any positive integers m, n we denote by $J_{m,n}$ the m by n matrix with all its entries 1; we write J_m for $J_{m,m}$ and $\mathbf{j}_m$ for $J_{1,m}$, so that $\mathbf{j}_m$ is a row vector. If the context is clear, we sometimes write J or $\mathbf{j}$ for $J_{m,n}$ or $\mathbf{j}_m$.

Exercise 1.2. If A is an incidence matrix for a structure $\mathscr{S}$, show that

(a) $J_{m,v}A$ is an m by b matrix whose ith column consists completely of the entry $|y_i|$ (where y_i is the ith block of $\mathscr{S}$);

(b) $\mathscr{S}$ is uniform if and only if $J_{m,v}A=kJ_{m,b}$ for a constant $k\neq 0$.

If we now define a structure to be *regular* if $|P|=r>0$ for all points P, then Exercise 1.2 has an obvious extension: $\mathscr{S}$ is regular if and only if $AJ_{b,m}=rJ_{v,m}$ for a constant $r\neq 0$.

Let $\mathscr{S}$ be a structure with $v>0$ points. If there exist integers λ, t with $0<\lambda$ and $0\leqslant t\leqslant v$ such that every subset of t points of $\mathscr{S}$ is incident with exactly λ common blocks then we say that $\mathscr{S}$ is a *t-structure for* λ, or merely a *t-structure*. (Note that a 0-structure is merely a non-empty structure.) A uniform t-structure with block size k is called a (uniform) t-(v, k, λ) *structure* or a t-*structure for* (v, k, λ), where v is the number of points and λ is the number of common blocks on t points. Note that, by definition, any uniform structure or t-structure must be non-empty. Also by talking about a t-structure for (v, k, λ)

we are automatically implying that it is uniform by referring to the 'k'. Hence we shall usually drop the word 'uniform' in this situation.

Uniform t-structures have the interesting, and very useful, property that they are also uniform s-structures for all s satisfying $0 \leqslant s \leqslant t$. We shall often refer to a set of j symbols as a j-set.

Theorem 1.2. If $\mathscr{S}$ is a t-structure for (v, k, λ) then, for any integer s satisfying $0 \leqslant s < t$, there are exactly λ_s blocks of $\mathscr{S}$ which are incident with any given s-set of points of $\mathscr{S}$, where

$$\lambda_s = \lambda \frac{(v-s)(v-s-1)\cdots(v-t+1)}{(k-s)(k-s-1)\cdots(k-t+1)}.$$

Proof. If $\mathscr{B}$ is any fixed subset of s points of $\mathscr{S}$, let m be the number of blocks of $\mathscr{S}$ which contain $\mathscr{B}$. We shall prove the theorem by computing m and showing that it depends only on s and is independent of the choice of $\mathscr{B}$. In order to compute m we define an *admissible pair* $(\mathscr{T}, y)$ to be a set $\mathscr{T}$ of t points which contains $\mathscr{B}$ and a block y which contains $\mathscr{T}$, and then compute the number of admissible pairs in two different ways.

Each of the m blocks which contain $\mathscr{B}$ contains $\binom{k-s}{t-s}$ t-sets which contain $\mathscr{B}$, so the number of admissible pairs is $m\binom{k-s}{t-s}$. On the other hand, there are $\binom{v-s}{t-s}$ ways to choose a set $\mathscr{T}$ of t points which contains $\mathscr{B}$, and each of these sets is on exactly λ blocks. Thus the number of admissible pairs is also $\lambda\binom{v-s}{t-s}$. Equating these two values for the number of admissible pairs, we see that

$$m = \lambda \frac{(v-s)\cdots(v-t+1)}{(k-s)\cdots(k-t+1)}$$

and the theorem is proved. □

Corollary 1.3. If $\mathscr{S}$ is a t-structure for (v, k, λ), $t > 0$, and s is any integer satisfying $0 < s < t$, then $\mathscr{S}$ is an s-structure for (v, k, λ_s), and, in particular, $\mathscr{S}$ is regular.

Proof. Obvious. □

One immediate consequence of Theorem 1.2 is the non-existence of uniform t-structures for most choices of v, k, λ. This is because each of the quantities λ_s

$(0 \leqslant s \leqslant t)$, where $\lambda_t = \lambda$, must be an integer. As a simple illustration the reader should solve the following exercise.

Exercise 1.3

(a) Show that there cannot exist a 4-structure for $(11, 7, 2)$.

(b) Show that if $t=5$, $v=24$, $k=8$ and $\lambda=1$ then $\lambda_4, \lambda_3, \lambda_2$ and λ_1 are all integers.

If $\mathscr{S}$ is a t-(v, k, λ) structure then, as always, we will denote the number of blocks by b. If $t \geqslant 1$ then, by Theorem 1.2, the number of blocks through a point is a constant which we denote by r, and if $t \geqslant 2$ then the integer $n = r - \lambda_2$ is called the *order* of $\mathscr{S}$. (Note that $b = \lambda_0$, $r = \lambda_1$, and $\lambda = \lambda_t$.) The integers $t, v, b, k, \lambda, r, n$ are called the *parameters* of $\mathscr{S}$. Clearly, they are not independent and we now restate certain special cases of Theorem 1.2 to emphasise some of the most important relations between them.

Corollary 1.4. If $\mathscr{S}$ is a t-structure for (v, k, λ) then:

(a) $$b = \lambda \frac{v(v-1)\cdots(v-t+1)}{k(k-1)\cdots(k-t+1)};$$

(b) if $t > 0$ then $bk = vr$;

(c) if $t > 1$ then $r(k-1) = \lambda_2(v-1)$. □

Exercise 1.4. Define a *flag* in a structure to be a pair (P, x), where P is a point and x is a block on P. By counting flags in two ways give a direct proof of (b) of Corollary 1.4.

Exercise 1.5. Define a *2-flag* (P, Q, x) in a structure to be a pair of distinct points P, Q and x a block incident with them both. By counting 2-flags in two ways give a direct proof of (c) of Corollary 1.4.

As we have already pointed out, any uniform t-structure with $t \geqslant 1$ is also a uniform 1-structure, which means that it is regular. The concept of *flag* introduced in Exercise 1.4 is extremely useful and arises again and again; the flag-counting of that exercise (and of Exercise 1.5) is one of the most common techniques in this book, and in the theory as a whole.

If a structure $\mathscr{S}$ has equally many points and blocks then any incidence matrix for $\mathscr{S}$ is square. For this reason we call a structure with $b = v$ a *square* structure.

Exercise 1.6. Show that a uniform 1-structure is square if and only if $k=r$.

We are now able to define the structures which are the central topic of this book. A *design* is a uniform, reduced structure, i.e. a uniform structure with no repeated blocks. (In terms of incidence matrices this means that the columns of an incidence matrix have equal sums but are all distinct.) Thus in a design a block is uniquely determined by the points on it and we may identify the blocks with distinguished subsets of points. The definition of a *t-design* is clear: namely a t-structure which is also a design. Since a design is a uniform structure, our results on uniform structures also apply to designs with the words 'uniform structure' replaced by 'design' throughout. We now introduce an important convention: if $\mathscr{S}$ is a t-design for (v, k, λ) then we say merely that $\mathscr{S}$ is a t-(v, k, λ). The point of this is simply that we shall be concerned with designs much more often than with general structures, and we want the most economical notation for them.

We have already seen that there do not exist t-(v, k, λ)s for arbitrary choices of t, v, k, λ. However, for any given t, v, k with $0 \leqslant t \leqslant k \leqslant v$ (note that these inequalities are necessary!) there is a t-$(v, k, \binom{v-t}{k-t})$. This is obtained by calling every possible k-subset of a given v-set a block. These designs are called trivial and we shall, in general, ignore them. More generally a uniform structure with block size k is called *trivial* if every k-set of points is incident with at least one block.

Exercise 1.7. If $\mathscr{S}$ is a design with block size k show that $\mathscr{S}$ is trivial if and only if $\mathscr{S}$ is a t-design for all t satisfying $0 \leqslant t \leqslant k$. (Note that a trivial uniform structure need not be a t-structure for any $t > 0$.)

We are now able to illustrate the vast difference between t-designs and uniform t-structures, and perhaps give some indication as to why we concentrate on designs. There are, at present, no known examples of non-trivial t-designs with $t \geqslant 7$. In Chapters 4 and 8 we construct some t-designs for $t=3, 4$ and 5; in [8] some 6-designs have been constructed (indeed, the only known examples at present). The situation for uniform t-structures could hardly be more different: as we shall prove in our next theorem, for any given t, v, k, with $0 < t < k \leqslant v/2$ there is a value of λ such that there exists a non-trivial structure for t-(v, k, λ). The restrictions $0 < t < k$ are obviously necessary for a non-trivial t-structure, but the reason for restricting outselves to $k \leqslant v/2$ will not become apparent until we discuss complementary structures in Section 1.4. In the proof (which should be left until later if the reader finds it difficult) the structure constructed always has repeated blocks and, consequently, is never a design.

Theorem 1.5. Given any positive integers t, v, k with $t<k\leqslant v/2$ there is a value of λ such that there exists a non-trivial t-(v, k, λ) structure.

Proof. Let $\mathscr{V}$ be a set of v elements. Let $\mathscr{X}_1, \ldots, \mathscr{X}_{\binom{v}{t}}$ be the t-subsets of $\mathscr{V}$, let $\mathscr{Y}_1, \ldots, \mathscr{Y}_{\binom{v}{k}}$ be the k-subsets and let A be the $\binom{v}{t}$ by $\binom{v}{k}$ matrix given by $a_{ij}=1$ if $\mathscr{X}_i \subset \mathscr{Y}_j$ and $a_{ij}=0$ otherwise. Since $t<k\leqslant v/2$, $\binom{v}{k}>\binom{v}{t}$. Thus the rank of A is at most $\binom{v}{t}$ and the columns of A must be linearly dependent. (This is true over any field.) Thus, if $\mathbf{c}_i$ is the ith column of A $(1\leqslant i\leqslant\binom{v}{k})$ there exist rationals $\alpha_1, \ldots, \alpha_{\binom{v}{k}}$ such that $\alpha_j\neq 0$ for some value of j and $\sum_{i=1}^{\binom{v}{k}} \alpha_i\mathbf{c}_i=\mathbf{0}$. Hence there exist integers $n_1, \ldots, n_{\binom{v}{k}}$ (not all zero) such that $\sum_{i=1}^{\binom{v}{k}} n_i\mathbf{c}_i=\mathbf{0}$. Since each $\mathbf{c}_i$ is a vector of 0s and 1s, each coordinate of $\sum_{i=1}^{\binom{v}{k}} n_i\mathbf{c}_i$ is the sum of some of the n_i. Consequently some of the integers n_i are negative. Let $-m$ be the least of the n_i and put $m_i=n_i+m$ for $i=1, \ldots, \binom{v}{k}$. Clearly, by the choice of m, $m_i\geqslant 0$ for all i and there is an integer l such that $m_l=0$. Also, since $\sum_{i=1}^{\binom{v}{k}} n_i\mathbf{c}_i=\mathbf{0}$,

$$\sum_{i=1}^{\binom{v}{k}} m_i\mathbf{c}_i=\sum_{i=1}^{\binom{v}{k}} n_i\mathbf{c}_i+\sum_{i=1}^{\binom{v}{k}} m\mathbf{c}_i=m\sum_{i=1}^{\binom{v}{k}} \mathbf{c}_i.$$

But the jth coordinate of $\sum_{i=1}^{\binom{v}{k}} \mathbf{c}_i$ is the sum of the entries in the jth row of A. This is equal to the number of k-subsets containing $\mathbf{X}_j$ which, of course, is $\binom{v-t}{k-t}$. Thus

$$\sum_{i=1}^{\binom{v}{k}} m_i\mathbf{c}_i=m\sum_{i=1}^{\binom{v}{k}} \mathbf{c}_i=m\binom{v-t}{k-t}\mathbf{j},$$

where $\mathbf{j}$ is the vector having each entry 1.

We now define a structure $\mathscr{S}$. The points of $\mathscr{S}$ are the elements of $\mathscr{V}$ and for each i $(1\leqslant i\leqslant\binom{v}{k})$, $\mathscr{Y}_i$ is a block of $\mathscr{S}$ m_i times. Clearly, since each block is a k-subset, $\mathscr{S}$ is uniform and, since $m_l=0$, it is non-trivial. The number of blocks through the t-subset $\mathscr{X}_h$ is merely the hth coordinate of $\sum_{i=1}^{\binom{v}{k}} m_i\mathbf{c}_i$ which, as we have seen, is $m\binom{v-t}{k-t}$ for every h. Thus $\mathscr{S}$ is a non-trivial t-structure for $(v, k, m\binom{v-t}{k-t})$. □

It is clear from the proof that at least one of the m_i is greater than one, so that $\mathscr{S}$ is definitely not a design. (One of the n_i is greater than zero and so is m.) However, we can also see this from the value of λ.

Exercise 1.8. If $\mathscr{S}$ is a t-(v, k, λ) show that $\lambda\leqslant\binom{v-t}{k-t}$. Show also that $\lambda=\binom{v-t}{k-t}$ if and only if $\mathscr{S}$ is trivial.

Although we know that the structure $\mathscr{S}$ of Theorem 1.5 is never a design, it is not inconceivable that $\mathscr{S}$ reduced is. However, it seems very difficult to determine whether or not this is true. Clearly, since the α_i are not unique, there are many possibilities for the structure $\mathscr{S}$ and also, note, even for the integer λ. If it were possible to choose the α_i so that they only assumed two values, then there would only be two values for the n_i and, hence, only one non-zero value for the m_i. If we denote this value by c then every k-subset of $\mathscr{V}$ would be a block of $\mathscr{S}$ either 0 or c times. Under these very exceptional (and presumably unlikely) conditions $\mathscr{S}/R$ would be a t-design.

Exercise 1.9. If $\mathscr{S}$ is a t-structure for (v, k, λ) in which every block has multiplicity c show that $\mathscr{S}/R$ is a t-$(v, k, \lambda/c)$.

Problem 1.1. If $\mathscr{S}$ is a t-structure for (v, k, λ) what, if anything, can you say about $\mathscr{S}/R$? Either try to prove that $\mathscr{S}/R$ is a t-design or find a counter-example.

Exercise 1.10. Show that a 2-design with $v=8$, $k=3$ is trivial.

Exercise 1.11.* Find a non-trivial 2-structure with $v=8, k=3$. (This exercise is somewhat more difficult than it might appear, and the reader might find it profitable to return to it later. But, in connection with Exercise 1.10, it emphasises the difference between structures and designs.)

We conclude this section with some definitions.

If $\mathscr{S}$ and $\mathscr{T}$ are two structures an *anti-isomorphism* α from $\mathscr{S}$ to $\mathscr{T}$ is a bijection from the points of $\mathscr{S}$ onto the blocks of $\mathscr{T}$ and the blocks of $\mathscr{S}$ onto the points of $\mathscr{T}$ such that P is on x in $\mathscr{S}$ if and only if x^α is on P^α in $\mathscr{T}$. An anti-isomorphism of a structure $\mathscr{S}$ onto itself is called a *correlation* and, as always, an *automorphism* of $\mathscr{S}$ is an isomorphism of $\mathscr{S}$ onto itself. (Note that a structure $\mathscr{S}$ cannot have any correlations unless $v=b$; i.e. unless it is square.)

Exercise 1.12. If α is an anti-isomorphism from $\mathscr{S}$ onto $\mathscr{T}$ and β is an anti-isomorphism from $\mathscr{T}$ onto $\mathscr{S}$, show that $\alpha\beta$ is an automorphism of $\mathscr{S}$.

From Exercise 1.12 it follows that the product of two correlations of $\mathscr{S}$ is an automorphism. If α is a correlation such that α^2 is the identity automorphism (i.e. the identity mapping on both the points and blocks of $\mathscr{S}$), then α is a *polarity*.

1.3 Examples

In this section we shall list a number of examples of structures and designs. These are to illustrate the definitions and results of the last section and also for future reference. In each case we state the basic properties of the given structure – usually in the form of an exercise. Most of these exercises are very similar and, once he has established that he understands the basic definitions, etc., the reader should skip those exercises which are merely straightforward verification of stated facts.

We begin by listing some structures that are so small we are able to write out the entire example. In these cases we represent the blocks by the set of points incident with them, so that incidence is given by set theoretic inclusion.

Example $\mathscr{A}$

The points of $\mathscr{A}$ are

$$\{1, 2, 3, 4, 5\}.$$

The blocks of $\mathscr{A}$ are

$$a_1=\{1, 2, 3\}, \quad a_2=\{1, 2, 3\}, \quad a_3=\{2, 3, 4\}, \quad a_4=\{3, 4, 5\}.$$

If we take $P_i=i$ $(i=1, \ldots, 5)$ then the incidence matrix for $\mathscr{A}$ is

$$A=\begin{pmatrix} 1 & 1 & 0 & 0 \\ 1 & 1 & 1 & 0 \\ 1 & 1 & 1 & 1 \\ 0 & 0 & 1 & 1 \\ 0 & 0 & 0 & 1 \end{pmatrix}.$$

Clearly, $\mathscr{A}$ is uniform and, since 1 is in two blocks but 5 is in only one, it is not regular and hence, by Corollary 1.3, is not a t-structure for any $t>1$. Also, since a_1 has multiplicity 2, it is not a design. Furthermore, since 5 is on only one block $\mathscr{A}$ is not standardised.

$\mathscr{A}/R$ has three blocks $a_1=\{1, 2, 3\}, a_3=\{2, 3, 4\}, a_4=\{3, 4, 5\}$. So $\mathscr{A}/R$ is an example of a design. Note that 5 is an isolated point and 3 is a full point of $\mathscr{A}/R$ so that $\mathscr{A}/R$ is not standardised. Removing 3 and 5 leaves us with three points $\{1, 2, 4\}$ and blocks $a_1 = \{1, 2\}, a_2 = \{2, 4\}, a_3 = \{3\}$. The resulting structure has a_3 as an isolated block and it is not difficult to see that $\overline{\mathscr{A}/R}$ is empty. Thus this example illustrates what a drastic effect total reduction can have.

Exercise 1.13. If $Q_1 = 2, Q_2 = 3, Q_3 = 5, Q_4 = 4, Q_5 = 1, y_1 = a_1, y_2 = a_4, y_3 = a_2, y_4 = a_3$, write down the incidence matrix B for $\mathscr{A}$, where $b_{ij} = 1$ if Q_i is on y_j. Find permutation matrices P and Q with $PAQ = B$.

Example $\mathscr{B}$

The points of $\mathscr{B}$ are

$\{1, 2, 3, 4, 5\}$.

The blocks are

$$b_1 = \{1, 2, 3\}, \quad b_2 = \{1, 4\}, \quad b_3 = \{2, 4\}, \quad b_4 = \{3, 4\}.$$

Clearly, in this example, 5 is an isolated point. The points of $\bar{\mathscr{B}}$ are $\{1, 2, 3, 4\}$ and the blocks of $\bar{\mathscr{B}}$ are the blocks of $\mathscr{B}$.

Obviously, $\bar{\mathscr{B}}$ is not uniform. However, straightforward verification shows that it is a 2-structure with $\lambda = 1$. Note that $\bar{\mathscr{B}}$ is not a 1-structure, so that Theorem 1.2 is not true for non-uniform structures.

Exercise 1.14. Verify that $\alpha: b_1 \leftrightarrow 4, b_2 \leftrightarrow 1, b_3 \leftrightarrow 2, b_4 \leftrightarrow 3$, is a polarity of $\bar{\mathscr{B}}$.

Exercise 1.15. Show that every automorphism of $\mathscr{B}$ must fix the block b_1 and the points 4 and 5. Hence show that $\mathscr{B}$ admits exactly six automorphisms (write them down), which form the symmetric group on three letters.

Our first two examples were not standardised. However, we will not usually be interested in non-standardised structures and so our remaining examples will have no full or isolated elements.

Example $\mathscr{C}$

The points of $\mathscr{C}$ are

$$\{1, 2, 3, 4, 5, 6\}.$$

The blocks are

$$d_1 = \{1, 2, 3\}, \quad d_2 = \{4, 5, 6\}, \quad d_3 = \{1, 4\}, \quad d_4 = \{1, 5\},$$
$$d_5 = \{1, 6\}, \quad d_6 = \{2, 4\}, \quad d_7 = \{2, 5\}, \quad d_8 = \{2, 6\},$$
$$d_9 = \{3, 4\}, \quad d_{10} = \{3, 5\}, \quad d_{11} = \{3, 6\}.$$

Exercise 1.16. Verify that $\mathscr{C}$ is a 2-structure with $\lambda = 1$ and a 1-structure with $r = 4$.

Clearly, $\mathscr{C}$ is not uniform. Thus, although Theorem 1.2 does not hold for non-uniform structures, a non-uniform 2-structure may also be a 1-structure, i.e. regular.

Example $\mathscr{D}$

The points of $\mathscr{D}$ are

$$\{1, 2, 3, 4, 5, 6\}.$$

The blocks are

$$d_1 = \{1, 2, 3\}, \quad d_2 = \{1, 4, 5\}, \quad d_3 = \{1, 2, 6\}, \quad d_4 = \{1, 3, 6\},$$
$$d_5 = \{2, 4, 5\}, \quad d_6 = \{2, 4, 6\}, \quad d_7 = \{3, 4, 5\}, \quad d_8 = \{3, 5, 6\}.$$

Exercise 1.17. Verify that $\mathscr{D}$ is a 1-design with parameters $v = 6$, $b = 8$, $r = 4$, $k = 3$ and determine the rank of $\mathscr{D}$.

Example $\mathscr{E}$

The points of $\mathscr{E}$ are

$$\{1, 2, 3, 4, 5, 6, 7, 8\}.$$

The blocks of $\mathscr{E}$ are

$$e_1=\{1,2,3,4\},\quad e_2=\{1,3,5,6\},\quad e_3=\{1,4,7,8\},$$
$$e_4=\{2,5,6,7\},\quad e_5=\{2,5,7,8\},\quad e_6=\{3,4,6,8\}.$$

Exercise 1.18. Verify that $\mathscr{E}$ is a 1-design with parameters $v=8$, $b=6$, $r=3$, $k=4$. Show that α: $i\to e_i$, $i=1,\ldots,6$, and $d_j\to j$, $j=1,\ldots,8$, is an anti-isomorphism from $\mathscr{D}$ to $\mathscr{E}$. Write down the incidence matrices for $\mathscr{D}$ and $\mathscr{E}$ with the given labellings and observe that one is the transpose of the other.

Examples $\mathscr{D}$ and $\mathscr{E}$ and Exercise 1.18 illustrate one of the standard methods of constructing a new structure from a given one. We shall be discussing this construction in the next section, but the reader might like to establish it for himself.

Exercise 1.19. If A is an incidence matrix for a structure $\mathscr{S}$ show that A^{T} is an incidence matrix for a structure which is anti-isomorphic to $\mathscr{S}$.

Example $\mathscr{F}$

The points of $\mathscr{F}$ are

$$\{1,2,3,4,5,6,7\}.$$

The blocks are

$$f_1=\{1,2,4\},\quad f_2=\{1,3,7\},\quad f_3=\{2,3,5\},\quad f_4=\{1,5,6\},$$
$$f_5=\{3,4,6\},\quad f_6=\{4,5,7\},\quad f_7=\{2,6,7\}.$$

Exercise 1.20. Verify that $\mathscr{F}$ is a 2-(7, 3, 1) and hence also, by Theorem 1.2, a 1-(7, 3, 3). Write out the incidence matrix A for $\mathscr{F}$ with the given labelling, observe that A is symmetric and find a polarity of $\mathscr{F}$.

Our next three examples will all be 2-structures for (7, 3, 2). Each will also contain the seven blocks of Example $\mathscr{F}$. However, as we shall easily see, no pair is isomorphic.

Example $\mathscr{G}$

The points are

$$\{1, 2, 3, 4, 5, 6, 7\}.$$

The blocks are

$$f_1, f_2, f_3, f_4, f_5, f_6, f_7,$$
$$g_1 = \{1, 3, 7\}, \quad g_2 = \{1, 2, 4\}, \quad g_3 = \{3, 4, 5\}, \quad g_4 = \{2, 3, 6\},$$
$$g_5 = \{2, 5, 7\}, \quad g_6 = \{1, 5, 6\}, \quad g_7 = \{4, 6, 7\}.$$

Example $\mathscr{H}$

The points are

$$\{1, 2, 3, 4, 5, 6, 7\}.$$

The blocks are

$$f_1, f_2, f_3, f_4, f_5, f_6, f_7,$$
$$h_1 = \{1, 3, 6\}, \quad h_2 = \{1, 4, 5\}, \quad h_3 = \{2, 3, 4\}, \quad h_4 = \{3, 5, 7\},$$
$$h_5 = \{2, 5, 6\}, \quad h_6 = \{1, 2, 7\}, \quad h_7 = \{4, 6, 7\}.$$

Example $\mathscr{I}$

The points are

$$\{1, 2, 3, 4, 5, 6, 7\}.$$

The blocks are

$$f_1, f_2, f_3, f_4, f_5, f_6, f_7,$$
$$i_1 = \{1, 2, 4\}, \quad i_2 = \{2, 3, 5\}, \quad i_3 = \{3, 4, 6\}, \quad i_4 = \{4, 5, 7\},$$
$$i_5 = \{1, 5, 6\}, \quad i_6 = \{2, 6, 7\}, \quad i_7 = \{1, 3, 7\}.$$

Exercise 1.21. Verify that each of these three examples is a 2-(7, 3, 2) structure but that only $\mathscr{H}$ is a 2-(7, 3, 2). Show that $\mathscr{I}/R$ is a 2-(7, 3, 1) and that $\mathscr{G}/R$ is not a 2-design. Hence show that no pair of these examples is isomorphic. Finally show $\mathscr{I}/R \cong \mathscr{F}$.

These examples illustrate the very important fact that structures with identical parameters need not be isomorphic. In this instance the non-isomorphism is established by looking at the multiplicities of the various blocks. (Note that we could also have looked at their ranks.) However, this simple method will not usually be available, and non-isomorphism can be very difficult to establish.

Our next example is also closely related to $\mathscr{F}$ and is a special case of another general construction.

Example $\mathscr{J}$

The points of $\mathscr{J}$ are

$$\{1, 2, 3, 4, 5, 6, 7\}.$$

The blocks are

$$j_1=\{3, 5, 6, 7\},\quad j_2=\{2, 4, 5, 6\},\quad j_3=\{1, 4, 6, 7\},\quad j_4=\{2, 3, 4, 7\},$$
$$j_5=\{1, 2, 5, 7\},\quad j_6=\{1, 2, 3, 6\},\quad j_7=\{1, 3, 4, 5\}.$$

Exercise 1.22. Show that $\mathscr{J}$ is a 2-(7, 4, 2) and note that this implies it is a 1-(7, 4, 4).

Exercise 1.23. Write out incidence matrices for $\mathscr{F}$ and $\mathscr{J}$ with the points and blocks indexed in the given orders, and note that the sum of these two matrices is J_7. What does this say about the two given designs?

Our final small example will be a 3-design.

Example $\mathscr{K}$

The points of $\mathscr{K}$ are

$$\{1, 2, 3, 4, 5, 6, 7, 8\}.$$

The blocks are

$$k_1=\{1, 3, 7, 8\},\quad k_2=\{1, 2, 4, 8\},\quad k_3=\{2, 3, 5, 8\},\quad k_4=\{3, 4, 6, 8\},$$
$$k_5=\{4, 5, 7, 8\},\quad k_6=\{1, 5, 6, 8\},\quad k_7=\{2, 6, 7, 8\},\quad k_8=\{1, 2, 3, 6\},$$
$$k_9=\{1, 2, 5, 7\},\quad k_{10}=\{1, 3, 4, 5\},\quad k_{11}=\{1, 4, 6, 7\},\quad k_{12}=\{2, 3, 4, 7\},$$
$$k_{13}=\{2, 4, 5, 6\},\quad k_{14}=\{3, 5, 6, 7\}.$$

Exercise 1.24. Show that $\mathscr{K}$ is a 3-(8, 4, 1) and note that this implies that it is also a 2-(8, 4, 3) and a 1-(8, 4, 7).

Before discussing some infinite families of 2-designs we have two exercises which are easy but illustrate very important points.

Exercise 1.25. (a) Show that there is a unique automorphism of Example $\mathscr{C}$ whose permutation action on the points is given by $1 \to 1, 2 \to 2, 3 \to 3, 4 \to 5, 5 \to 6, 6 \to 4$. Show also that there is a unique automorphism of $\mathscr{C}$ whose permutation on the points is the identity (hence note that this implies that an automorphism of $\mathscr{C}$ is uniquely determined by its action on points).

(b) Show that there is an automorphism of Example $\mathscr{I}$ which is the identity on points but which is not the identity on blocks. (So this implies that an automorphism of a structure is not necessarily uniquely determined by its action of the points.)

Exercise 1.26. Show that an automorphism of a structure is uniquely determined by its permutation action on the points if and only if the structure has no repeated blocks.

There are, of course, many infinite families of structures and designs. Among the most important of these are those obtained from finite projective and affine geometries. We conclude this section by defining these geometries and establishing some of their basic combinatorial properties. We shall not, however, attempt to give a comprehensive treatment of these geometries. (More detailed accounts of their combinatorial properties are given in Dembowski [4] or Hirschfeld [6], while excellent studies of them considered as a branch of linear algebra can be found in Artin [1], Baer [2] or Gruenberg & Weir [5].)

A *projective geometry* $\mathscr{P}$ is a (not necessarily finite) structure of points and lines such that

(P1) every pair of points are on a unique common line;
(P2) every line contains at least three points;
(P3) $\mathscr{P}$ contains a set of three points which are not on a common line;
(P4) if X_1, X_2, X_3 are distinct points and l_1, l_2, l_3 are distinct lines with X_i on l_j for all $i \neq j$, and if l is any line intersecting l_2 and l_3 but not containing X_1, then l intersects l_1. (In more familiar language: if a line intersects two sides of a triangle but does not contain their common vertex then it intersects the third side.)

Among these projective geometries there is a special subclass which we want to single out. A *projective plane* is a projective geometry satisfying

(P5) every pair of distinct lines contain a common point.

Note, by the way, that (P5) implies (P4) and that the point common to a pair of distinct lines is, by (P1), necessarily unique.

There is a very simple way to construct projective geometries. For K a skewfield, let V be an $(n+1)$-dimensional vector space over K, and define $\mathscr{P}(n, K)$ to be the collection of all subspaces of V, with an 'incidence relation' given by ordinary inclusion. For $i=0, 1, \ldots, n$, we say that the subspace U of V has *g-dimension* i if U has dimension $i+1$, and that $\mathscr{P}(n, K)$ itself has g-dimension n. We define $\mathscr{P}_{n,i}(K)$ to be the structure whose points are the subspaces of g-dimension 0 and whose blocks are the subspaces of g-dimension i; so, in particular, if the subspaces of g-dimension 1 are called *lines*, then the following is easy:

Exercise 1.27. Show that $\mathscr{P}_{n,1}(K)$ is a projective geometry if $n>1$.

With an appropriate finiteness condition (and in particular if the number of points is finite), it turns out that any projective geometry is either a projective plane or is isomorphic to $\mathscr{P}_{n,1}(K)$ for a unique n and a unique K (see [1, 2] for this important result). *Hence a common attitude is to refer to all of* $\mathscr{P}(n, K)$ *as being a projective geometry*, and we shall adopt this usage from now on if no ambiguity results. So a projective geometry $\mathscr{P}(n, K)$ with $n>2$ has many substructures $\mathscr{P}_{n,i}(K)$ which are proper objects of study; among these we have seen that $\mathscr{P}_{n,1}(K)$ is important, and perhaps the other one of greatest interest is $\mathscr{P}_{n,n-1}(K)$, whose blocks, the subspaces of g-dimension $n-1$, are the *hyperplanes* of $\mathscr{P}(n, K)$. We write $\mathscr{P}_n(K)$ for this structure $\mathscr{P}_{n,n-1}(K)$, and shall meet it again in the book.

If K is finite then it has q elements, where q is a prime power, and K is even unique up to isomorphism once q is specified. So we can write $\mathscr{P}(n, q)$, $\mathscr{P}_{n,i}(q)$, etc., instead of $\mathscr{P}(n, K)$, $\mathscr{P}_{n,i}(K)$, etc.

Projective planes are singled out for consideration because they are projective geometries but are not necessarily of the form $\mathscr{P}(n, K)$, as we shall see in Chapter 3.

Exercise 1.28. Show that $\mathscr{P}_{n,i}(K)$ is a projective plane if and only if $n=2$ and $i=1$.

Exercise 1.29. Show that $\mathscr{P}_{2,1}(2)$ is isomorphic to Example $\mathscr{F}$.

Note that $\mathscr{P}_2(K)$ only contains points and lines and so it is unambiguous to write $\mathscr{P}_{2,1}(K)=\mathscr{P}_2(K)=\mathscr{P}(2, K)$.

Exercise 1.30. Show that $\mathscr{P}_2(q)$ is a 2-$(q^2+q+1, q+1, 1)$ with $b=q^2+q+1$, $r=q+1$ and order $n=q$.

We establish a useful counting result:

Lemma 1.6. Let V be a vector space of dimension n over $GF(q)$ and let U be an m-dimensional subspace of V, $m\geqslant 0$. Then the number of $(m+h)$-dimensional subspaces containing U is

$$\frac{(q^{n-m}-1)(q^{n-m}-q)\cdots(q^{n-m}-q^{h-1})}{(q^h-1)(q^h-q)\cdots(q^h-q^{h-1})}.$$

Proof. Let $\mathbf{a}_1, \ldots, \mathbf{a}_m$ be any fixed basis of U. If W is any $(m+h)$-dimensional subspace of V containing U, the $\mathbf{a}_1, \ldots, \mathbf{a}_m$ can be extended to form a basis of W. We will prove the lemma by counting the number of linearly independent ordered sets $\mathbf{a}_1, \ldots, \mathbf{a}_m, \mathbf{b}_1, \ldots, \mathbf{b}_h$ in V, and dividing this number by the number of these sets which generate the same $(m+h)$-dimensional subspace (i.e. dividing by the number of bases of an $(m+h)$-dimensional subspace with $\mathbf{a}_1, \ldots, \mathbf{a}_m$ as the first m basis vectors).

If we have a set $\mathbf{a}_1, \ldots, \mathbf{a}_m, \mathbf{b}_1, \ldots, \mathbf{b}_j$ of $(m+j)$ independent vectors and if $U_j=\langle \mathbf{a}_1, \ldots, \mathbf{a}_m, \mathbf{b}_1, \ldots, \mathbf{b}_j\rangle$ then a vector in V is independent of $\mathbf{a}_1, \ldots, \mathbf{a}_m, \mathbf{b}_1, \ldots, \mathbf{b}_j$ if and only if it is not in U_j. Thus, since there are q^x vectors in an x-dimensional vector space over $GF(q)$, there are q^n-q^{m+j} choices for $\mathbf{b}_{j+1}, j=0, \ldots, h-1$. Hence the number of sets of independent vectors $\mathbf{a}_1, \ldots, \mathbf{a}_m, \mathbf{b}_1, \ldots, \mathbf{b}_h$ is $(q^n-q^m)(q^n-q^{m+1}) \cdots (q^n-q^{m+h-1})$.

But the number of bases of an $(m+h)$-dimensional subspace W containing U is obtained by letting W replace V in the above argument, i.e. $(q^{m+h}-q^m)(q^{m+h}-q^{m+1}) \cdots (q^{m+h}-q^{m+h-1})$.

The number of $(m+h)$-dimensional subspaces containing U is the quotient of these two numbers. □

Lemma 1.7 (Grassman's Identity). Let U and W be subspaces of $\mathscr{P}(n, q)$; then $gd(U+W)+gd(U\cap W)=gd(U)+gd(W)$.

Any two points A, B of $\mathscr{P}(n, q)$ are contained in a unique line l and a subspace of g-dimension i contains both A and B if and only if it contains l. Thus, the number of subspaces of g-dimension i through two points is the number of subspaces containing a subspace of g-dimension 1 which, by Lemma 1.6, is a constant, i.e. independent of the two points. Clearly, any two subspaces of g-dimension i contain the same number of points and hence, by taking the subspaces of g-dimension i as blocks, we have a 2-design.

Exercise 1.31. If q is a prime power and i, n are positive integers with $n>i>1$ show that $\mathscr{P}_{n,i}(q)$ is a

$$2-\left(\frac{q^{n+1}-1}{q-1}, \frac{q^{i+1}-1}{q-1}, \prod_{j=0}^{j=i-2}\left(\frac{q^{n-1-j}-1}{q^{j+1}-1}\right)\right),$$

and show that $\mathscr{P}_{n,1}(q)$ is a 2-$((q^{n+1}-1)/q-1, q+1, 1)$. In each case find b and r.

When we defined a projective geometry we used points and lines and, since they have to be distinct, we only dealt with $\mathscr{P}(n, q)$ for $n \geqslant 2$. However, it is convenient to define $\mathscr{P}(1, q)$ to be the set of one-dimensional subspaces of a two-dimensional vector space over $GF(q)$. We call $\mathscr{P}(1, q)$ a *projective line.* Clearly, since there are no subspaces to take as lines, $\mathscr{P}(1, q)$ is not, properly speaking, a projective geometry. Nevertheless it occurs frequently in the study of designs and we shall often need it.

An *affine geometry* arises by deleting a fixed hyperplane and all its subspaces from a projective geometry. Just as projective geometries can be axiomatically defined by properties of points and lines, so can affine geometries, but we shall not develop that point of view. And also, just as projective geometries have not only points and lines, but also (in general) subspaces of higher g-dimension, so do affine geometries have subspaces other than points and lines (in general). And finally, affine planes are exceptional in the theory of affine geometry exactly as projective planes are exceptional in the theory of projective geometry.

An *affine plane* $\mathscr{A}$ is a (not necessarily finite) structure of points and lines such that

(A1) every pair of distinct points are on a unique common line;

(A2) if P, y is a non-incident point-line pair, then there is a unique line z containing P which does not meet y;

(A3) $\mathscr{A}$ contains a set of three points not on a common line.

Next, let K be a skewfield and V a vector space of dimension $n>1$ over K. Let $\mathscr{A}(n, K)$ be the collection of all the co-sets $U+\mathbf{a}$, where U is any subspace of V and $\mathbf{a}$ is a vector in V. If U has dimension m, then we say that $U+\mathbf{a}$ is a *subspace* of $\mathscr{A}(n, K)$ of *g-dimension m* (note the difference from the projective case). The *points* of $\mathscr{A}(n, K)$ are the subspaces of g-dimension 0; i.e. the points are the vectors of V. The *lines* of $\mathscr{A}(n, K)$ are the subspaces of g-dimension 1, and the *hyperplanes* are the subspaces of g-dimension $n-1$. For any i $(1 \leqslant i \leqslant n-1)$, we define $\mathscr{A}_{n,i}(K)$ to be the structure whose points are the points of $\mathscr{A}(n, K)$ and whose blocks are the subspaces of g-dimension i. As before, we write $\mathscr{A}_n(K)$ for $\mathscr{A}_{n,n-1}(K)$. It is then true that an affine geometry is either an affine plane or is isomorphic to an $\mathscr{A}_{n,1}(K)$ for a unique n and K, in complete analogy with the situation for projective geometries.

We can write $\mathscr{A}(n, q)$, etc., instead of $\mathscr{A}(n, K)$, etc., if K is a finite field with q elements. As in the case of projective geometries, we shall refer to $\mathscr{A}(n, K)$ as being an affine geometry, even though strictly speaking an affine geometry consists only of the points and lines of $\mathscr{A}(n, K)$. See [5] for further details, and proofs, of all the results and definitions we have given here. (See Exercise 1.43 as well.)

Finally, if $n=2$, then a hyperplane of $\mathscr{P}(2, K)$ is a line, and the subspaces of $\mathscr{A}(2, K)$ all have g-dimension 0 or 1, so we may refer to $\mathscr{A}(2, K)$ or $\mathscr{A}_{2,1}(K)$ or $\mathscr{A}_2(K)$ as the same objects (and they are all affine planes).

Exercise 1.32. Show that $\mathscr{A}(2, 2)$ is trivial and that $\mathscr{A}_3(2)$ is isomorphic to Example $\mathscr{K}$.

Exercise 1.33. Show that $\mathscr{A}_{n,i}(q)$ is a

$$2\text{-}\left(q^n, q^i, \prod_{j=0}^{j=i-2} \frac{q^{n-1-j}-1}{q^{j+1}-1}\right)$$

if $i \geqslant 2$ and that $\mathscr{A}_{n,1}(q)$ is a $2\text{-}(q^n, q, 1)$. Find b and r.

In Exercise 1.32 we saw that $\mathscr{A}_3(2)$ is actually a 3-design. This, of course, poses a natural question: when, if ever, are $\mathscr{P}_{n,i}(q)$ or $\mathscr{A}_{n,i}(q)$ 3-designs? From Exercise 1.31 we know that in $\mathscr{P}_{n,i}(q)$ every line contains $q+1 \geqslant 3$ points, so that three points may be collinear or not. If three points are collinear the number of blocks through them is equal to the number of subspaces of g-dimension i containing a line, whereas the number of blocks through three non-collinear points is equal to the number of subspaces of g-dimension i containing a plane. Since these two numbers are never equal, $\mathscr{P}_{n,i}(q)$ is never a 3-design. The same argument proves that $\mathscr{A}_{n,1}(q)$ cannot possibly be a 3-design if a line has three or more points on it. Since a line of $\mathscr{A}_{n,i}(q)$ contains q points, $\mathscr{A}_{n,i}(q)$ cannot possibly be a 3-design unless $q=2$. However, if $q=2$ then any 3-set of points lies in a unique plane and hence, since the number of subspaces of g-dimension i through any plane is a constant, $\mathscr{A}_{n,i}(2)$ is always a 3-design (provided, of course, that i is large enough for a block to contain three points, i.e. $i \geqslant 2$).

Exercise 1.34. Show that $\mathscr{A}_{n,i}(2)$ is a

$$3\text{-}\left(2^n, 2^i, \prod_{j=0}^{i-3} \frac{2^{n-2-j}-1}{2^{j+1}-1}\right)$$

provided $n > i \geqslant 3$, and that $\mathscr{A}_{n,2}(2)$ is a 3-$(2^n, 4, 1)$.

In the study of affine geometries the concept of parallelism plays a central role. Two lines in an affine plane are said to be *parallel* if they are either equal or do not meet, while two blocks $U_1 + \mathbf{a}$, $U_2 + \mathbf{b}$ of $\mathscr{A}_{n,i}(q)$ are *parallel* if and only if $U_1 = U_2$. Clearly, two parallel blocks in any $\mathscr{A}_{n,i}(q)$ are either equal or have no common point. However, the converse is not always true.

Exercise 1.35. Show that two distinct blocks in $\mathscr{A}_n(q)$ are parallel if and only if they are disjoint, but that $\mathscr{A}_{n,i}(q)$ always has two non-intersecting, non-parallel blocks if $i \neq n-1$.

1.4 Related structures

If $\mathscr{S}$ is a structure then there are many ways in which we can use $\mathscr{S}$ to construct other related structures. Perhaps the simplest of these is the *dual* of $\mathscr{S}$, which we denote by $\mathscr{S}^{\mathrm{T}}$, in which the roles of points and blocks are interchanged: for every point P of $\mathscr{S}$ there is a block P' of $\mathscr{S}^{\mathrm{T}}$, and for each block x of $\mathscr{S}$ there is a point x' of $\mathscr{S}^{\mathrm{T}}$, while incidence in $\mathscr{S}^{\mathrm{T}}$ is given by the rule that x' is on P' in $\mathscr{S}^{\mathrm{T}}$ if and only if P is on x in $\mathscr{S}$. Clearly, $(\mathscr{S}^{\mathrm{T}})^{\mathrm{T}} = \mathscr{S}$. The incidence matrices of $\mathscr{S}$ and $\mathscr{S}^{\mathrm{T}}$ are obviously closely related.

Lemma 1.8. If A is an incidence matrix for a structure $\mathscr{S}$, then the transpose A^{T} of A is an incidence matrix for $\mathscr{S}^{\mathrm{T}}$.

Proof. Let the points and blocks of $\mathscr{S}$ be labelled $P_1, \ldots, P_v, x_1, \ldots, x_b$, so that

P_i is on x_j if and only if $a_{ij}=1$, and let $B=(b_{ij})$ be the incidence matrix of $\mathscr{S}^{\mathrm{T}}$ given by the labelling $x'_1, \ldots, x'_b$ of the points and $P'_1, \ldots, P'_v$ of the blocks. Then $b_{ij}=1$ if and only if x'_i is on P'_j. Thus, from the definition of $\mathscr{S}^{\mathrm{T}}$, $b_{ij}=1$ if and only if P_j is on x_i, i.e. if and only if $a_{ji}=1$. Hence $b_{ij}=a_{ji}$ or $B=A^{\mathrm{T}}$. □

Corollary 1.9. $\mathscr{S}$ is regular if and only if $\mathscr{S}^{\mathrm{T}}$ is uniform.

Proof. This is an immediate consequence of Lemma 1.9 and Exercises 1.2 and 1.6. □

Exercise 1.36. Give an example of a 1-design whose dual is not a 1-design.

Clearly, the mapping $\alpha : \mathscr{S} \to \mathscr{S}^{\mathrm{T}}$, given by $P^{\alpha}=P'$, $x^{\alpha}=x'$ for all points P and blocks x, is an anti-isomorphism. So the dual of $\mathscr{S}$ is always anti-isomorphic to $\mathscr{S}$. It is also possible for $\mathscr{S}$ to be isomorphic to $\mathscr{S}^{\mathrm{T}}$. As an illustration let us consider Example $\mathscr{F}$. The points and blocks of $\mathscr{F}^{\mathrm{T}}$ may be labelled $f'_1, f'_2, \ldots, f'_7, 1', \ldots, 7'$ and the rules for incidence in $\mathscr{F}$ enable us to write out the blocks of $\mathscr{F}^{\mathrm{T}}$ as point sets. Thus

$1'=\{f'_1, f'_2, f'_4\}$, $2'=\{f'_1, f'_3, f'_7\}$, $3'=\{f'_2, f'_3, f'_5\}$, $4'=\{f'_1, f'_5, f'_6\}$, $5'=\{f'_3, f'_4, f'_6\}$, $6'=\{f'_4, f'_5, f'_7\}$, $7'=\{f'_2, f'_6, f'_7\}$.

Straightforward verification now shows that the mapping θ: $i \to f'_i$, $f_i \to i'$ is an isomorphism from $\mathscr{F}$ to $\mathscr{F}^{\mathrm{T}}$. If a structure $\mathscr{S}$ has the property that $\mathscr{S} \cong \mathscr{S}^{\mathrm{T}}$ then we say that $\mathscr{S}$ is *self-dual*. It is an immediate consequence of the definitions that if $\mathscr{S}$ is self-dual then an anti-isomorphism $\mathscr{S} \to \mathscr{S}^{\mathrm{T}}$ followed by an isomorphism $\mathscr{S}^{\mathrm{T}} \to \mathscr{S}$ gives a correlation of $\mathscr{S}$, and that $\mathscr{S}$ cannot have a correlation unless it is self-dual. Thus, for the above example, the mapping $\beta=\alpha\theta^{-1}$ is a correlation of $\mathscr{F}$. In fact since β interchanges i with f_i it has order two and is a polarity.

Since the points of $\mathscr{S}^{\mathrm{T}}$ are the blocks of $\mathscr{S}$ a structure cannot be self-dual unless it is square. (Note, as we shall see later, there exist square structures which are not self-dual.) Thus we already know many examples of non-self-dual designs. We have already seen that the dual of a 1-design is again a 1-design. We have also seen (see Example $\mathscr{F}$) an example of a 2-design whose dual is also a 2-design. In general, however, the dual of a 2-design is usually only a 1-design. (Recall that, by Corollary 1.3, a 2-design is a 1-design.) To illustrate this consider Example $\mathscr{K}$. The blocks k_1 and k_2 have two points in common, while k_1 and k_{13} do not intersect. So, in $\mathcal{K}^{\mathrm{T}}$, k'_1 and k'_2 are on two blocks while k'_1 and k'_{13} are on no common block.

We end this brief discussion on dual structures by observing that the self-duality of Example $\mathcal{F}$ could have been established more quickly by using Lemmas 1.8 and 1.1.

Exercise 1.37. Use incidence matrices to show that Examples $\mathcal{F}$ and $\mathcal{J}$ are both self-dual.

The next related structure which we discuss is constructed from $\mathcal{S}$ by complementing the incidence relation: it is called the *complement* of $\mathcal{S}$ and is represented by $\mathcal{C}(\mathcal{S})$. To be precise, the points of $\mathcal{C}(\mathcal{S})$ are the points of $\mathcal{S}$, and for each block x in $\mathcal{S}$ we let x^* be a block of $\mathcal{C}(\mathcal{S})$ with the incidence rule that P is on x^* if and only if P is not on x. Once again the incidence matrices of $\mathcal{S}$ and $\mathcal{C}(\mathcal{S})$ are closely related.

Exercise 1.38. If A is an incidence matrix for $\mathcal{S}$, where $\mathcal{S}$ has v points and b blocks, show that $J_{v,b}-A$ is an incidence matrix for $\mathcal{C}(\mathcal{S})$.

By referring back to Exercise 1.23 we see that Example $\mathcal{F}$ is the complement of Example $\mathcal{J}$. Clearly, for any $\mathcal{S}$, $\mathcal{C}(\mathcal{C}(\mathcal{S}))=\mathcal{S}$ so $\mathcal{J}$ is also the complement of $\mathcal{F}$. As in the case of the dual, many properties of $\mathcal{S}$ are shared by $\mathcal{C}(\mathcal{S})$. For instance, as an immediate consequence of Exercise 1.38, we have

Lemma 1.10

(a) $\mathcal{S}$ is uniform if and only if $\mathcal{C}(\mathcal{S})$ is.

(b) $\mathcal{S}$ is regular if and only if $\mathcal{C}(\mathcal{S})$ is.

(c) $\mathcal{S}$ is a design if and only if $\mathcal{C}(\mathcal{S})$ is.

Proof. Use Exercises 1.38, 1.2 and 1.6. □

In fact we can prove even more.

Theorem 1.11. If $\mathcal{S}$ is a 2-structure for (v, k, λ) with $2 \leqslant k \leqslant v-2$ then $\mathcal{C}(\mathcal{S})$ is a 2-structure for $(v, v-k, b-2\lambda_1+\lambda)$.

Proof. Clearly, $\mathscr{C}(\mathscr{S})$ has v points and any block of $\mathscr{C}(\mathscr{S})$ contains $v-k$ points. Let A and B be any two distinct points of $\mathscr{C}(\mathscr{S})$ and, hence, of $\mathscr{S}$. By Theorem 1.2, A is on $\lambda_1=\lambda(v-1)/(k-1)$ blocks of $\mathscr{S}$. Similarly B is on λ_1 blocks and A, B are on λ common blocks. Thus the number of blocks of $\mathscr{S}$ which contain at least one of A or B is $2\lambda_1-\lambda$. (Note that if $k\geqslant v-1$ then this number is b.) But every block of $\mathscr{S}$ which contains neither A nor B gives a block of $\mathscr{C}(\mathscr{S})$ which contains them both. Hence there are $b-2\lambda_1+\lambda$ blocks of $\mathscr{C}(\mathscr{S})$ through A and B. □

Corollary 1.12. If $\mathscr{S}$ is a 2-(v, k, λ) with $2\leqslant k\leqslant v-2$ then $\mathscr{C}(\mathscr{S})$ is a 2-$(v, v-k, b-2\lambda_1+\lambda)$. □

As well as b and v a third parameter, n (the order), is also common to $\mathscr{S}$ and $\mathscr{C}(\mathscr{S})$. (Geometrically if A and B are a pair of distinct points then n is the number of blocks which contain A but not B.)

Lemma 1.13. If $\mathscr{S}$ is a 2-structure for (v, k, λ) with $2\leqslant k\leqslant v-2$ then the order of $\mathscr{S}$ is equal to the order of $\mathscr{C}(\mathscr{S})$.

Proof. By Theorem 1.11, $\mathscr{C}(\mathscr{S})$ is a 2-structure for $(v, v-k, v-2r+\lambda)$. Since there are r blocks of $\mathscr{S}$ through a given point A there are $b-r$ blocks of $\mathscr{S}$ which do not contain A, i.e. there are $b-r$ blocks of $\mathscr{C}(\mathscr{S})$ through A. Hence the order of $\mathscr{C}(\mathscr{S})$ is $b-r-(b-2r+\lambda)=r-\lambda=$ order of $\mathscr{S}$. □

There is an analogue to Theorem 1.11 for t-structures with $t\geqslant 3$. However, it is harder to prove and we will postpone its proof until later. The reader might like to establish it for himself.

Exercise 1.39.* Generalise Theorem 1.11 to the case where $\mathscr{S}$ is a t-structure for (v, k, λ) with $t\geqslant 2$ and $t\leqslant k\leqslant v-t$.

Whenever we are studying a structure $\mathscr{S}$ then, obviously, we are also studying its complement $\mathscr{C}(\mathscr{S})$. Furthermore, if $\mathscr{S}$ is uniform with block size k then $\mathscr{C}(\mathscr{S})$ has block size $v-k$ and either $k\leqslant v/2$ or $v-k\leqslant v/2$. Hence we shall often restrict outselves to considering t-structures for (v, k, λ) with $t\leqslant k\leqslant v/2$. (This is the justification for the hypothesis $k\leqslant v/2$ in Theorem 1.5.)

Now suppose that $\mathscr{S}$ is a structure and that P is a point of $\mathscr{S}$. We define a new structure $\mathscr{S}_P$, called the *internal structure* (or the *contraction*) of $\mathscr{S}$ at P, to be the set of all blocks of $\mathscr{S}$ which contain P and the set of all points of $\mathscr{S}$, except P, which lie on at least one of those blocks. Similarly the *external structure* of $\mathscr{S}$ at P, written $\mathscr{S}^P$, is the set of all blocks not on P and the set of all points on at least one of those blocks. In the same way we define the internal and external structures $\mathscr{S}_x$ and $\mathscr{S}^x$, where x is a block: thus $\mathscr{S}_x$ consists of all the points on x and all the blocks ($\neq x$) which meet x, while $\mathscr{S}^x$ consists of all the points not on x and the set of all blocks which contain at least one point not on x. (Of course, incidence in one of these new structures is that induced by $\mathscr{S}$.) In general, even though $\mathscr{S}$ itself may have 'nice' properties, we cannot always say very much about these new structures. To illustrate we will look at two of our examples.

Example $\mathscr{A}$ is a uniform structure with five points and four blocks. However, $\mathscr{A}^3$ has four points and no blocks while $\mathscr{A}^4$ has points $\{1, 2, 3, 5\}$ and blocks $a_1 = \{1, 2, 3\}$, $a_2 = \{1, 2, 3\}$ (note that 5 is an isolated point). This small example shows $\mathscr{S}^P$ and $\mathscr{S}^Q$ need not be isomorphic if $P \neq Q$. It also shows that the 'nice' properties of $\mathscr{S}$ need not be inherited by these new structures.

Example $\mathscr{K}$ is a 3-(8, 4, 1). $\mathscr{K}^{k_1}$ has points $\{2, 4, 5, 6\}$. The blocks of $\mathscr{K}^{k_1}$ are

$$k_2 = \{2, 4\}, \quad k_3 = \{2, 5\}, \quad k_4 = \{4, 6\}, \quad k_5 = \{4, 5\}, \quad k_6 = \{5, 6\},$$
$$k_7 = \{2, 6\}, \quad k_8 = \{2, 6\}, \quad k_9 = \{2, 5\}, \quad k_{10} = \{4, 5\}, \quad k_{11} = \{4, 6\},$$
$$k_{12} = \{2, 4\}, \quad k_{13} = \{2, 4, 5, 6\}, \quad k_{14} = \{5, 6\}.$$

Thus $\mathscr{K}^{k_1}$ has 13 blocks. There are 12 with two points and each of these has multiplicity two, and k_{13} is a full block. If we totally reduce $\mathscr{K}^{k_1}$ we get a design with six blocks of size two. Since $6 = \binom{4}{2}$, this is the trivial design with $v = 4$, $k = 2$.

A similar analysis shows that $\mathscr{K}_{k_1}$ has four points and 12 blocks, since k_{13} contains no points of k_1 and hence is not in $\mathscr{K}_{k_1}$. The 12 blocks all have multiplicity two, and if we totally reduce $\mathscr{K}_{k_1}$ we will get again the trivial 2-(4, 2, 1).

$\mathscr{K}_8$ has points $\{1, 2, 3, 4, 5, 6, 7\}$. The blocks of $\mathscr{K}_8$ are

$$k_1 = \{1, 3, 7\}, \quad k_2 = \{1, 2, 4\}, \quad k_3 = \{2, 3, 5\}, \quad k_4 = \{3, 4, 6\},$$
$$k_5 = \{4, 5, 7\}, \quad k_6 = \{1, 5, 6\}, \quad k_7 = \{2, 6, 7\}.$$

(Note that since the blocks $k_8, \ldots, k_{14}$ of $\mathscr{K}$ do not contain the point 8 they are not blocks of $\mathscr{K}_8$.) So here we have an example where the new structure is again a design. The same will be true of $\mathscr{K}^8$.

Exercise 1.40. Show that $\mathscr{K}_8 \cong \mathscr{F}$ and that $\mathscr{K}^8 \cong \mathscr{J}$.

We now want to consider how certain properties of $\mathscr{S}$ are reflected in the internal and external structures. Notice that if $\mathscr{S}$ is a 2-structure, then $\mathscr{S}_P$ will contain all the points of $\mathscr{S}$ except P, and that if $\mathscr{S}$ is a design, so will $\mathscr{S}_P$ and $\mathscr{S}^P$ be designs.

(In the following theorems the reader should recall that $\lambda_0 = b$.)

Theorem 1.14. Let $\mathscr{S}$ be a structure and let P be a point of $\mathscr{S}$. Then

(a) if $\mathscr{S}$ is a t-structure with $t \geqslant 2$, $\mathscr{S}_P$ *is a* $(t-1)$-structure;

(b) if $\mathscr{S}$ is a uniform t-structure for (v,k,λ) with $t \geqslant 2$, $\mathscr{S}_P$ is a uniform $(t-1)$-structure for $(v-1,k-1,\lambda)$;

(c) if $\mathscr{S}$ is a t-(v,k,λ) with $t \geqslant 2$ then $\mathscr{S}_P$ is a $(t-1)-(v-1,k-1,\lambda)$.

Proof. First we note that (b) implies (c) and we have merely stated (c) as this is the form in which the theorem is most often used.

Any $(t-1)$-set of points in $\mathscr{S}_P$, together with P, form a t-set of points in $\mathscr{S}$. Thus if every t-set of points in $\mathscr{S}$ is on λ common blocks of $\mathscr{S}$, every $(t-1)$-set of points of $\mathscr{S}_P$ is on λ common blocks in $\mathscr{S}_P$. Clearly, $\mathscr{S}_P$ has $v-1$ points and, if each block of $\mathscr{S}$ contains k points, each block of $\mathscr{S}_P$ contains $(k-1)$ points. □

Theorem 1.15. Let $\mathscr{S}$ be a structure and let P be a point of $\mathscr{S}$. Then

(a) if $\mathscr{S}$ is a uniform t-structure for (v, k, λ) with $t \geqslant 2$ then $\mathscr{S}^P$ is a uniform $(t-1)$-structure for $(v-1, k, \lambda_{t-1} - \lambda)$;

(b) if $\mathscr{S}$ is a t-(v, k, λ) with $t \geqslant 2$ then $\mathscr{S}^P$ is a $(t-1)-(v-1, k, \lambda_{t-1} - \lambda)$.

Proof. We need only prove (a).

Any $(t-1)$-set of points in $\mathscr{S}^P$ are on λ_{t-1} blocks of $\mathscr{S}$. But, of these λ_{t-1} blocks, exactly λ will contain P (because P, together with the given $(t-1)$-set, forms a t-set of points in $\mathscr{S}$). Hence there are $\lambda_{t-1} - \lambda$ blocks of $\mathscr{S}^P$ containing the given $(t-1)$-set of points. □

The contraction or internal structure $\mathscr{S}_P$ is important, and is also called the *restriction* of $\mathscr{S}$ at P. We have already seen that the contraction of a t-design $(t \geqslant 2)$ at any point is a $(t-1)$-design. If $\mathscr{T}$ is a t-design and if $\mathscr{S}$ is a $(t-1)$-design such that $\mathscr{S} \cong \mathscr{T}_X$ for some point X of $\mathscr{T}$ then $\mathscr{T}$ is said to be an *extension* of $\mathscr{S}$. (Note that if X, Y are two points of an incidence structure $\mathscr{D}$ there is no reason for supposing $\mathscr{D}_X \cong \mathscr{D}_Y$. In fact we have already seen an example with $\mathscr{D}_X \not\cong \mathscr{D}_Y$, and, as we shall see later, there are many examples where $\mathscr{D}_X \not\cong \mathscr{D}_Y$.)

Exercise 1.41. (a) If $\mathscr{S}$ is a structure and α is an automorphism of $\mathscr{S}$ show that, for any point A of $\mathscr{S}$, $\mathscr{S}_A \cong \mathscr{S}_B$ and $\mathscr{S}^A \cong \mathscr{S}^B$, if $B = A^\alpha$.

(b) If $\mathscr{S}$ is a structure and A, B are distinct points of $\mathscr{S}$ show that $(\mathscr{S}_A)_B = (\mathscr{S}_B)_A$ and $(\mathscr{S}^A)^B = (\mathscr{S}^B)^A$.

Note that if $\mathscr{S}$ is a structure such that, for some point P of $\mathscr{S}$, $\mathscr{S}_P$ *is a* $(t-1)$-design, then $\mathscr{S}$ need not be a t-design and, consequently, need not be an extension of $\mathscr{S}_P$. As a simple illustration, consider the structure $\mathscr{S}$ whose points are $\{1, 2, \ldots, 8\}$ and whose blocks are blocks $k_1, \ldots, k_7$ of example $\mathscr{K}$ plus the extra block $\{1, 2, 3, 4, 5, 6\}$. Clearly, $\mathscr{S}$ is not uniform so it is certainly not a 3-design but $\mathscr{S}_8 = \mathscr{K}_8 \cong \mathscr{F}$ which is a 2-(7, 3, 1).

In fact, given any structure $\mathscr{S}$ it is always possible to construct a structure $\mathscr{T}$ with $\mathscr{T}_P \cong \mathscr{S}$ for some point P in $\mathscr{T}$. (We merely add P to the point set of $\mathscr{S}$, define $x \cup \{P\}$ to be a block of $\mathscr{T}$ for every block x of $\mathscr{S}$, and then make sure there are no further blocks of $\mathscr{T}$ containing P.)

Since the extension of a t-design is a $(t+1)$-design, a 'natural' way of constructing t-designs for large t suggests itself: namely, start with a 1-design and then repeatedly extend it. However, it is the exception rather than the rule for a t-design to have an extension (and, as we have already noted, there are no known non-trivial t-designs with $t \geqslant 7$). If we start with a trivial design then an extension exists but it is again trivial.

Exercise 1.42. Show that any extension of a trivial design is trivial.

There is one very simple criterion for the non-existence of an extension of a given t-(v, k, λ). It is given by computing the number of blocks which the extension would have to have and observing that, if this number is not an integer, then the extension cannot exist.

Theorem 1.16. Let $\mathscr{S}$ be a t-(v, k, λ) with b blocks. If $\mathscr{T}$ is an extension of $\mathscr{S}$ then $\mathscr{T}$ is a $(t+1)$-$(v+1, k+1, \lambda)$ with $b(v+1)/(k+1)$ blocks.

Proof. From Theorem 1.2

$$b = \lambda_0 = \lambda \frac{v(v-1)\cdots(v+1-t)}{k(k-1)\cdots(k+1-t)}.$$

By (c) of Theorem 1.14 $\mathcal{T}$ is a $(t+1)$-$(v+1, k+1, \lambda)$ and so, by Theorem 1.2 again, the number of blocks of $\mathcal{T}$ is

$$\lambda \frac{(v+1)v\cdots(v+1-t)}{(k+1)k\cdots(k+1-t)} = \frac{b(v+1)}{k+1}. \quad \square$$

Corollary 1.17. A necessary condition for a t-(v, k, λ) to have an extension is $k+1 \mid b(v+1)$.

Proof. The number of blocks of the extension is, by Theorem 1.16, $b(v+1)/(k+1)$ and this must be an integer. $\square$

Of course, if $(k+1) \mid b(v+1)$ there is still no guarantee that a t-(v, k, λ) with b blocks has an extension. We will be returning to this problem later.

The internal and external structures of a block are not so easy to deal with since, even if $\mathcal{S}$ is a 2-design, these structures need not even be designs. From our discussion of Example $\mathcal{K}$ we have already seen some of the many difficulties which may arise. For instance, if x is a block of $\mathcal{S}$ other than y, then (regarding x and y as point sets) $x \cap y$ is empty or is a block of $\mathcal{S}_y$. But even if $\mathcal{S}$ is a t-design for large t there is no reason why $x \cap y$ and $z \cap y$ should not both be non-empty but of different sizes, and then $\mathcal{S}_y$ is not even uniform. Similar considerations apply to $\mathcal{S}^y$. Furthermore, for given blocks x and y there may exist a third block w, say, such that $x \cap y$ and $w \cap y$ are non-empty and equal as point sets; then $\mathcal{S}_y$ will have repeated blocks. All of these possibilities arise and can be of considerable importance. We often call $\mathcal{S}^y$ the *residual* of $\mathcal{S}$ with respect to the block y. If a design $\mathcal{T}$ is isomorphic to a residual of a structure $\mathcal{S}$ then we say that $\mathcal{T}$ can be *embedded* in $\mathcal{S}$.

As an illustration of internal and external structures with respect to blocks we will consider projective and affine geometries. If $\mathcal{P} = \mathcal{P}_n(q)$ with $n \geq 2$ then the points of $\mathcal{P}$ have g-dimension 0 and the blocks have g-dimension $n-1$. For any distinct blocks x and y, $x \cap y$ is a subspace of g-dimension $n-2$ (see Grassman's Identity: Lemma 1.7). Thus $|x \cap y|$ is a constant for all x and y. This immediately says that $\mathcal{P}_x$ (and hence $\mathcal{P}^x$) is uniform. Furthermore, since $x \cap y \neq \phi$ for any x and y, $\mathcal{P}_x$ and $\mathcal{P}^x$ have no isolated elements.

Since $\mathcal{P}$ is a 2-$((q^{n+1}-1)/(q-1), (q^n-1)/(q-1), (q^{n-1}-1)/(q-1))$ we know that $\mathcal{P}^x$ has $(q^{n+1}-1)/(q-1)-(q^n-1)/(q-1)=q^n$ points. Further, since a subspace of g-dimension $n-2$ has $(q^{n-1}-1)/(q-1)$ points, each block of $\mathcal{P}^x$ has $(q^n-1)/(q-1)-(q^{n-1}-1)/(q-1)=q^{n-1}$ points. The only block of $\mathcal{P}^x$ which is not a block of $\mathcal{P}$ is x and so, for any pair of points A, B in $\mathcal{P}^x$, the

$(q^{n-1}-1)/(q-1)$ blocks of $\mathscr{P}$ through A and B are all blocks of $\mathscr{P}^x$. Thus $\mathscr{P}^x$ is certainly a 2-$(q^n, q^{n-1}, (q^{n-1}-1)/(q-1))$ structure. Is it a design? Clearly, $\mathscr{P}^x$ is a design if and only if it has no repeated blocks. If y and z are two distinct blocks of $\mathscr{P}$ then $|y \cap z| = (q^{n-1}-1)/(q-1)$ in $\mathscr{P}$. But, by Grassman's Identity, $x \cap y \cap z$ is a subspace of g-dimension either $n-1$ or $n-2$. So $|x \cap y \cap z| = (q^{n-2}-1)/(q-1)$ or $(q^{n-1}-1)/(q-1)$. But this means $|y \cap z \backslash x \cap y \cap z| = q^{n-2}$ or 0. However, $|y \cap z \backslash x \cap y \cap z|$ is the size of the intersection of y and z when considered as blocks of $\mathscr{P}^x$. Thus, since this number is never q^{n-1} (the size of a block in $\mathscr{P}^x$), y and z can never be incident with the same set of points in $\mathscr{P}^x$, i.e. $\mathscr{P}^x$ cannot have a repeated block. This establishes that $\mathscr{P}^x$ is a 2-$(q^n, q^{n-1}, (q^{n-1}-1)/(q-1))$.

We have now shown that $\mathscr{P}^x$ is a 2-design with the same parameters as $\mathscr{A}_n(q)$. Since $\mathscr{P}^x$ is obtained by deleting a block x from $\mathscr{P}$, and since an affine geometry is obtained from a projective geometry by deleting a hyperplane, it is reasonable to guess that $\mathscr{P}^x \cong \mathscr{A}_n(q)$. This is, in fact, true, but we must stress again that we have not proved it. We leave the proof as an exercise for the reader with sufficient knowledge of algebra and geometry.

Exercise 1.43.* Show that, for any hyperplane x of $\mathscr{P}_n(q)$ with $n \geqslant 2$, $\mathscr{P}_n(q))^x \cong \mathscr{A}_n(q)$.

Since $x \cap y$ is a subspace of g-dimension $n-2$, the number of blocks through $x \cap y$ is the number of subspaces of g-dimension $n-1$ through a fixed subspace of g-dimension $n-2$. But, by Lemma 1.6, this is $q+1$. Thus there are $q-1$ (and $q-1 \geqslant 1$) blocks z such that $x \cap y = x \cap z$ but $y \neq z$. So $\mathscr{P}_x$ always has repeated blocks. In fact, every block has multiplicity q. Similar calculations to these just carried out for $\mathscr{P}^x$ now give that $\mathscr{P}_x$ is a 2-$[(q^n-1)/(q-1), (q^{n-1}-1)/(q-1), ((q^{n-1}-1)/(q-1))-1]$ structure with every block having multiplicity q. (Note that the number of blocks of $\mathscr{P}_x$ through two points is only $(q^{n-1}/(q-1))-1$ because x is not a block of $\mathscr{P}_x$.) Thus $\mathscr{P}_x$ is a 2-$((q^n-1)/(q-1), (q^{n-1}-1)/(q-1), q(q^{n-2}-1)/(q-1))$ structure. But this means, by Exercise 1.9, that $\mathscr{P}_x/R$ is a 2-$((q^n-1)/(q-1), (q^{n-1}-1)/(q-1), (q^{n-2}-1)/(q-1))$, so that, although $\mathscr{P}_x$ itself is not a 2-design, $\mathscr{P}_x$ reduced is.

Exercise 1.44. Show that, for any hyperplane x of $\mathscr{P}_n(q)$ with $n \geqslant 2$, $(\mathscr{P}_n(q))_x$ reduced is isomorphic to $\mathscr{P}_{n-1}(q)$.

As with $\mathscr{P}_n(q)$, if $\mathscr{A} = \mathscr{A}_n(q)$ then $\mathscr{A}_x$ may have repeated blocks and need not be a design. However, $\mathscr{A}_x$ reduced will, of course, be a design and the interested reader should verify that $\mathscr{A}_x$ reduced is isomorphic to $\mathscr{A}_{n-1}(q)$.

If $\mathscr{S}$ is any structure and x is a block of $\mathscr{S}$ then it is frequently true that it is $\mathscr{S}_x$ or $\mathscr{S}^x$ reduced, rather than $\mathscr{S}_x$ or $\mathscr{S}^x$ themselves, which have interesting properties. We shall frequently abuse our notation and write $\mathscr{S}_x$ or $\mathscr{S}^x$ for $\mathscr{S}_x$ or $\mathscr{S}^x$ reduced but only, of course, when it is clear that we are doing so. We shall, for instance, say that if $\mathscr{P}=\mathscr{P}_n(q)$ or $\mathscr{A}=\mathscr{A}_n(q)$ then, for any block x of the respective design, $\mathscr{P}_x\cong\mathscr{P}_{n-1}(q)$ and $\mathscr{A}_x\cong\mathscr{A}_{n-1}(q)$.

These internal and external structures of a structure $\mathscr{S}$ all consist of subsets of the points and blocks of $\mathscr{S}$ and have the same incidence as $\mathscr{S}$. In general, if $\mathscr{S}$ is any structure then $\mathscr{C}$ is a *substructure* of $\mathscr{S}$ if $\mathscr{C}$ is a subset of the points and blocks of $\mathscr{S}$ such that two elements of $\mathscr{C}$ are incident in $\mathscr{C}$ if and only if they are incident in $\mathscr{S}$. A *subdesign* of a design is defined analogously. Clearly, as in the restriction at a point, if x is a block of a substructure $\mathscr{C}$ of $\mathscr{S}$ the number of points of $\mathscr{C}$ on x may be fewer than the number of points of $\mathscr{S}$ on it. We say that a substructure $\mathscr{C}$ of $\mathscr{S}$ is *complete* if, for every block x in $\mathscr{C}$, every point of $\mathscr{S}$ on x is also in $\mathscr{C}$.

If $\mathscr{S}$ is a structure and if we construct $\mathscr{S}/R$ by letting one block represent an equivalence class, then $\mathscr{S}/R$ is a complete substructure of $\mathscr{S}$.

The observant reader may have noticed that we have not considered the possibility of two distinct points being incident with the same block sets, and have not defined a repeated point, etc. This is because we shall mainly be concerned with 2-designs (or at least uniform 2-structures) and here the situation cannot arise.

Lemma 1.18. If $\mathscr{S}$ is a 2-structure for (v, k, λ) with $v>k$ then no two distinct points are incident with the same set of blocks.

Proof. If two points in $\mathscr{S}$ are incident with the same set of blocks then the number of blocks through two points of $\mathscr{S}$ (which is a constant since $\mathscr{S}$ is a 2-structure), must be the same as the number of blocks through one point, i.e. $\lambda=r$. But, by Theorem 1.2, $\lambda=r$ implies $v=k$. □

If $\mathscr{S}$ is a uniform and regular structure then we have already seen that $\mathscr{S}^{\mathrm{T}}$, the dual of $\mathscr{S}$, has the same properties. But a repeated block in $\mathscr{S}$ 'is' a repeated point in $\mathscr{S}^{\mathrm{T}}$ (see Exercise 1.36). So whenever one studies uniform 1-structures which are not 2-structures it no longer makes sense to regard the points as more 'basic' objects than the blocks. But for various reasons (e.g. in the next section we are mainly interested in uniform 2-structures, at other points in the book the designs we study have close connections with geometry, etc.) it is often convenient to consider the points as being in fact the basic objects.

1.5 The incidence matrix

Suppose $\mathcal{S}$ is a 2-structure for λ. Let $P_1, \ldots, P_v$ be the points of $\mathcal{S}$, let $x_1, \ldots, x_b$ be the blocks and let $r_i = |P_i|$ (the number of blocks on P_i). If A is the incidence matrix of $\mathcal{S}$ with this labelling then:

Lemma 1.19. $AA^{\mathrm{T}} = N + \lambda J_v$, where N is a diagonal v by v matrix with $n_i = r_i - \lambda$ in the ith position on the diagonal.

Proof. The (i, j)th entry of AA^{T} is the inner product of the ith and jth rows of A. If $i \neq j$, this inner product counts a 1 whenever there is a block on both P_i and P_j and a 0 otherwise; so the entry is λ. When $i = j$, the inner product of row i with itself counts the number of blocks on P_i, so the entry is $|P_i| = r_i = n_i + \lambda$. □

We shall only be interested in 2-structures for which each of these numbers n_i is non-zero. For this reason we define a structure to be *proper* if there is at least one block in it which contains more than one point but does not contain all points.

Lemma 1.20. If $\mathcal{S}$ is a proper 2-structure then $n_i > 0$ for all i.

Proof. Clearly, $n_i \geqslant 0$ for all i in any 2-structure. Suppose $n_i = 0$ for some i. Then $r_i = \lambda$ and there are only λ blocks on P_i. If P_j is any other point then, since $\mathcal{S}$ is a 2-structure, P_j is on λ blocks through P_i, i.e. P_j is on every block through P_i. Thus every block through P_i contains every other point of $\mathcal{S}$. But this means that for any pair of points P_j, P_l $(j \neq l)$ the λ blocks through them all contain P_i and, consequently, all contain every point. Hence each block which has at least two points contains every point. This is impossible since $\mathcal{S}$ is proper. □

Lemma 1.21. If $\mathscr{S}$ is a proper 2-structure for λ and A is as above, then

$$\det(AA^{\mathrm{T}}) = \prod_{i=1}^{v} n_i\left(1+\lambda\sum_{j=1}^{v}\frac{1}{n_j}\right).$$

Proof. To compute $\det(AA^{\mathrm{T}})$ we first subtract the first row of AA^{T} $(=N+\lambda J_v)$ from every other row to obtain

$$\begin{bmatrix} n_1+\lambda & \lambda \cdots & & \lambda \\ -n_1 & n_2 & 0 & \\ -n_1 & & n_3 & \\ \vdots & & & \ddots \\ & & 0 & \\ -n_1 & & & n_v \end{bmatrix}$$

For $i=2, 3, \ldots, v$, we multiply the ith column by n_1/n_i ($n_i \neq 0$ by Lemma 1.20) and add to the first column. This gives

$$\begin{bmatrix} x & \lambda & \lambda \cdots & \lambda \\ & n_2 & & \\ & & n_3 & \\ & & & \ddots \\ & 0 & & \\ & & & n_v \end{bmatrix}$$

where

$$x = n_1+\lambda+\lambda\left[\frac{n_1}{n_2}+\frac{n_1}{n_3}+\cdots+\frac{n_1}{n_n}\right].$$

So

$$\det(AA^{\mathrm{T}}) = \prod_{i=2}^{v} n_i\cdot x = \prod_{i=1}^{v} n_i\cdot\left(1+\lambda\sum_{j=1}^{v}\frac{1}{n_j}\right). \quad \square$$

Corollary 1.22. If $\mathscr{S}$ is a proper 2-structure with v points then rank $\mathscr{S}$ is v.

Proof. By Lemmas 1.20 and 1.21, $\det(AA^{\mathrm{T}}) \neq 0$. Thus rank $AA^{\mathrm{T}} = v$. But, clearly, rank $AA^{\mathrm{T}} \leqslant$ rank A which means that rank $A \geqslant v$. However, since A

only has v rows, rank $A \leqslant v$. Thus rank $A = v$, and, since by definition rank $\mathscr{S} =$ rank A, the corollary is proved. □

Corollary 1.23. If $\mathscr{S}$ is a proper 2-structure with v points and b blocks then $b \geqslant v$.

Proof. Since A only has b columns, rank $A \leqslant b$. Corollary 1.23 now follows from Corollary 1.22. □

This last corollary is a slight generalisation of the famous Fisher's Inequality.

Corollary 1.24 (Fisher's Inequality). If $\mathscr{S}$ is a 2-(v, k, λ), with $v > k$, then $b \geqslant v$.

Proof. The condition $v > k$ implies that $\mathscr{S}$ is proper. (Note that $\lambda > 0$ implies $k \geqslant 2$.) □

Exercise 1.45. Show that if $\mathscr{S}$ is a proper 2-structure then, for any pair of distinct points A and B, $\langle A \rangle \neq \langle B \rangle$.

Corollary 1.25. A square proper 2-structure is reduced.

Proof. If a structure $\mathscr{S}$ has repeated blocks then any incidence matrix for $\mathscr{S}$ will have two identical columns. Thus rank $\mathscr{S} < b$. But if $\mathscr{S}$ is a square proper 2-structure, rank $\mathscr{S} = v$ (by Corollary 1.22) and $v = b$ (since $\mathscr{S}$ is square). This contradiction establishes the corollary. □

In view of Corollary 1.25, we know that a square uniform 2-structure is either a design or has $k = v$. We now define a *symmetric* design to be a square 2-design for (v, k, λ) with $v > k$. For symmetric designs we have $k = r$ (see Corollary 1.4(b)) and so we know, from Exercise 1.2, that, for any incidence matrix A, $AJ_v = J_v A = kJ_v$. Also since A is a v by v matrix of rank v, A^{-1} exists.

Exercise 1.46. Verify that $(1/n)A^{\mathrm{T}} - (\lambda/nk)J_v = A^{-1}$.

Symmetric designs have many interesting properties and we shall discuss them in considerable detail in Chapter 2. Meanwhile we prove a useful theorem to illustrate their importance.

Theorem 1.26. If $\mathscr{S}$ is a proper uniform 2-structure for (v, k, λ) then the following are equivalent:

(a) $b = v$;
(b) $k = r$;
(c) $\mathscr{S}^{\mathrm{T}}$ is a 2-structure;
(d) $\mathscr{S}^{\mathrm{T}}$ is a 2-structure for λ;
(e) both $\mathscr{S}$ and $\mathscr{S}^{\mathrm{T}}$ are symmetric designs for (v, k, λ).

Proof. Clearly, (e) implies all the other conditions, and (d) implies (c). By Corollary 1.4(b), (a) and (b) are equivalent. By Corollary 1.25, (a) and (d) together imply (e). We shall show that (c) implies (a) and that (a) implies (d), which finishes the proof.

Since $\mathscr{S}$ is a proper uniform 2-structure it is also regular (see Theorem 1.2) and so, by Corollary 1.9, $\mathscr{S}^{\mathrm{T}}$ is both uniform and regular. The number of points and blocks of $\mathscr{S}^{\mathrm{T}}$ are b and v respectively, so, by Corollary 1.23, if $\mathscr{S}^{\mathrm{T}}$ is a 2-structure either $v \geqslant b$ or $\mathscr{S}^{\mathrm{T}}$ is improper. But, since $\mathscr{S}$ is a proper 2-structure, Corollary 1.23 also gives $b \geqslant v$. Thus either (c) implies (a) or $\mathscr{S}^{\mathrm{T}}$ is improper. Suppose $\mathscr{S}^{\mathrm{T}}$ is improper. Then each block of $\mathscr{S}^{\mathrm{T}}$ contains either 1 or all points and thus, dualising, in $\mathscr{S}$ we have $r = 1$ or $r = b$. But since $\mathscr{S}$ is proper, Lemma 1.20 shows that $r > \lambda$ which, because $\lambda > 0$, means $r \neq 1$. Similarly, since $\mathscr{S}$ is proper, we cannot have the situation where every block contains every point. Thus $r \neq b$ and we have shown that (c) implies (a).

To show that (a) implies (d) we consider the equation of Lemma 1.19. Since $\mathscr{S}$ is regular $n_i = n = r - \lambda$ for each i, so $AA^{\mathrm{T}} = nI_v + \lambda J_v$ (where I_v is the v by v identity matrix). Multiply both sides on the right by A which gives

$$
\begin{aligned}
(AA^{\mathrm{T}})A &= (nI_v + \lambda J_v)A \\
&= nI_vA + \lambda J_vA \\
&= nA + \lambda kA \quad \text{(by Exercise 1.2)} \\
&= nA + \lambda AJ_v \quad \text{(since } \mathscr{S} \text{ is a symmetric 2-structure it is regular and uniform with } k = r\text{).}
\end{aligned}
$$

So $(AA^{\mathrm{T}})A = A(nI + \lambda J_v)$. But $(AA^{\mathrm{T}})A = A(A^{\mathrm{T}}A)$, so $A(A^{\mathrm{T}}A) = A(nI + \lambda J_v)$ and, since A is non-singular, $A^{\mathrm{T}}A = nI + \lambda J_v$. But this implies that the inner product of two distinct columns of A is λ, i.e. that two distinct blocks of $\mathscr{S}$ intersect in λ points. Thus $\mathscr{S}^{\mathrm{T}}$ is a (proper) 2-structure for λ. □

Thus symmetric designs are precisely those proper uniform 2-structures whose duals are also 2-structures or, equivalently, those proper uniform 2-structures such that every pair of distinct blocks intersect in a constant number of points. (And then this constant is forced to be λ.) By Theorem 1.2 any square uniform t-structure with $t>2$ is also a square uniform 2-structure, so the results of this discussion all apply to proper uniform t-structures with $t\geqslant 3$. We now show that any square uniform t-structure with $t\geqslant 3$ is trivial, so the only really interesting square uniform t-structures with $t\geqslant 2$ are our symmetric designs.

Theorem 1.27. Let $\mathscr{S}$ be a square proper uniform t-structure with $t\geqslant 3$. Then the blocks of $\mathscr{S}$ are the $(v-1)$-subsets of the v points of $\mathscr{S}$, i.e. $\mathscr{S}$ is trivial.

Proof. By Theorem 1.2 we may assume $t=3$.

If P is any point of $\mathscr{S}$ then, by Theorem 1.14, $\mathscr{S}_p$ is a uniform 2-structure with $v-1$ points, r blocks and $k-1$ points on each block. If $\mathscr{S}_p$ is improper then $k-1=1$ or $v-1$. But $k-1=1$ implies $k=2$, which is impossible since $\mathscr{S}$ is a 3-structure, while $k-1=v-1$ implies $k=v$, which is impossible since $\mathscr{S}$ is proper. Thus $\mathscr{S}_p$ is proper and, by Fisher's Inequality, $r\geqslant v-1$. But, since $\mathscr{S}$ is square, $k=r$ (see Corollary 1.4(b)). Thus $k\geqslant v-1$. However, since $\mathscr{S}$ is proper, we know $k\neq v$ and so $k=v-1$. Finally since $b=v$ we know that every subset of size $v-1$ must be a block. □

Note, of course, that trivial square proper t-$(v, v-1, v-t)$ designs exist for all v and t with $t\leqslant v-1$.

Exercise 1.47.* Find all square proper t-structures with $t>2$.

Exercise 1.48.* Find all square t-structures with $t>2$.

There are so many examples of square 1-structures (even uniform and proper) that one cannot say much about them. However, we shall consider some of the more interesting 1-designs in Chapter 7. As we have already mentioned, we discuss symmetric designs in considerable detail in Chapter 2. There are many unsolved problems about square proper 2-structures which are not designs (i.e. not uniform), but most of the known partial results are difficult.

To conclude this section we will establish a number of interesting combinatorial properties of 2-structures. As we have already seen (Theorem 1.2) any uniform 2-structure is regular. Our first result concerns regular 2-structures.

Theorem 1.28. Let $\mathscr{S}$ be a regular 2-structure for λ and let X, Y, Z be any three distinct points of $\mathscr{S}$. If c_0 is the number of blocks which contain none of X, Y and Z while c_{XYZ} is the number containing all three, then c_0+c_{XYZ} is a constant independent of X, Y and Z.

Proof. Let c_{XY} be the number of blocks containing X, Y but not Z, c_X be the number of blocks containing X but not Y or Z, etc. Then, since the number of blocks on X and Y is $c_{XY}+c_{XYZ}$, we have $c_{XY}+c_{XYZ}=c_{YZ}+c_{XYZ}=c_{ZX}+c_{XYZ}=\lambda$. Since $\mathscr{S}$ is also regular and since the number of blocks through X is $c_X+c_{XY}+c_{XZ}+c_{XYZ}$ we also have $c_X+c_{XY}+c_{XZ}+c_{XYZ}=r$, with two similar expressions involving c_Y and c_Z. Thus $c_X=c_Y=c_Z=r-2\lambda+c_{XYZ}$. Finally $c_0+c_X+c_Y+c_Z+c_{XY}+c_{YZ}+c_{ZX}+c_{XYZ}=b$, since the left-hand side is merely the sum of all the blocks of $\mathscr{S}$.

Simplifying these expressions we get $c_0+c_{XYZ}=b-3r+3\lambda$. □

Exercise 1.49. If $\mathscr{S}$ is a uniform 2-structure for (v, k, λ), show that c_0+c_{XYZ} of Theorem 1.28 is equal to λ if and only if $v=k$ or $v=2k+1$.

If $\mathscr{S}$ is a regular 2-structure and P is any point of $\mathscr{S}$ then we define another related structure $\mathscr{S}^*(P)$ with the same point set as $\mathscr{S}$. For any y, a block of $\mathscr{S}$, we define a block y^* of $\mathscr{S}^*(P)$ as follows: if $P\in y$ then y^* contains precisely these points not on y in $\mathscr{S}$, while if $P\notin y$ then y^* contains the points of y in $\mathscr{S}$ and P itself.

Lemma 1.29. $\mathscr{S}^*(P)$ is a 2-structure for $\lambda^*=r-\lambda$.

Proof. Consider two points P, X of $\mathscr{S}^*(P)$ with $P\neq X$. A block y^* contains them both if and only if y is a block of $\mathscr{S}$ containing X but not P. Since there are r blocks through X and λ of these contain P, there are $r-\lambda$ blocks of $\mathscr{S}^*(P)$ containing P and X.

Now consider two points X, Y of $\mathscr{S}^*(P)$ with $X \neq Y$ and neither equal to P. There are now two kinds of blocks of $\mathscr{S}^*(P)$ containing X and Y: (i) blocks y^* where y contains X and Y but not P and (ii) blocks y^* where y contains P but neither X nor Y. So to count the blocks of $\mathscr{S}^*(P)$ through X and Y we must (in the notation of Theorem 1.27 with P replacing Z) evaluate $c_{XY} - c_P$. But $c_{XY} = \lambda - c_{XYP}$ and $c_P = r - 2\lambda + c_{XYP}$, so $c_{XY} + c_P = r - \lambda$. □

Exercise 1.50. If $\mathscr{S}$ is a symmetric design show that, for any point P of $\mathscr{S}$, $\mathscr{S}^*(P)$ is a square proper 2-structure which is not uniform. What are the block sizes of $\mathscr{S}^*(P)$?

Although regular 2-structures are, in general, not uniform the extra condition of being square is sufficient to imply uniformity.

Lemma 1.30. Let $\mathscr{S}$ be a square regular 2-structure for λ, with r blocks on a point. Then $\mathscr{S}$ is uniform with block size r.

Proof. Let $\mathscr{S}$ have v points and, as earlier, put $n = r - \lambda$.

If $n = 0$ then every block contains every point so $\mathscr{S}$ is improper with block size v. But $\mathscr{S}$ is square, so $\mathscr{S}$ has v blocks and, since each contains every point, $r = v$.

If $n > 0$, then, by Corollary 1.22, $\mathscr{S}$ has rank v. If A is the incidence matrix from a labelling $P_1, \ldots, P_v$ of the points and $y_1, \ldots, y_v$ of the blocks then $AA^{\mathrm{T}} = nI_v + \lambda J_v$ (see Lemma 1.19). In order to solve Exercise 1.46 it was only really necessary to know that $AJ_v = rJ_v$, i.e. that $\mathscr{S}$ was regular. Thus a similar argument to the solution of Exercise 1.46 gives $A^{-1} = (1/n)A^{\mathrm{T}} - (\lambda/nr)J_v$. Multiplying on the right by A gives $I_v = A^{-1}A = (1/n)A^{\mathrm{T}}A - (\lambda/nr)J_vA$ or, on rearranging,

$$A^{\mathrm{T}}A = nI_v + \frac{\lambda}{r} J_vA.$$

The diagonal entries on the left-hand side are the block sizes $k_1, k_2, \ldots, k_v$, where $k_i = |y_i|$. The diagonal entries of J_vA are also $k_1, k_2, \ldots, k_v$ so, comparing each side, we have $k_i = n + (\lambda k_i/r)$ for each i. Thus $k_i(1 - (\lambda/r)) = n$ or $k_i = rn/(r - \lambda) = r$ for all i. □

In Exercise 1.50 we showed how to construct a square proper 2-structure which is not uniform. In fact, the situation in that exercise is almost characteristic of such structures and it is often conjectured that those examples are the only square proper 2-structures which are not designs. Our next theorem is a possible step towards establishing the conjecture.

Theorem 1.31. Let $\mathscr{S}$ be a square proper 2-structure which is not a design. Then there are exactly two block sizes in $\mathscr{S}$ and their sum is $v+1$.

Proof. Let $\mathscr{S}$ be a 2-structure for λ. If $P_1, \ldots, P_v$ are the points and $y_1, \ldots, y_v$ are the blocks, let A be the corresponding incidence matrix, let $r_i = |P_i|$ and let $k_i = |y_i|$, $i = 1, \ldots, v$. If $n_i = r_i - \lambda$ (so, by Lemma 1.20, $n_i > 0$ for all i) let N be the diagonal v by v matrix with the n_i as diagonal entries. Then

$$AA^{\mathrm{T}} = N + \lambda J_v \tag{1}$$

(by Lemma 1.19).

Each entry in the ith row of AJ_v is the sum of the entries in the ith row of A, i.e. r_i. Thus

$$AJ_v = (N + \lambda I_v)J_v. \tag{2}$$

We now observe that

$$(N + \lambda J_v)^{-1} = N^{-1} - cN^{-1}J_vN^{-1} \tag{3}$$

where

$$c = \lambda \bigg/ \left(1 + \lambda \sum_{i=1}^{v} n_i^{-1}\right). \tag{4}$$

In order to see (3) and (4) note that $J_vN^{-1}J_vN^{-1}$ is $(\sum_{i=1}^{v} 1/n_i)J_vN^{-1}$. Now, from (1), $AA^{\mathrm{T}}(N + \lambda J_v)^{-1} = I_v$ so, multiplying on the left by A^{-1}, we have $A^{\mathrm{T}}(N + \lambda J_v)^{-1} = A^{-1}$, and now, multiplying on the right by A gives

$$A^{\mathrm{T}}(N + \lambda J_v)^{-1}A = I_v. \tag{5}$$

Using (3), we multiply (5) out, to obtain

$$A^{\mathrm{T}}N^{-1}A = I + cA^{\mathrm{T}}N^{-1}J_vN^{-1}A. \tag{6}$$

Now the entry in the (i, i)-position in (6) gives us

$$\sum_{P_j \in y_i} \frac{1}{n_j} = 1 + c\left[\sum_{P_j \in y_i} \frac{1}{n_j}\right]^2.$$

Writing $s_i = \sum_{P_j \in y_i} 1/n_j$, this is

$$s_i = 1 + cs_i^2. \tag{7}$$

But if we multiply (5) on the right by J_v, we find

$$A^{\mathrm{T}}(J_v - c(v-1)N^{-1}J_v) = J_v, \tag{8}$$

and the entry in row i on the left in (8) is $k_i - c(v-1)s_i$, so

$$k_i - c(v-1)s_i = 1, \quad \text{or} \quad s_i = \frac{k_i - 1}{c(v-1)}. \tag{9}$$

If we now use (9) to substitute in (7) we see that k_i must satisfy the following quadratic

$$(k_i - 1)^2 - (v-1)(k_i - 1) + c(v-1)^2 = 0. \tag{10}$$

Thus there are just two possible values for $k_i - 1$ and their sum is $v-1$. In other words, there are two possible values for k_i and their sum is $v+1$. □

Problem 1.2. Find all square proper 2-structures. (Start by trying to prove that they are all given by the examples in Exercise 1.50.)

The incidence matrix permits us to prove other results. For instance, suppose $\mathscr{S}$ is a square 2-structure for λ, with v points, and σ is a polarity of $\mathscr{S}$. If the points of $\mathscr{S}$ are $P_1, P_2, \ldots, P_v$, we can label the blocks $y_1 = P_1^\sigma$, $y_2 = P_2^\sigma, \ldots,$ $y_v = P_v^\sigma$, and the corresponding incidence matrix A will satisfy $A = A^{\mathrm{T}}$, since P_i is on y_j if and only if P_i^σ is on y_j^σ, i.e. if and only if P_j is on y_i. Then the equation of Lemma 1.19 becomes

$$A^2 = N + \lambda J_v.$$

Now, since $\det(A^2 - xI) = \det[(A - \sqrt{x}I)(A + \sqrt{x}I)] = \det(A - \sqrt{x}I) \cdot \det(A + \sqrt{x}I)$, we have:

Lemma 1.32. If A is the incidence matrix, as above, of the square 2-structure $\mathscr{S}$, possessing a polarity σ, then the eigenvalues of A are square-roots of the eigenvalues of $N + \lambda J_v$.

Proof. If a is an eigenvalue of A^2, then one of $\det(A - \sqrt{a}I)$ and $\det(A + \sqrt{a}I)$ is zero, so one of $\pm\sqrt{a}$ is an eigenvalue of A. □

But it is easy to compute the eigenvalues of $N + \lambda J_v$.

Lemma 1.33. Let $\mathscr{S}$ be a symmetric design for (v, k, λ). Then the eigenvalues of AA^{T} are k^2, with multiplicity 1, and $n = k - \lambda$, with multiplicity $v-1$.

Proof. In this case, $AA^{\mathrm{T}}=nI_v+\lambda J_v$. Clearly, $(1, 1, 1, \ldots, 1)(AA^{\mathrm{T}})=(n+\lambda v)(1, 1, \ldots, 1)$, so $n+\lambda v$ is an eigenvalue. But $n+\lambda v=k-\lambda+\lambda v=k+\lambda(v-1)=k+k(k-1)=k^2$ since, by Corollary 1.4, $b=v$ is equivalent to $\lambda(v-1)=k(k-1)$. We now note that each of the $v-1$ linearly independent vectors $(0, 0, \ldots, 0, 1, -1, 0, \ldots, 0)$ is an eigenvector of $nI_v+\lambda J_v$, with eigenvalue n (this is because each of these vectors when multiplied by J_v gives the all zero vector). Since we have now exhibited v linearly independent eigenvectors and AA^{T} is a v by v matrix, we have found all the eigenvalues. □

If σ is a polarity of a square structure $\mathscr{S}$ we define an *absolute element* of σ to be one which is incident with its image. If A is the incidence matrix of $\mathscr{S}$ associated with σ as above, then P_i is an absolute point if and only if the (i, i)-entry of A is 1. Thus

Lemma 1.34. If A is an incidence matrix of a square structure $\mathscr{S}$ corresponding to a polarity σ, then the number of absolute points of σ is tr (A), the trace of A. □

But the trace of a matrix is the sum of its eigenvalues. So

Lemma 1.35. Let $\mathscr{S}$ be a symmetric design for (v, k, λ) with a polarity σ. Then the number of absolute points of σ is of the form $k+s\sqrt{n}-(v-1-s)\sqrt{n}$, where s is an integer satisfying $\frac{1}{2}(v-1-(k/\sqrt{n}))\leqslant s\leqslant v-1$.

Proof. From Lemmas 1.32 and 1.33 we know that A^2 has eigenvalues k^2 and n so that A can only have k, $-k$, $\sqrt{n}$ or $-\sqrt{n}$ for its eigenvalues. Clearly, $(1, 1, 1, \ldots, 1)$ is an eigenvector of A with eigenvalue k. So, as k^2 only has multiplicity 1 as an eigenvalue of A^2, A has k as an eigenvalue with multiplicity 1 and $-k$ is not an eigenvalue of A. If $\sqrt{n}$ is an eigenvalue with multiplicity s then, since n has multiplicity $v-1$ as an eigenvalue of A^2, $-\sqrt{n}$ must have multiplicity $v-1-s$. Hence the sum of the eigenvalues is $k+s\sqrt{n}-(v-1-s)\sqrt{n}$.

Clearly, the number of absolute points must be at least zero. Thus $k+s\sqrt{n}-(v-1-s)\sqrt{n}\geqslant 0$ or $2s\geqslant v-1-(k/\sqrt{n})$. □

Corollary 1.36. If n is not a square, then v must be odd and σ has exactly k absolute points.

Proof. Clearly, tr (A) must be an integer. So if $\sqrt{n}$ is not integral $s\sqrt{n}-(v-1-s)\sqrt{n}=0$, i.e. $v=2s+1$. Thus v is odd and σ has k absolute points. □

Corollary 1.37. If σ has no absolute points then n is a square and $\sqrt{n}$ divides λ.

Proof. If $k+s\sqrt{n}-(v-1-s)\sqrt{n}=0$ then, clearly $\sqrt{n}$ is an integer and $k=(v-1-2s)\sqrt{n}$. Thus $\sqrt{n}$ divides k. But $k=n+\lambda$ and hence $\sqrt{n}$ divides λ. □

As we shall see later, the situations described in these last two corollaries can both occur.

We now use the incidence matrix to obtain another result, which is part of a theorem to be exploited more fully in Chapter 2. Lemma 1.20 yields:

Theorem 1.38. If $\mathscr{S}$ is a proper square 2-structure for λ, with v points, then

$$\prod_{i=1}^{v} n_i\left(1+\lambda\sum_{i=1}^{v}\frac{1}{n_i}\right)$$

is the square of an integer.

Proof. Since $\det(AA^{\mathrm{T}})=\det A\cdot\det A^{\mathrm{T}}=(\det A)^2$ the result follows. (Note that since A only has integer entries, $\det(A)$ must be an integer.) □

Corollary 1.39. If $\mathscr{S}$ is a symmetric design for (v, k, λ), and if v is even, then $n=k-\lambda$ is a square.

Proof. In this case $n_i=n$ for each i. So the expression in Theorem 1.38 becomes:

$$\prod_{i=1}^{v} n_i\left(1+\lambda\prod_{i=1}^{v}\frac{1}{n_i}\right)=n^v\left(1+\lambda\frac{v}{n}\right)=n^{v-1}(n+\lambda v)$$
$$=n^{v-1}k^2.$$

This number has to be a square, so n^{v-1} is a square. But if v is even then, since $v-1$ is then odd, n^{v-1} is a square only if n is itself a square. □

1.6 Block's Lemma and the Orbit Theorem

In this section we prove an elementary lemma about matrices but then deduce from it some surprisingly strong results about various types of structures and their automorphism groups. Let A be a v by b matrix over a field K. A *decomposition* of A is any partition $\mathscr{P}_1,\ldots,\mathscr{P}_{v_1}$ of the rows of A and a partition $\mathscr{X}_1,\ldots,\mathscr{X}_{b_1}$ of the columns of A. If $|\mathscr{P}_i|=p_i$ and $|\mathscr{X}_j|=x_j$ then the p_i by x_j matrices M_{ij}, which consist of all the entries in the rows of $\mathscr{P}_i$ and the columns of x_j are the *decomposition matrices* of the decomposition. Obviously an arbitrary decomposition is much too general a concept to be useful, and we shall usually be only interested in decompositions which satisfy some extra conditions. If, for each i and j, the sum of the entries of each row of M_{ij} is a constant, which we will denote by r_{ij}, then we say that the decomposition is *row-tactical*. Similarly we say the decomposition is *column-tactical* if the sum of each column of M_{ij} is a constant c_{ij}, and that it is *tactical* if it is both row- and column-tactical. If a decomposition is row-tactical then we define the v_1 by b_1 matrix $R=(r_{ij})$ and if it is column-tactical we define $C=(c_{ij})$.

Theorem 1.40 (Block's Lemma). Suppose that A is a v by b matrix with a decomposition $\mathscr{P}_1,\ldots,\mathscr{P}_{v_1}$ of the rows and $\mathscr{X}_1,\mathscr{X}_2,\ldots,\mathscr{X}_{b_1}$ of the columns. For any matrix, X, let $\rho(X)$ denote the rank of X; then

(a) if the decomposition is row-tactical then $\rho(A)-\rho(R)\leqslant b-b_1$;

(b) if the decomposition is column-tactical then $\rho(A)-\rho(C)\leqslant v-v_1$.

Proof. We will first prove (b). Suppose the decomposition is column-tactical. If $\mathscr{R}$ is a set of $\rho(A)$ linearly independent rows of A then the remaining $v-\rho(A)$ rows of A fall in, at most, $v-\rho(A)$ different point classes $\mathscr{P}_i$. Thus at least $v_1-(v-\rho(A))$ of the point classes of the decomposition are completely contained in the rows of $\mathscr{R}$. Of course $v_1-(v-\rho(A))$ might be negative or zero but, if this is the case, then $v_1-(v-\rho(A))\leqslant 0\leqslant\rho(C)$, which gives $\rho(A)-\rho(C)\leqslant v-v_1$, as required. So suppose $v_1-(v-\rho(A))$ is positive and consider the corresponding $v_1-(v-\rho(A))$ rows of C. If some linear combination of these rows were equal to zero then the same linear combination of the rows of the corresponding $v_1-(v-\rho(A))$ row classes of A (where each row inside a given class has the same coefficient) would also be zero. This is an immediate consequence of the fact that the decomposition is column-tactical since multiplying each row of M_{ij} by a constant h and adding the rows gives a row vector with each entry equal to hc_{ij}. However, the $v_1-(v-\rho(A))$ point classes are completely contained in $\mathscr{R}$

and the rows of $\mathscr{R}$ were chosen to be linearly independent. Thus the $v_1-(v-\rho(A))$ rows of C must be linearly independent, which gives $v_1-(v-\rho(A))\leqslant\rho(C)$ or $\rho(A)-\rho(C)\leqslant v-v_1$. This establishes (b).

The proof of (a) is given by similar considerations on the columns. □

Corollary 1.41. If the rank of A is v and the decomposition of Theorem 1.40 is tactical then $0\leqslant b_1-v_1\leqslant b-v$.

Proof. Clearly, $\rho(R)\leqslant v_1$. So, from Block's Lemma, $v=\rho(A)\leqslant\rho(R)+b-b_1\leqslant v_1+b-b_1$ or $b-v\geqslant b_1-v_1$. Also, since $\rho(C)\leqslant b_1$, we have $v=\rho(A)\leqslant\rho(C)+v-v_1\leqslant b_1+v-v_1$ or $b_1-v_1\geqslant 0$. Hence $0\leqslant b_1-v_1\leqslant b-v$. □

So far all that we appear to have done in this section is prove a couple of simple results about matrices. However, tactical decompositions of the incidence matrices of structures occur in very many natural ways, and these last results can then be applied to give some interesting results about structures. If A is an incidence matrix of a structure $\mathscr{S}$ then, clearly, any decomposition of A gives rise to a partitioning of the points and blocks of $\mathscr{S}$. Similarly any partitioning of the points and blocks of $\mathscr{S}$ gives, in a natural way, a decomposition of A. We shall say that a partitioning of the points and blocks is a *point-tactical*, *block-tactical* or *tactical* decomposition of $\mathscr{S}$ if it is equivalent to a row-tactical, column-tactical or tactical decomposition of A. We can, in fact, recognise the various decompositions of $\mathscr{S}$ without considering incidence matrices.

Lemma 1.42. Let $\mathscr{P}_1,\ldots,\mathscr{P}_{v_1}$ and $\mathscr{x}_1,\ldots,\mathscr{x}_{b_1}$ be a partitioning of the points and blocks of a structure $\mathscr{S}$. Then

(a) it is point-tactical if and only if, for any i and j, each point of class i is incident with a constant number of blocks from class j. (We denote this constant by $(\mathscr{x}_j,\mathscr{P}_i)$.);

(b) it is block-tactical if and only if, for any i and j, each block in class i is incident with a constant number of points from class j. (We denote this constant by $(\mathscr{P}_j,\mathscr{x}_i)$.).

Proof. We shall only prove (b). By definition the decomposition of $\mathscr{S}$ is block-tactical if and only if it is equivalent to a column-tactical decomposition of A, i.e. if and only if, for any i and j, the entries in any column of M_{ij} add up to a constant c_{ij}. But, since A is a (0, 1)-matrix, this is the same as saying that each column of M_{ij} has c_{ij} non-zero entries. However, the number of non-zero entries in a given column of M_{ij} is equal to the number of points of class i incident with that particular block of class j. This establishes (b) and a similar argument will prove (a). □

Exercise 1.51. If $\mathscr{P}_1, \ldots, \mathscr{P}_{v_1}, \mathscr{x}_1, \ldots, \mathscr{x}_{b_1}$ is a tactical decomposition of a structure $\mathscr{S}$ show, for any i and j, $(\mathscr{P}_i, \mathscr{x}_j)|\mathscr{x}_j| = (\mathscr{x}_j, \mathscr{P}_i)|\mathscr{P}_i|$.

As a first illustration of a tactical decomposition of a structure we consider the orbits of any automorphism group. But first we set some exercises on automorphisms.

Exercise 1.52. Show that, for any structure $\mathscr{S}$, the set of all automorphisms of $\mathscr{S}$ form a group whose binary operation is the usual product of mappings.

The group of all automorphisms of a structure $\mathscr{S}$ is called the *full automorphism group* of $\mathscr{S}$ and is denoted by Aut $\mathscr{S}$. An *automorphism group* of $\mathscr{S}$ is merely a subgroup of Aut $\mathscr{S}$.

If α is an automorphism on a structure $\mathscr{S}$ then α acts as a permutation on the points of $\mathscr{S}$ and as another permutation on the blocks of $\mathscr{S}$. Clearly, these two permutations are very closely related. Let $\mathscr{S}$ be any structure and let $\alpha \in \text{Aut } \mathscr{S}$. If the points of $\mathscr{S}$ are labelled $P_1, \ldots, P_v$ and the blocks $y_1, \ldots, y_b$ then the action of α on the points is represented by a v by v permutation matrix P_α, while the action on the blocks is given by a b by b permutation matrix Q_α.

Exercise 1.53. If A is the incidence matrix of $\mathscr{S}$ corresponding to the above labelling show that $P_\alpha A = AQ_\alpha$. (Note: this is very similar to the proof of Lemma 1.1.)

Exercises 1.52 and 1.53 are both very easy. However, Exercise 1.53 has a very interesting consequence when A is square and non-singular. In this case $P_\alpha = AQ_\alpha A^{-1}$, i.e. P_α and Q_α are similar. But this means that P_α and Q_α have the same trace or, in other words, the same numbers of fixed elements. Thus we have proved:

Lemma 1.43. If $\mathscr{S}$ is any square structure with v points and rank v and if $\alpha \in$ Aut $\mathscr{S}$ then α fixes equal numbers of points and blocks. □

Corollary 1.44. An automorphism of a symmetric design fixes equal numbers of points and blocks. □

When we were discussing our examples in Section 1.3 we gave an example of an automorphism which fixed every point of a structure but was not the identity (see Exercise 1.26). This means, of course, that two different automorphisms can have the same action on points. If $\mathscr{S}$ is a structure and $\alpha, \beta \in \text{Aut}\,\mathscr{S}$ are such that $P_\alpha = P_\beta$ then, clearly, $P_{\alpha\beta^{-1}} = P_I = I_v$. So a structure $\mathscr{S}$ has two different automorphisms which act the same way on its points if and only if there is a non-identity automorphism which fixes every point. One fortunate consequence of Exercise 1.26 is that automorphism of designs are completely determined by the permutations they induce on the points.

If G is a permutation group on a set $\mathscr{X}$ then, for any $x \in \mathscr{X}$, the *orbit* of x under G is $\{x^\gamma \mid \gamma \in G\}$. We denote the orbit of x under G by x^G. Clearly, $y \in x^G$ if and only if $x \in y^G$ and the orbits under G partition $\mathscr{X}$. If $x^G = \mathscr{X}$ for some x (and hence for all $x \in \mathscr{X}$) then we say that G is *transitive* on $\mathscr{X}$. An automorphism group of a structure $\mathscr{S}$ will have two kinds of orbits: point orbits and block orbits.

Lemma 1.45. Let G be an automorphism group of a structure $\mathscr{S}$. Then the point and block orbits of G form a tactical decomposition of $\mathscr{S}$.

Proof. Let $\mathscr{P}_1, \ldots, \mathscr{P}_{v_1}$ be the point orbits of G and let $\mathscr{X}_1, \ldots, \mathscr{X}_{b_1}$ be the block orbits. If $|\mathscr{P}_i| = 1$ then, trivially, for every j every point of $\mathscr{P}_i$ is on the same number of blocks in $\mathscr{X}_j$. If $|\mathscr{P}_i| \geqslant 2$ let X, Y be two distinct points in $\mathscr{P}_i$ and let $a_1, \ldots, a_m$ be the blocks of $\mathscr{X}_j$ through X. Since X and Y are in the same orbit of G, there exists γ in G with $X^\gamma = Y$ and then $a_1^\gamma, \ldots, a_m^\gamma$ are the blocks of $\mathscr{X}_j$ through Y. Thus each point of $\mathscr{P}_i$ is on m blocks of $\mathscr{X}_j$ and, by Lemma 1.44, the decomposition is point-tactical. A similar argument shows it is also block-tactical and proves the result. □

Now that we know that the orbits of an automorphism group form a tactical decomposition, we can use our results on decompositions to obtain a result about the number of orbits.

Theorem 1.46 (The Generalised Orbit Theorem). Let $\mathscr{S}$ be a structure with v points, b blocks and having rank v. If G is an automorphism group of $\mathscr{S}$ with v_1 point orbits and b_1 block orbits then $0 \leqslant b_1 - v_1 \leqslant b - v$.

Proof. By Lemma 1.45 the point and block orbits form a tactical decomposition and so the result is an immediate consequence of Corollary 1.41. □

If we put $b=v$ in Theorem 1.46 we see that $b_1=v_1$. Thus, in particular, an automorphism group of a symmetric design has the same number of point and block orbits. This special case of Theorem 1.46 was proved earlier and is often called the Orbit Theorem, and it is for this reason that Theorem 1.46 is known as the Generalised Orbit Theorem. Theorem 1.46 has many interesting consequences. For instance, a uniform 2-structure $\mathscr{S}$ with $v>k$ has rank v, thus if an automorphism group is transitive on the blocks of $\mathscr{S}$ it must also be transitive on the points. (This is easily seen since $b_1=1$ and $0 \leqslant b_1-v_1$ forces $v_1=1$.) Note, however, that the converse of this is false and that an automorphism group can be transitive on points without being transitive on blocks. It is also worth noting that Corollary 1.44 does not extend to *automorphism groups* of symmetric designs. Thus, while an automorphism group of a symmetric design has the same number of point and block orbits, it can fix different numbers of points and blocks.

Exercise 1.54. Since Example $\mathscr{F}$ is a symmetric design an element of Aut $\mathscr{F}$ is uniquely determined by its action on the points. Let $\alpha, \beta, \gamma \in \text{Aut } \mathscr{F}$ all fix the points 1, 2 and 4 and let $\alpha: 3 \leftrightarrow 7, 5 \leftrightarrow 6, \beta: 3 \leftrightarrow 5, 6 \leftrightarrow 7$ and $\gamma: 3 \leftrightarrow 6, 5 \leftrightarrow 7$. First check that, as claimed, α, β and γ are automorphisms. Then show that $H=\{I, \alpha, \beta, \gamma\}$ is an automorphism group of $\mathscr{F}$. Show that H has four point orbits and four block orbits, but has three fixed points and one fixed block.

Exercise 1.55. Let $\mathscr{A}=\mathscr{A}_n(q)$ and let H be the automorphism group of $\mathscr{A}$ consisting of all mappings $\mathbf{x} \to \mathbf{x}+\mathbf{a}$ for all $\mathbf{a} \in V_n(q)$. Show

(a) H is transitive on the points of $\mathscr{A}$;

(b) H has $\dfrac{q^{n-1}-1}{q-1}$ block orbits.

This exercise also gives an illustration of a case where $b_1-v_1=b-v$. Theorem 1.46 can be generalised still further.

Exercise 1.56. Let $\mathscr{S}$ be a structure with v points, b blocks and having rank v. If $G \leqslant H \leqslant \mathrm{Aut}\,\mathscr{S}$ are such that G has b_2 block orbits and v_2 point orbits while H has b_1 block orbits and v_1 point orbits, show that $0 \leqslant b_1 - v_1 \leqslant b_2 - v_2$. (Note Theorem 1.46 is a special case of this exercise where G is the identity.)

As well as the orbits of automorphism groups there are many other interesting tactical decompositions. We conclude this section by mentioning two examples which we will discuss more fully in Chapter 5. A *resolution* is a tactical decomposition with exactly one point class. (One example of a design with a resolution is given by Exercise 1.56 where the resolution is given by the orbits of H.) We call the block classes of the tactical decomposition the *resolution classes.*

If $\mathscr{S}$ is a structure with a resolution then the points of $\mathscr{S}$ together with the blocks of any one of the resolution classes form a 1-structure. We call these various 1-structures the *resolution structures.* We will not say much about resolutions here except to observe that it is an interesting problem to try to construct 2-structures with given 1-structures as resolution structures.

Before concluding this section we will mention one special kind of resolution. A *parallelism* of structure is a resolution such that each point is on exactly one block from each class. In this case we call the resolution classes the *parallel classes.*

Exercise 1.57. Let $\mathscr{S}$ be a structure of rank v with v points and b blocks. Show
(a) if $\mathscr{S}$ has a resolution with x classes then $b - v \geqslant x - 1$, and
(b) if $\mathscr{S}$ has a parallelism then $b - v \geqslant r - 1$.

1.7 Graphs

There are various definitions of a graph, but for our purposes a particularly simple one will suffice. A *graph* $\Gamma = \Gamma(\mathscr{V}, \mathscr{E})$ consists of a finite set $\mathscr{V}$ of *vertices* together with a set $\mathscr{E}$ of *edges*, where the edges are subsets of $\mathscr{V}$ consisting of two distinct vertices; no edge is repeated, so a graph has neither 'loops' nor 'multiple edges'. Graph theory has considerable importance in a number of fields of mathematics, including the study of incidence structures. In this section our aim is merely to introduce some terminology and some of the standard graphs associated with structures.

If Γ is a graph then the *degree* or *valency* of a vertex X is the number of edges on X or, equivalently, the number of vertices 'adjacent to' X. If each vertex has the same degree d, then the graph is said to be *regular*, of degree d; a regular graph of degree 2 is said to be *divalent*.

A graph is *complete* if every vertex is adjacent to every other. A graph is *bipartite* if the vertex set can be partitioned into two subsets $\mathscr{V}_1$ and $\mathscr{V}_2$ such that no two vertices of $\mathscr{V}_1$ are adjacent and no two vertices of $\mathscr{V}_2$ are adjacent (so all edges join a vertex in $\mathscr{V}_1$ with a vertex in $\mathscr{V}_2$). A graph is *complete bipartite* if, in addition, every vertex in $\mathscr{V}_1$ is adjacent to every vertex in $\mathscr{V}_2$.

If $\Gamma = \Gamma(\mathscr{V}, \mathscr{E})$ is a graph, we can consider the edges as blocks and construct an incidence matrix for the resulting structure. But a more useful matrix is the *adjacency matrix* $N = N(\Gamma)$: index the vertices $P_1, P_2, \ldots, P_v$, and let the (i, j)-entry in N be 1 or 0 according as P_i and P_j are adjacent or not. Clearly, N is symmetric and has 0s on its main diagonal. So $NN^{\mathrm{T}} = N^2$ and the (i, j)-entry of N^2 is the inner product of the ith and jth rows of N. Thus the (i, i)-entry of N^2 is the degree of P_i and the (i, j)-entry $(i \neq j)$ is the number of vertices adjacent to both P_i and P_j. In particular, Γ is regular if and only if N^2 has the same entry everywhere on its main diagonal.

For any graph Γ, the *complement* Γ^{c} of Γ is the graph defined with the same vertex set as Γ, but with adjacency and non-adjacency interchanged: so if $X \neq Y$, then X and Y are adjacent in Γ^{c} if and only if X and Y are not adjacent in Γ.

If $\mathscr{S}$ is a structure, there are several graphs that can be constructed from $\mathscr{S}$. The *point-adjacency* graph of $\mathscr{S}$, written $\Gamma_{\mathscr{P}}(\mathscr{S})$, has the points of $\mathscr{S}$ for vertices and two are adjacent if they are on a common block of $\mathscr{S}$; the *block-adjacency* graph $\Gamma_{\mathscr{B}}(\mathscr{S})$ is defined similarly. If $\mathscr{S}$ is a 2-structure, then $\Gamma_{\mathscr{P}}(\mathscr{S})$ is complete and not very interesting, but there can still be considerable interest in the block-adjacency graph, as we shall see in later chapters. (Sometimes the complement of the block-adjacency graph is the more natural setting for study.) Another standard graph associated with $\mathscr{S}$ is the *incidence* graph $\Gamma(\mathscr{S})$: the vertices of $\Gamma(\mathscr{S})$ are the points and the blocks of $\mathscr{S}$ and two elements are adjacent if and only if they are incident in $\mathscr{S}$. Clearly, $\Gamma(\mathscr{S})$ is bipartite and the structure $\mathscr{S}$ can be recovered (up to isomorphism) from it.

Exercise 1.58. Show that $\Gamma(\mathscr{S})$ is regular if and only if $\mathscr{S}$ is regular and uniform with $k = r$.

Two vertices X, Y of a graph Γ are *connected* if there is a sequence $X = X_0, X_1, \ldots, X_n = Y$ of vertices X_i such that X_i is adjacent to X_{i+1} for each i. The graph Γ is *connected* if every pair of its vertices is connected. The *distance* $d(X, Y)$

between the vertices X and Y is the length n of the minimal sequence of X_i as above, if such a sequence exists. If Γ is connected, then the *diameter* of Γ is the maximal distance between any two of its vertices. If $\mathscr{S}$ is a structure, then we define terms such as distance, connected, etc., to reflect the corresponding properties in $\Gamma(\mathscr{S})$. We will give each definition using the terminology of points, blocks and incidence in $\mathscr{S}$, but the reader should translate into graph theoretic terms and verify that they do reflect the corresponding properties of $\Gamma(\mathscr{S})$.

For any incidence structure a *chain* of length n is a sequence of elements a_0, $a_1, \ldots, a_n$ such that a_i is incident with a_{i+1} for $0 \leqslant i \leqslant n-1$. The elements a_0 and a_n are called the *supports* of the chain, and a chain is *reduced* if it contains no element more than once. Two elements of a structure are *connected* if there is a chain which has them as supports. If a and b are connected, then the distance $d(a, b)$ from a to b is the length of the shortest reduced chain which has a and b as supports. So the distance (if it exists) between two points is always even and the distance between a point and a block is odd (again, if it exists). We define the distance from an element to itself to be 0. The *diameter* of a connected structure is the maximum distance between two of its elements.

Exercise 1.59. Show that in a connected structure $\mathscr{S}$

(a) $d(x, y) + d(y, z) \geqslant d(x, z)$ for any elements x, y, z of $\mathscr{S}$;

(b) if $d(x, y) = m$ and $d(y, z) = 1$, then $d(x, z) = m-1$ or $m+1$.

An *ordinary m-gon* is a chain of length $2m$ with identical support elements but no other element repeated, and an *ordinary polygon* is an ordinary m-gon for some m.

Exercise 1.60. Show that a connected divalent graph is the point-incidence graph of an ordinary polygon.

Exercise 1.61. Show that if a chain is not reduced then it must contain an ordinary polygon.

Exercise 1.62. If $\mathscr{S}$ and $\mathscr{S}^{\mathrm{T}}$ are both non-trivial 2-structures determine the diameter of $\Gamma(\mathscr{S})$.

Exercise 1.63. Find the point-incidence graph, the block-incidence graph and the incidence graph, for the 11 structures $\mathscr{A}, \mathscr{B}, \ldots, \mathscr{K}$ of Section 1.3 (many of these will be complete and hence require little description!).

The reader interested not only in more about graphs, but in particular their connections with structures and designs, may consult Cameron & van Lint [3].

References

References [1] to [7] are for text books which provide excellent coverage of the background material of this chapter. [8] is the research paper exhibiting the first (and at present the only) non-trivial 6-designs.

[1] Artin, E. *Geometric Algebra.* New York, Interscience, 1957.

[2] Baer, R. *Linear Algebra and Projective Geometry.* New York, Academic Press, 1952.

[3] Cameron, P. & van Lint, J. H. *Graph Theory, Coding Theory and Block Designs.* London Mathematical Society Lecture Note Series 19, Cambridge University Press.

[4] Dembowski, H. P. *Finite Geometries.* Berlin–Heidelberg–New York, Springer, 1968.

[5] Gruenberg, K. & Weir, J. *Linear Geometry.* New York–London, Van Nostrand, 1967.

[6] Hirschfeld, J. *Projective Geometries over Finite Fields.* Oxford University Press, 1983.

[7] Hughes, D. R. & Piper, F. C. *Projective Planes.* Berlin–Heidelberg–New York, Springer, 1973.

[8] Magliveras, S. S. & Leavitt, D. W. 'Simple 6-(33, 8, 36) Designs from $P\Gamma L_2(32)$.' *Proceedings of the 1982 Durham Symposium on Computational Group Theory.*

2 Symmetric designs

2.1 Introduction

In Section 2.2 we investigate residual designs of symmetric designs and discuss the (difficult) problem of embedding a 'quasi-residual' of a symmetric design in a symmetric design. However, we do not get too deeply involved in this problem as the main results in this area are too difficult to include here. Section 2.3 contains M. Hall's elementary proof of the celebrated Bruck–Ryser–Chowla Theorem; the only known non-existence result about symmetric designs. In Section 2.4, Singer groups and difference sets are introduced and, in particular, it is shown that the symmetric designs defined by finite projective geometries, i.e. $\mathscr{P}_n(q)$, all have Singer groups. Then, in Section 2.5, we define multipliers and include a proof of M. Hall's original 'Multiplier Theorem'. This section involves much more group theory than the rest of this chapter, most of which is only relevant to the proof and not to an understanding of the theorem, and so the reader with less background in groups might skip the proofs and the technical lemmas, going straight to Theorem 2.15 and the examples that follow it. Section 2.6 deals with some simple relations between the parameters v and n of a symmetric design and this leads naturally to a brief discussion of projective planes and Hadamard 2-designs (as extreme cases of certain inequalities). Finally in Section 2.7 we give a proof of the famous Dembowski–Wagner Theorem, which gives a powerful set of characterisations of the designs $\mathscr{P}_n(q)$.

In this chapter we introduce the abbreviated terminology of a 'symmetric design for (v, k, λ)', since a symmetric design is always a 2-design. (If the context makes it clear, we can also sometimes omit the word 'symmetric'.)

2.2 Residual structures

If y is a block of an arbitrary 2-design $\mathscr{D}$ then, as we saw in Chapter 1, neither $\mathscr{D}_y$ nor $\mathscr{D}^y$ need be designs. However, if $\mathscr{D}$ is a symmetric 2-(v, k, λ) then, by

Theorem 1.25, any two blocks of $\mathscr{D}$ intersect in λ points. Thus when $\mathscr{D}$ is symmetric both $\mathscr{D}_y$ and $\mathscr{D}^y$ must be uniform structures; $\mathscr{D}_y$ has block size λ and $\mathscr{D}^y$ has block size $k-\lambda$. But in general they may still have repeated blocks and, consequently, need not be designs. However, if $2k<v$ then $\mathscr{D}^y$ is a 2-design for every block y.

Theorem 2.1. If $\mathscr{D}$ is a symmetric 2-(v, k, λ) with $2k<v$ then, for any block y of $\mathscr{D}$, $\mathscr{D}^y$ is a 2-$(v-k\ k-\lambda, \lambda)$.

Proof. Clearly, $\mathscr{D}^y$ contains $v-k$ points. Since any two blocks of $\mathscr{D}$ intersect in λ points, each block of $\mathscr{D}^y$ contains $k-\lambda$ points and, since the blocks of $\mathscr{D}^y$ through two points are merely the blocks of $\mathscr{D}$ through the same points, every pair of points of $\mathscr{D}^y$ are on λ common blocks. Thus $\mathscr{D}^y$ is a 2-structure for $(v-k, k-\lambda, \lambda)$. In order to show that $\mathscr{D}^y$ is a 2-$(v-k, k-\lambda, \lambda)$ we have to show that $\mathscr{D}^y$ has no repeated blocks. But if two blocks of $\mathscr{D}^y$ were incident with the same set of points then $\mathscr{D}$ would have to contain two blocks with at least $k-\lambda$ common points. The restriction $2k<v$ gives, when substituted in Corollary 1.4(c), $k(k-1)>\lambda(2k-1)$, which implies $2\lambda<k$ or $\lambda<k-\lambda$. Hence, since any pair of blocks of $\mathscr{D}$ intersect in λ points, $\mathscr{D}^y$ has no repeated blocks and is a 2-$(v-k, k-\lambda, \lambda)$. □

The converse of Theorem 2.1 is also true, i.e. if a 2-design $\mathscr{D}$ has the property that, for every block y, $\mathscr{D}^y$ is a 2-design then $\mathscr{D}$ is symmetric. This is an immediate corollary of the following even stronger result.

Theorem 2.2. If a non-trivial 2-(v, k, λ) design $\mathscr{D}$ has a block y such that $\mathscr{D}^y$ is a 2-design then $\mathscr{D}$ is symmetric.

Proof. For any block x of $\mathscr{D}$ label the remaining blocks $x_1, x_2, \ldots, x_{b-1}$ and let $n_{x,i}$ be the number of points in $x \cap x_i$ $(i=1, \ldots, b-1)$. We will now count in two ways the number of flags (P, z) with P on x and $z \neq x$. There are k choices for P and, given any of these choices, then $r-1$ choices for z. So the number of such flags is $k(r-1)$. However, for each x_i there are $n_{x,i}$ choices of P, so there are also $\sum_{i=1}^{b-1} n_{x,i}$ such flags. Hence

$$k(r-1)=\sum_{i=1}^{b-1} n_{x,i}. \tag{1}$$

Similarly, counting 2-flags (P, Q, z) with P, Q in $x \cap z$, $P \neq Q$, $z \neq x$ gives

$$k(k-1)(\lambda-1) = \sum_{i=1}^{b-1} n_{x,i}(n_{x,i}-1). \tag{2}$$

If we now put $w_x = \sum_{i=1}^{b-1} n_{x,i}/(b-1)$, i.e. let w_x be the average value of $n_{x,i}$, then we can get an expression for $\sum_{i=1}^{b-1} (n_{x,i}-w_x)^2$.

$$\begin{aligned}
\sum_{i=1}^{b-1} (n_{x,i}-w_x)^2 &= \sum_{i=1}^{b-1} n_{x,i}^2 - 2w_x \sum_{i=1}^{b-1} n_{x,i} + (b-1)w_x^2 \\
&= \sum_{i=1}^{b-1} n_{x,i}(n_{x,i}-1) - (2w_x-1) \sum_{i=1}^{b-1} n_{x,i} + (b-1)w_x^2 \\
&= k(k-1)(\lambda-1) - \left(\frac{2k(r-1)}{b-1} - 1\right)k(r-1) + (b-1)\frac{k^2(r-1)^2}{(b-1)^2}.
\end{aligned}$$

For the purpose of this proof there is no need to simplify this expression any further but merely to note that its value is independent of the original block x. Thus if it is zero for one particular block it will be zero for every block.

Since $\mathscr{D}^y$ is a 2-design, it is uniform and each block has the same number of points, k' say. Thus for any block c of $\mathscr{D}$, $c \neq y$, $|c \cap y| = k - k'$. But this means $n_{y,i} = k - k' = w_y$ for $i = 1, \ldots, b-1$, or $\sum_{i=1}^{b-1} (n_{y,i}-w_y)^2 = 0$. Hence for every block x, $\sum_{i=1}^{b-1} (n_{x,i}-w_x)^2 = 0$. Since $\mathscr{D}$ is non-trivial, $2 < k < v$ and so, by Fisher's Inequality and Corollary 1.4(b), $r > 1$. If x is any block and P is on x then $r > 1$ implies that there is another block through P, i.e. that $w_x > 0$. Hence, since $n_{x,i} = w_x$ for all i, every other block meets x in $w_x > 0$ points. But y also meets x in w_y points, so $w_x = w_y$ for all blocks y and any two blocks of $\mathscr{D}$ intersect in w_y points. Theorem 2.22 now completes the proof. □

If $\mathscr{S}$ is a 2-$(v-k, k-\lambda, \lambda)$, with $v-1$ blocks, then $\mathscr{S}$ is called a *quasiresidual* 2-design; a quasiresidual is a candidate to be of the form $\mathscr{D}^y$ for a symmetric design $\mathscr{D}$, i.e. to be embeddable. The following example shows that a quasiresidual is not always embeddable.

Example 2.1 (Bhattacharya [2]). Let $\mathscr{S}$ be the structure whose points are the 16 integers 1, 2, ..., 16 and whose blocks are the 24 subsets x_i below:

$x_1 = \{1,2,7,8,14,15\}$ $x_2 = \{3,5,7,8,11,13\}$ $x_3 = \{2,3,8,9,13,16\}$
$x_4 = \{3,5,8,9,12,14\}$ $x_5 = \{1,6,7,9,12,13\}$ $x_6 = \{2,5,7,10,13,15\}$
$x_7 = \{3,4,7,10,12,16\}$ $x_8 = \{3,4,6,13,14,15\}$ $x_9 = \{4,5,7,9,12,15\}$
$x_{10} = \{2,4,9,10,11,13\}$ $x_{11} = \{3,6,7,10,11,14\}$ $x_{12} = \{1,2,3,4,5,6\}$
$x_{13} = \{1,4,7,8,11,16\}$ $x_{14} = \{2,4,8,10,12,14\}$ $x_{15} = \{5,6,8,10,15,16\}$

$x_{16}=\{1,6,8,10,12,13\}$ $x_{17}=\{1,2,3,11,12,15\}$ $x_{18}=\{2,6,7,9,14,16\}$
$x_{19}=\{1,4,5,13,14,16\}$ $x_{20}=\{2,5,6,11,12,16\}$ $x_{21}=\{1,3,9,10,15,16\}$
$x_{22}=\{4,6,8,9,11,15\}$ $x_{23}=\{1,5,9,10,11,14\}$ $x_{24}=\{11,12,13,14,15,16\}$

Then $\mathscr{S}$ is a 2-(16, 6, 3) and is a quasiresidual for a possible symmetric design with parameters (25, 9, 3). But x_5 and x_{16} above have four points in common, and in a symmetric design with $\lambda=3$, two blocks meet in three points and hence in three or fewer points in any substructure. So $\mathscr{S}$ is not embeddable.

Exercise 2.1. Check that $\mathscr{S}$ in Example 2.1 is a 2-(16, 6, 3).

In fact if $\lambda=1$ or 2, then a quasiresidual 2-design is always embeddable, and even uniquely. For $\lambda=1$ the proof is easy and we give it in Section 2.6. The proof for $\lambda=2$ is more difficult and can be found in [3]. In fact, the general situation is that for arbitrary λ there is a function $f(\lambda)$ such that, if $v>f(\lambda)$, then a quasiresidual 2-design for $(v-k,\ k-\lambda,\ \lambda)$ is always embeddable; for this considerably deeper result see [5].

We shall continue to see that the order n $(=r-\lambda)$ of a 2-design plays an important role. For quasiresidual 2-designs it has a precise value.

Exercise 2.2. Show that if $\mathscr{S}$ is a quasiresidual 2-(v, k, λ) then $n=k$, i.e. $r-k-\lambda=0$. Show also that this implies $b+1=v+r$.

2.3 The Bruck–Ryser–Chowla Theorem

Since, in a symmetric (v, k, λ) design, $b=v$ and $r=k$, Corollary 1.4(c) gives $(v-1)\lambda=k(k-1)$ or $v=[k(k-1)+\lambda]/\lambda$. However, if v, k and λ are integers satisfying this relation this certainly does not guarantee the existence of a symmetric 2-(v, k, λ). We now prove a remarkable theorem which, in fact, proves the non-existence of a symmetric 2-(v, k, λ) unless v, k and λ satisfy some other, very powerful, restrictions. For completeness we include a proof of this result. However, the proof uses some elementary results and techniques from number theory and is of a completely different nature to the other proofs in the book. The reader should be prepared to omit reading the proof if he finds it difficult. Failure to understand the proof will in no way hinder him at any later stage of the book.

Theorem 2.3 (The Bruck–Ryser–Chowla Theorem). If v, k, λ are integers satisfying $(v-1)\lambda = k(k-1)$ then for the existence of a symmetric 2-(v, k, λ) $\mathscr{D}$ it is necessary that

(a) if v is even then $k-\lambda$ is a square;

(b) if v is odd, then $z^2 = (k-\lambda)x^2 + (-1)^{(v-1)/2}\lambda y^2$ has a non-trivial solution in integers x, y and z.

Proof. (a) See Corollary 1.38.

(b) We first recall a number theoretic result (for proof the reader should consult [4]):

Every positive integer is the sum of four integral squares. (1)

We shall also use the following elementary identity.

$$(b_1^2+b_2^2+b_3^2+b_4^2)(x_1^2+x_2^2+x_3^2+x_4^2) = y_1^2+y_2^2+y_3^2+y_4^2, \tag{2}$$

where

$$y_1 = b_1x_1 - b_2x_2 - b_3x_3 - b_4x_4,$$
$$y_2 = b_2x_1 + b_1x_2 - b_4x_3 + b_3x_4,$$
$$y_3 = b_3x_1 + b_4x_2 + b_1x_3 - b_2x_4,$$
$$y_4 = b_4x_1 - b_3x_2 + b_2x_3 + b_1x_4.$$

Let $\mathscr{D}$ be a symmetric 2-(v, k, λ) and write $n = k-\lambda$. Let $P_1, \ldots, P_v$ be a labelling of the points of $\mathscr{D}$, let $c_1, \ldots, c_v$ be a labelling of the blocks and let $A = (a_{ij})$ be the incidence matrix given by these labellings. We now take $x_1, x_2, \ldots, x_v$ as independent variables and determine v linear forms $L_i = \sum_{j=1}^{v} a_{ij}x_j$ $(i = 1, \ldots, v)$. If we now form $L_1^2 + \cdots + L_v^2$ then, since each x_i occurs in exactly k distinct L_j, each x_i^2 will occur k times. Similarly each x_i and x_j (with $i \neq j$) will occur together exactly 2λ times. This gives

$$L_1^2 + \cdots + L_v^2 = n(x_1^2 + \cdots + x_v^2) + \lambda(x_1 + \cdots + x_v)^2 \equiv Q. \tag{3}$$

If we write n as $b_1^2+b_2^2+b_3^2+b_4^2$, which we can do by (1), then we may use (2) to write

$$n(x_i^2 + x_{i+1}^2 + x_{i+2}^2 + x_{i+3}^2) = y_i^2 + y_{i+1}^2 + y_{i+2}^2 + y_{i+3}^2. \tag{4}$$

By assumption, v is odd and we will now suppose $v \equiv 1 \pmod 4$. In this case if we apply (4) to the right-hand side of (3) with four variables at a time and if we put $w = \sum_{i=1}^{v} x_i$ we get

$$L_1^2 + \cdots + L_v^2 = y_1^2 + \cdots + y_{v-1}^2 + nx_v^2 + \lambda w^2. \tag{5}$$

However, using (2) and the fact that the linear transformation from the xs to the ys is non-singular we can solve rationally for the xs in terms of the ys and so, with $w = x_1 + \cdots + x_v$ expressed in terms of $y_1, \ldots, y_{v-1}, x_v$, we have an identity in independent variables $y_1, \ldots, y_{v-1}, x_v$, where the Ls and w are rational linear forms in these variables.

If L_1 (as a linear form in $y_1, \ldots, y_{v-1}, x_v$) does not have the coefficient $+1$ for y_1 we put $L_1 = y_1$, but if the coefficient is $+1$ we put $L_1 = -y_1$. (Note that (5) is an identity for all rational forms in the given variables. Thus (5) must hold for any choice of the y_i.) In both cases we may use this relation to solve for y_1 as a rational linear combination of $y_2, \ldots, y_{v-1}, x_v$ and we also have the identity $L_1^2 = y_1^2$. Thus (5) becomes an identity in $y_2, \ldots, y_{v-1}, x_v$.

$$L_2^2 + \cdots + L_v^2 = y_2^2 + \cdots + y_{v-1}^2 + nx_v^2 + \lambda w^2. \tag{6}$$

In the same way we can put $L_2 = \pm y_2, \ldots, L_{v-1} = \pm y_{v-1}$ to solve for $y_2, \ldots, y_{v-1}$ in terms of later variables. This finally gives

$$L_v^2 = nx_v^2 + \lambda w^2, \tag{7}$$

with x_v an independent variable and L_v and w rational multiples of x_v. Taking x_v as an integer which is a multiple of the denominators appearing in L_v, and w, we have an equation in integers $z^2 = (k-\lambda)x^2 + \lambda y^2$.

If $v \equiv 3 \pmod 4$ we take a new variable x_{v+1} and add nx_{v+1}^2 to both sides of (3) to obtain

$$L_1^2 + \cdots + L_v^2 + nx_{v+1}^2 = n(x_1^2 + \cdots + x_{v+1}^2) + \lambda w^2, \tag{8}$$

to which we apply (4) and get

$$L_1^2 + \cdots + L_v^2 + nx_{v+1}^2 = y_1^2 + \cdots + y_{v+1}^2 + \lambda w^2. \tag{9}$$

This can be reduced in the same way as above to the form

$$nx_{v+1}^2 = y_{v+1}^2 + \lambda w^2, \tag{10}$$

where y_{v+1} and w are rational multiples of the independent variable x_{v+1} and in this case we have integer solutions to $(k-\lambda)x^2 = z^2 + \lambda y^2$. However, since $v \equiv 3 \pmod 4$, $(-1)^{(v-1)/2} = -1$, and so we have integer solutions to $z^2 = (k-\lambda)x^2 + (-1)^{(v-1)/2}\lambda y^2$. □

We shall repeatedly quote this theorem and will often refer to it as the BRC Theorem. Although it may appear somewhat difficult to use, there are various techniques available which allow its immediate application. Before giving examples to illustrate this it is worth pointing out that the BRC Theorem has a particularly simple form if $\lambda = 1$.

Corollary 2.4. If a symmetric design with $\lambda = 1$ and order n exists and if $n \equiv 1$ or $2 \pmod 4$ then n can be expressed as a sum of two integral squares.

Proof. Let $\mathscr{D}$ be a symmetric 2-$(v, k, 1)$ and write $n = k-1$. Then $v = n^2+n+1$, which is always odd (since $n^2+n = n(n+1)$ is even). If $v \equiv 1 \pmod 4$ then $n \equiv 0$ or $3 \pmod 4$, while if $v \equiv 3 \pmod 4$ then $n \equiv 1$ or $2 \pmod 4$.

If $v \equiv 1 \pmod 4$, then $(-1)^{(v-1)/2} = 1$, so the BRC Theorem says that we want a solution to $nx^2 = z^2 - y^2$. But, clearly, $x=0$, $y=z=1$ is a non-trivial solution, and hence no values of k are excluded. But if $v \equiv 3 \pmod 4$, then we want solutions to $nx^2 = y^2 + z^2$, or $n = a^2 + b^2$, where a, b are rational numbers. But an integer is a sum of two rational squares if and only if it is a sum of two integral squares (again, see [4]), proving the corollary. □

An equivalent formulation of Corollary 2.4 is:

Corollary 2.5. If a symmetric design with $\lambda = 1$ and of order n exists and if $n \equiv 1$ or $2 \pmod 4$, then the square-free part of n has no prime divisor $\equiv 3 \pmod 4$.

Proof. It is well known that an integer n can be expressed as a sum of two squares if and only if the square-free part of n has no prime divisor $\equiv 3 \pmod 4$ (see [4]). □

We now give some examples of the use of the BRC Theorem. Before we do, however, note that by Exercise 1.30 we know that there is a symmetric design with $\lambda = 1$ of order n for every prime power n.

Example 2.2. There is no symmetric design with $\lambda = 1$ and $n = 6$, i.e. there is no 2-(43, 7, 1).

Proof. Since $n = 6$, $n \equiv 2 \pmod 4$. Thus, since 6 is square free but $3 \mid 6$, Corollary 2.5 establishes the result. (Alternatively, merely by trying all possibilities, we could check that 6 is not the sum of two squares and use Corollary 2.3.) □

The next smallest integer which is not a prime power is 10. Since $10 = 3^2 + 1^2$ the BRC Theorem does not exclude the possibility of such a design existing. Despite the concerted efforts of many mathematicians it is still not known whether a 2-(111, 11, 1) exists or not.

Example 2.3. There are no symmetric designs with parameters (22, 7, 2) or (29, 8, 2).

Proof. A design for (22, 7, 2) would have even v and yet $n=7-2$ is not a square.

For the existence of a symmetric design with parameters (29, 8, 2), we need a non-trivial solution in integers for $z^2=6x^2+2y^2$. We may assume that x, y, z have no common factors, and we see that z must be even, so we write $z=2z_1$. Then substituting and simplifying, we want non-trivial solutions to $2z_1^2=3x^2+y^2$. If we reduce this equation modulo 3, we obtain: $2z_1^2\equiv y^2 \pmod 3$. If $z_1\not\equiv 0 \pmod 3$, this implies that $2\equiv(y/z_1)^2 \pmod 3$, i.e. that 2 is a square modulo 3. This is not so, hence 3 must divide z_1. But writing $z_1=3z_2$, substituting and simplifying, this gives $18z_2^2=3x^2+y^2$, so 3 divides y. Repeat the process, to find that 3 divides x as well: that is, 3 divides x, y and z, contrary to hypothesis. □

The techniques of Example 2.3 are, in fact, always sufficient to examine the Diophantine equations arising from the BRC Theorem.

Exercise 2.3. Examine all possible parameter sets for symmetric designs with $\lambda=1$ and $k\leqslant 24$. With Exercise 1.30 in mind, decide for each k whether (a) a design exists, (b) a design does not exist, or (c) no information is available.

Exercise 2.4. Examine all possible parameter sets for symmetric designs with $\lambda=2$ and $k\leqslant 16$. We do not have many examples yet, so merely decide whether (a) a design cannot exist or (b) the BRC Theorem gives no information.

It is an interesting fact that at this time no parameter set for a symmetric design has ever been shown to be impossible except for those rejected by the BRC Theorem. Whether this is a consequence of our ignorance or conceals a deeper theorem is a central question.

2.4 Singer groups and difference sets

In this section we study certain very special automorphism groups which some symmetric designs possess. We will also show how certain groups may be used to construct symmetric designs. Before doing this, however, we introduce some notation and elementary results on permutation groups. (The group theory results can be found in any standard introduction to that subject.)

If G is a permutation group on a set $\mathscr{X}$ then, as in Chapter 1, for any $x \in \mathscr{X}$ we denote the orbit of x under G by x^G. If we let $G_x = \{g \in | x^g = x\}$ then G_x is a subgroup of G (G_x is the *stabiliser* of x) and:

Result 2.1. If G is a permutation group on a finite set $\mathscr{X}$ then, for any $x \in \mathscr{X}$, $|G| = |G_x| \cdot |x^G|$.

If G is transitive on $\mathscr{X}$ and $G_x = I$ for some $x \in \mathscr{X}$ then G is said to be *regular* on $\mathscr{X}$. As an immediate corollary of Result 2.1 we have.

Result 2.2. If G is regular on a finite set $\mathscr{X}$ then $|G| = |\mathscr{X}|$.

Note that if G is regular on $\mathscr{X}$ then each permutation, other than the identity, has no fixed element. Thus for any $x, y \in \mathscr{X}$ there is a unique $g \in G$ with $x^g = y$.

If G is an abelian permutation group on a set $\mathscr{X}$ and if $y \in x^G$ then $G_x = G_y$. This leads to:

Result 2.3. If an abelian permutation group G is transitive on a set $\mathscr{X}$ then it is regular on $\mathscr{X}$.

If $\mathscr{D}$ is a symmetric 2-(v, k, λ) and if $G \leqslant \text{Aut } \mathscr{D}$ is regular on the points of $\mathscr{D}$ then, by Result 2.2, $|G| = v$. By the Orbit Theorem, G must also be transitive, and consequently regular, on the blocks of $\mathscr{D}$. We call such a group a *Singer group* of $\mathscr{D}$.

Exercise 2.5. Let g be the cyclic permutation $g = (1\ 2\ 3\ 4\ 5\ 6\ 7)$ on the points of example $\mathscr{F}$. Show that the cyclic group G generated by g is a Singer group of $\mathscr{F}$; determine the action of g on the blocks of $\mathscr{F}$.

If G is a Singer group for a symmetric 2-$(v, k, \lambda)\mathscr{D}$ then, for any given point P and block x, the subset $D = \{g \text{ in } G \mid P^g \text{ is on } x\}$ is called a *difference set* of G, and the point P and block x are called the *base elements* of D. We will often signify that D has base elements P and x by writing it as $D(P, x)$. (The terminology is historical and, as we shall see, makes more sense when G is abelian and is written additively.)

Exercise 2.6. Show that $D=\{1, g, g^3\}$ is a difference set for the Singer group G of Exercise 2.5 and determine a pair of base elements.

We now prove two simple lemmas.

Lemma 2.6. If G is a Singer group of a symmetric 2-(v, k, λ) design $\mathscr{D}$ and if $D(P, x)$ is a difference set, then $|D(P, x)|=k$.

Proof. Label the points of x as $A_1, A_2, \ldots, A_k$. Since G is regular, for each i $(1 \leqslant i \leqslant k)$ there is a unique $g_i \in G$ with $P^{g_i}=A_i$. Then $D(P,x)=\{g_1, g_2, \ldots, g_k\}$ and $|D(P,x)|=k$. □

Lemma 2.7. If G is a Singer group of a symmetric 2-(v, k, λ) design $\mathscr{D}$ and if $D(P, x)$ is a difference set, then

(a) for any $a, b \in G$, $a^{-1}(D(P, x))b=D(P^a, x^b)$;

(b) if D' is any other difference set then there exist c, d in G with $D'=c^{-1}(D(P, x))d$.

Proof. (a) If $d \in D(P^a, x^b)$ then $P^{ad} \in x^b$, i.e. $P^{adb^{-1}} \in x$. Thus $adb^{-1} \in D(P, x)$ or $d \in a^{-1}(D(P, x))b$. This shows $D(P^a, x^b) \subseteq a^{-1}(D(P, x))b$. The opposite inclusion in (a) follows similarly.

(b) Let R, z be the base elements of D'. Since G is a Singer group there are (unique) elements c, d with $P^c=R$, $x^d=z$. Then $D'=D(P^c, x^d)= c^{-1}(D(P, x))d$. □

Our next theorem establishes the fundamental properties of difference sets.

Theorem 2.8. Let $\mathscr{D}$ be a symmetric 2-(v, k, λ) design with a Singer group G and let D be a difference set of G. Then for every $g \neq 1$ in G there exist exactly λ distinct pairs c_i, d_i in D such that $g=c_i, d_i^{-1}$; also there exist exactly λ distinct pairs e_i, f_i in D such that $g=e_i^{-1}f_i$.

Proof. Let P and x be the base elements of D. Since G is regular on the points of $\mathscr{D}$, P and P^g are distinct points. Thus there are λ blocks of $\mathscr{D}$ containing them both, and, by the transitivity of G, these blocks are x^{h_i} for λ elements h_i in G ($i = 1, 2, \ldots, \lambda$). But now $P^{h_i^{-1}}$ and $P^{gh_i^{-1}}$ are on x, for each i, and so h_i^{-1} and gh_i^{-1} are in D. Letting $c_i = gh_i^{-1}$ and $d_i = h_i^{-1}$, this gives us λ distinct pairs $c_i, d_i \in D$ such that $g = c_i d_i^{-1}$. We now want to show that these are the only such pairs of elements in D. Suppose $g = ab^{-1}$, where a and b are in D; then $gb = a$ is in D and so P^{gb} is on x. Hence P^g is on $x^{b^{-1}}$. Also, since b is in D, P is on $x^{b^{-1}}$. Thus $x^{b^{-1}}$ is one of the blocks x^{h_i}, and by the regularity of G, $b^{-1} = h_i$. Then it is straightforward that $a = c_i$ and $b = d_i$, and thus there are exactly λ pairs $c_i, d_i \in D$ such that $g = c_i d_i^{-1}$.

By considering the blocks x and x^g, similar considerations tell us that $g = e_i^{-1} f_i$ for exactly λ pairs $e_i, f_i \in D$. □

Theorem 2.8 has a very interesting converse which shows that any group with the properties of that theorem is a Singer group for a symmetric design. If G is a finite group, then a subset D of G such that every non-identity element of G can be represented exactly λ times as $c_i d_i^{-1}$ and exactly λ times as $e_i^{-1} f_i$, for $c_i, d_i, e_i, f_i \in D$, is called a *λ-difference set* for G. The *parameters* of the group G with a λ-difference set D are v, k, λ, where $v = |G|$, $k = |D|$. A *difference set* is a λ-difference set for some λ.

Theorem 2.9. Let G be a group of order v with a λ-difference set D of $k < v$ elements. Then there is a symmetric design $\mathscr{D}$ for (v, k, λ), unique up to isomorphism, such that G is a Singer group for $\mathscr{D}$ with difference set D

Proof. First we must construct $\mathscr{D}$. For each element $g \in G$ let (g) be a point of $\mathscr{D}$ and $[g]$ a block of $\mathscr{D}$; the incidence rule is: (g) is on $[h]$ if and only if $g \in Dh$. (Thus we could think of the block $[h]$ as the point set of elements Dh in G.)

If (g_1) and (g_2) are two distinct points of $\mathscr{D}$ then they lie on the block $[h]$ if and only if $g_1 h^{-1}$ and $g_2 h^{-1}$ are in D. Since $g_1 \neq g_2$ it follows that $g_1 g_2^{-1} \neq 1$, and so there exist exactly λ distinct pairs $c_i, d_i \in D$ ($i = 1, 2, \ldots, \lambda$) such that $g_1 g_2^{-1} = c_i d_i^{-1}$. For a fixed choice of i let $f = d_i^{-1} g_2$; then $g_1 f^{-1} = c_i$ and $g_2 f^{-1} = d_i$, and this gives us λ blocks $[d_i^{-1} g_2]$ ($i = 1, 2, \ldots, \lambda$) containing the two points (g_1), (g_2). Now we must show that these λ blocks are the only blocks of $\mathscr{D}$ which contain the two given points. Suppose (g_1) and (g_2) are on the block $[h]$. There is a unique $y \in G$ such that $h = y^{-1} g_2$ and since (g_2) is on $[h] = [y^{-1} g_2]$, y is in D. But (g_1) is also on $[h] = [y^{-1} g_2]$ and thus $g_1 (y^{-1} g_2)^{-1}$ is in

D; i.e. there is an element $m \in D$ such that $g_1 g_2^{-1} y = m \in D$, so $g_1 g_2^{-1} = my^{-1}$. But then m, y must be one of the pairs c_i, d_i and $[h]$ is one of the blocks $[d_i^{-1} g_2]$. So D is a 2-structure and, since every block clearly contains k points, $\mathscr{D}$ is uniform. Finally, $\mathscr{D}$ is square since it has v points and v blocks, and is proper since $k < v$. So $\mathscr{D}$ is a symmetric design for (v, k, λ).

For any $b \in G$, the mapping which sends the point (g) to (gb) and the block $[h]$ to the block $[hb]$ is easily seen to be an automorphism of $\mathscr{D}$, and so G induces an automorphism group of $\mathscr{D}$ which is obviously regular on both points and blocks, and so G becomes a Singer group of $\mathscr{D}$.

Exercise 2.7, which follows, finishes the proof of the uniqueness of $\mathscr{D}$.

Exercise 2.7. Show that the symmetric design $\mathscr{D}$ constructed in Theorem 2.9 is unique, up to isomorphism, with the property that G acts as a Singer group on $\mathscr{D}$ with difference set D. □

We call $\mathscr{D}$ the design *associated with* the Singer group G and its difference set D.

Exercise 2.8. Let $\mathscr{D}$ be the design, with parameters (v, k, λ), associated with the abelian Singer group G and its difference set D. Define σ by

$$\sigma : (g) \to [g^{-1}]$$
$$[h] \to (h^{-1}).$$

Show that σ is a polarity of $\mathscr{D}$, and if v is odd then σ has exactly k absolute points.

(Exercise 2.8 is important: it tells us that designs with abelian Singer groups have polarities and are self-dual, and this is very useful in a number of ways.)

Exercise 2.9. Let G be the additive group of integers modulo v, with D as given. Show that D is a difference set for G, find the parameters of the associated symmetric design, and determine the absolute points of the polarity σ of Exercise 2.8.

(a) $v = 7, \quad D = \{0, 1, 3\}$,

(b) $v = 11, \quad D = \{1, 3, 4, 5, 9\}$.

Exercise 2.10. Let G be the additive group of $V_4(2)$ (the 4-dimensional vector space over $GF(2)$, or note that G is merely the direct product of four cyclic groups of order 2). Show that the set

$$D=\{(0,0,0,0),(1,0,0,0),(0,1,0,0),(0,0,1,0),(0,0,0,1),(1,1,1,1)\}$$

is a difference set for G, and find the parameters of the associated symmetric design. Show that the design has a polarity in which all points are absolute. Show that the design also has a polarity in which no point is absolute (hint: replace D by one of its translates and apply Exercise 2.8).

Exercise 2.11. Show that in the additive group of integers modulo 19 the set $\{1,4,5,6,7,9,11,16,17\}$ is a difference set and find the parameters of the associated symmetric design.

As the exercises show, difference sets give a very simple and convenient way of representing certain symmetric designs. The entire design is determined by the Singer group and the difference set. If a symmetric design has a cyclic Singer group then it also has a very special type of incidence matrix. As an illustration let us return to Exercise 2.9(a). The blocks of the associated design $\mathscr{D}$ are the point sets $D+i$ (where addition is modulo 7). If we write x for the point (x) then the blocks are $D=\{0,1,3\}$, $D+1=\{1,2,4\}$, $D+2=\{2,3,5\}$, $D+3=\{3,4,6\}$, $D+4=\{4,5,0\}$, $D+5=\{5,6,1\}$, $D+6=\{6,0,2\}$.

Clearly, $\mathscr{D}$ has as an incidence matrix

$$\begin{bmatrix} 1&0&0&0&1&0&1\\ 1&1&0&0&0&1&0\\ 0&1&1&0&0&0&1\\ 1&0&1&1&0&0&0\\ 0&1&0&1&1&0&0\\ 0&0&1&0&1&1&0\\ 0&0&0&1&0&1&1 \end{bmatrix},$$

where the ith row represents the point $i-1$ and the jth column represents the block $D+j-1$. This is an example of a *cyclic matrix*: each row (or column) is obtained from the previous one by cycling each entry one place to the right (or down).

Exercise 2.12. Show that a symmetric design $\mathscr{D}$ has a cyclic incidence matrix if and only if it has a cyclic Singer group.

Certain families of symmetric designs admit Singer groups. We will now show that, for any $n \geqslant 2$ and any prime power q, $\mathscr{P}_n(q)$ has a Singer group. (This fact was first proved by Singer: hence the term Singer group.) If $K = GF(q)$ then there is a (unique) field $F = GF(q^{n+1})$, and F contains K as a subfield. We can think of F as a vector space over K in the following way: the sum of two 'vectors' f_1 and f_2 in F is simply $f_1 + f_2$, while the product of the 'vector' $f \in F$ and the 'scalar' $k \in K$ is kf. It is easy to see that these definitions satisfy all the rules of a vector space:

$$k(f_1 + f_2) = kf_1 + kf_2,$$
$$(k_1 k_2)f = k_1(k_2 f)$$
$$(k_1 + k_2)f = k_1 f + k_2 f$$
$$1 \cdot f = f,$$

for k, k_1, $k_2 \in K$ and f, f_1, $f_2 \in F$.

Since $|F| = q^{n+1}$, F has dimension $n+1$ over K, so we can consider the points and blocks of $\mathscr{P}_n(q)$ as the set of 1- and n-dimensional subspaces of F. For each $f \in F$, $f \neq 0$, define M_f to be the mapping of F onto F given by $M_f : x \to xf$. Then M_f is a linear transformation of F, for:

$$(x + y)M_f = xM_f + yM_f,$$
$$(kx)m_f = k(xM_f),$$

for $x, y \in F, k \in K$. M_f is non-singular since it is onto, and the set $G = \{M_f\}$ of all such mappings is a (cyclic) group of linear transformations of F, transitive on its non-zero elements. So G induces a group H on the subspaces of F and, since G is transitive on non-zero vectors, H is certainly transitive on 1-dimensional subspaces, i.e. on the points of $\mathscr{P}_n(q)$. Since H is a homomorphic image of G it is also cyclic, and hence, by Result 2.3, regular; in fact $H \cong G/N$, where $N = \{M_k \mid k \in K,\ k \neq 0\}$. Thus we have established:

Theorem 2.10. The symmetric design $\mathscr{P}_n(q), n \geqslant 2$, has a cyclic Singer group. □

Notice that although Theorem 2.10 guarantees the existence of a cyclic Singer group H it does not help us much to find a difference set for H. In fact, the practical problem of finding difference sets is very difficult. In the next section we will prove the Multiplier Theorem which, in certain cases, will help us decide if a group has a difference set and, when one exists, help to construct it.

Although all of the Singer groups given so far have been abelian, there are examples of non-abelian ones. However, less is known about them. Finally, note that if a Singer group G is cyclic of order v then we can represent it as the set of integers modulo v. A λ-difference set D of G is then a set of integers such that $x \not\equiv 0 \pmod{v}$ implies $x \equiv d_1 - d_2 \pmod{v}$ for exactly λ pairs $d_1, d_2 \in D$. This, by the way, is the reason for the term 'difference set'.

2.5 Multipliers

In this section we discuss 'multipliers' of abelian Singer groups: that is, an automorphism ϕ of the Singer group G which sends the difference set D onto Dg, for some g in G (and if G is cyclic and written as the additive group of integers modulo v, this just means we are looking for natural numbers t, prime to v, such that $tD = D + g$ for some g in G). The proofs involve rather more group theory than has been required so far, and a reader with insufficient group theory might skip the general results and go straight to the statements of Theorems 2.15 and 2.18, and the examples and exercises which follow.

If G is a Singer group of a symmetric design $\mathscr{D}$ we denote the normaliser of G in Aut $\mathscr{D}$ by N. For any $g \in N$ we let ϕ_g be the mapping ϕ_g: $y \to g^{-1}yg$, which is an inner automorphism of Aut $\mathscr{D}$, and, since $g \in N$ and G is normal in N, ϕ_g is an automorphism of G; this is the automorphism *induced* by g on G.

Lemma 2.11. If g is in N and ϕ_g is induced by g on G, then for a difference set D of G, we have $D^{\phi_g} = b_1^{-1}Db_2$ for some choice of elements $b_1, b_2 \in G$, and D^{ϕ_g} is a difference set for G.

Proof. If d is in D, then P^d is on x, and since $d^{\phi_g} = g^{-1}dg$, we have $(P^g)^{d^{\phi_g}} = P^{dg}$, which is on x^g. Thus D^g is a difference set with base point P^g and base block x^g. There exist elements $b_1, b_2 \in G$ such that $P^g = P^{b_1}$ and $x^g = x^{b_2}$, and so (see Lemma 2.7) $D^{\phi_g} = b_1^{-1}Db_2$. □

Lemma 2.12. If ϕ is an (abstract) automorphism of G and if there exist elements $b_1, b_2 \in G$ such that $D^{\phi} = b_1^{-1}Db_2$, then there exists an element $g \in N$ such that $\phi = \phi_g$.

Proof. For any $c \in G$, the point P^c is in $\mathscr{D}$ and we define $(P^c)^g = P^{b_1c^{\phi}}$ and, similarly, $(x^c)^g = x^{b_2c^{\phi}}$. This defines g on all of $\mathscr{D}$, and we must show that g is in

Aut $\mathscr{D}$, and then in N. Now P^c is on x^f if and only if $cf^{-1} \in D$. On the other hand, $(P^c)^g$ is on $(x^f)^g$ if and only if $(b_1 c^\phi)(b_2 f^\phi)^{-1} = b_1 c^\phi (f^\phi)^{-1} b_2^{-1} \in D$, which is equivalent to $(cf^{-1})^\phi \in b_1^{-1} D b_2$. But $cf^{-1} \in D$ if and only if $(cf^{-1})^\phi \in D^\phi = b_1^{-1} D b_2$. So $g \in \text{Aut } \mathscr{D}$.

For any $h \in G$, g^{-1} maps $P^{b_1 h\phi}$ onto P^h. Thus $(P^{b_1 h\phi})^{g^{-1}cg} = P^{hcg} = (P^{b_1 h\phi})^{c\phi}$, for any $c \in G$. Thus $g^{-1}cg = c^\phi$ for any $c \in G$, and so g normalises G. So $g \in N$. □

If $g \in G$ then, of course, $g \in N$, and ϕ_g is merely conjugation by g. Similarly if ϕ is an inner automorphism of G, then the automorphism ϕ_g of Lemma 2.12 is right multiplication by an element of G, and is already in G. But it is certainly possible for automorphisms of G to give rise to new automorphisms of $\mathscr{D}$.

Since G is regular on the points of $\mathscr{D}$ and is normal in N we know that $N = N_P G$ for any point P of $\mathscr{D}$. We define the *multiplier group* M of G to be N_p. Thus the multiplier group of G is a group of automorphisms of $\mathscr{D}$ whose non-identity elements act as outer automorphisms of G. Note also that, as an abstract group, the multiplier group is independent of the base point since, by transitivity of G, $N_P \cong N_Q$ for any Q in $\mathscr{D}$. One obvious problem immediately arises: which automorphisms of G give multipliers?

Lemma 2.13. If a, b are in N_P then $\phi_a = \phi_b$ implies $a = b$.

Proof. It $\phi_a = \phi_b$ then, clearly, $\phi_{ab^{-1}} = \phi_1 = 1$. Thus we may suppose, without any loss of generality, that $b = 1$. Hence ϕ_a is the identity automorphism of G so that $ah = ha$ for all h in G. If Q is any point of $\mathscr{D}$ then $Q = P^c$ for some c in G. Thus $Q^a = P^{ca} = P^{ac}$. But, since $a \in N_P$, $P^{ac} = P^c = Q$ and so $Q^a = Q$ for all Q in $\mathscr{D}$, i.e. $a = 1$. □

Theorem 2.14. The automorphism ϕ of G is induced by a multiplier a of M if and only if there is an element g of G with $D^\phi = Dg$.

Proof. Suppose $D^\phi = Dg$. Then define a by $(P^c)^a = P^{c\phi}$, $(x^c)^a = x^{gc^\phi}$ (or, equivalently, by $(c) \to (c^\phi)$, $[c] \to [gc^\phi]$). Then if P^c is on x^h we have $ch^{-1} \in D$. Hence $c^\phi h^{-\phi} \in D^\phi = Dg$ so that $c^\phi (gh^\phi)^{-1} \in D$ or $P^{c\phi}$ is on $x^{gh\phi}$. Clearly, a fixes P and normalises G and hence a is a multiplier.

Conversely, suppose $a \in N_P$. Since $a \in \text{Aut } \mathscr{D}$ and G is transitive on $\mathscr{D}$ there exists a g in G such that $x^a = x^g$. Now d is in D if and only if P^d is on x and so, for any d in D, $P^{da} = P^{a(a^{-1}da)} = P^{ad\phi_a}$ is on $x^a = x^g$. Thus d^ϕ is in Dg or $D^\phi = Dg$. □

Notice that the effect of a multiplier is now precisely given. If we know that ϕ is an automorphism of the Singer group G, then ϕ is a multiplier if and only if ϕ sends a difference set D onto a *translate* Dg; in this case ϕ acts on the points and blocks of $\mathscr{D}$ by the simple rule $(c) \to (c^{\phi})$ and $[c] \to [gc^{\phi}]$. Furthermore, Lemma 2.13 implies that the two representations of the multiplier group (i.e. as permutations of the points of $\mathscr{D}$ or as automorphisms of G) are permutation isomorphic.

As we have just seen, if a Singer group G has multipliers then the full automorphism group of the design $\mathscr{D}$ constructed from G must be larger than G. For any multiplier ϕ of G we shall refer to the corresponding automorphism b of $\mathscr{D}$ as the automorphism *associated* with ϕ. Note, by the way, that b will usually depend on D as well as ϕ.

Exercise 2.13. If ϕ is a multiplier of a Singer group G with difference set D and associated automorphism b, what is the automorphism associated with ϕ if D is replaced by $e^{-1}Df$, for some e, f in G?

If the group G is abelian then for each integer s prime to v the mapping $g \to g^s$ is an automorphism of G which we will denote by μ_s.

Theorem 2.15 (The Multiplier Theorem). If G is an abelian group with a difference set D having parameters (v, k, λ) then, for any prime p with $p > \lambda$, $(p, v) = 1$ and p dividing n, μ_p is a multiplier.

Proof. As in the case of the BRC Theorem, this proof has nothing to do with our geometric or combinatorial considerations. It is included for completeness and because we consider it important. However, the reader should not be afraid to omit it if he finds it difficult. Failure to read the proof will not hinder the student in any way.

We shall assume that G is an abelian group of order v, D is a subset of k elements and every g in G $(g \neq 1)$ can be represented in exactly λ ways as $g = d_1 d_2^{-1}$ for some elements d_1, d_2 in D. Let F be the rational field and let FG be the group algebra of G over F, where we identify the identities of F, G and FG, and write a typical element of FG as $\sum_{x \in G} a_x x$. We define two special elements:

$$\delta = \sum_{d \in D} d, \quad \sigma = \sum_{g \in G} g.$$

We note that if $\alpha = \sum a_x x$ then $\alpha\sigma = (\sum a_x)\sigma$.

If we define * by $(\sum a_x x)^* = \sum a_x x^{-1}$ then it is well known (and easy to prove) that * is an automorphism of FG. The condition that D is the difference set expresses itself in FG as:

$$\delta\delta^* = \delta^*\delta = k + \lambda(\sigma - 1) = n + \lambda\sigma. \tag{1}$$

If p is any prime divisor of n with $(p, v) = 1$ then $x \to x^p$ is an automorphism of G so that the mapping $\phi: \sum a_x x \to \sum a_x x^p$ is an automorphism of FG. If we expand $(\sum_{d \in D} d)^p$ by the 'binomial theorem' we get

$$\delta^\phi = \sum d^p = (\sum d)^p + p\alpha = \delta^p + p\alpha, \tag{2}$$

where α is an element $\sum a_x x$ with each a_x an integer (we call such an element an *integral element* of FG). Thus, using (1) and (2), we have

$$\begin{aligned}\delta^\phi\delta^* &= \delta^p\delta^* + p\alpha\delta^* = \delta^{p-1}(n + \lambda\sigma) + p\alpha\delta^* \\ &= \lambda\delta^{p-1}\sigma + p\left(\alpha\delta^* + \left(\frac{n}{p}\right)\right)\delta^{p-1} = \lambda\delta^{p-1}\sigma + p\alpha_1,\end{aligned}$$

where α_1 is an integral element.

But $\delta\sigma = k\sigma = (n + \lambda)\sigma$, and so, using the fact that n is a multiple of p, $\delta^{p-1}\sigma$ can be reduced successively, finally leading to

$$\delta^\phi\delta^* = \lambda^p\sigma + p\alpha_2, \tag{3}$$

where α_2 is integral. But, since λ is an integer, $\lambda^p \equiv \lambda (\text{mod } p)$ and so

$$\delta^\phi\delta^* = \lambda\sigma + p\alpha_3, \tag{4}$$

where α_3 is integral (since, of course, σ is certainly integral).

Now suppose $\alpha_3 = \sum a_x x$. Then, for any particular x in G, the coefficient of x on the left-hand side of (4) is certainly not negative. Hence we must have $0 \leqslant \lambda + pa_x$, from which we have $a_x \geqslant -\lambda/p$. But $p > \lambda$, so $a_x \geqslant -\lambda/p > -1$, and since α_3 is integral each of its coefficients is non-negative.

Now we compute: $(\delta^\phi\delta^*)(\delta^\phi\delta^*)^* = \delta^\phi\delta^*\delta\delta^{*\phi} = (\delta^*\delta)(\delta\delta^*)^\phi = (n + \lambda\sigma)(n + \lambda\sigma)^\phi = (n + \lambda\sigma)^2$. Thus

$$(\delta^\phi\delta^*)(\delta^\phi\delta^*)^* = n^2 + 2n\lambda\sigma + \lambda^2\sigma^2. \tag{5}$$

But, on the other hand, the left-hand side of (5) is equal to $(\lambda\sigma + p\alpha_3)(\lambda\sigma + p\alpha_3)^* = \lambda^2\sigma^2 + \lambda\sigma p(\alpha_3 + \alpha_3^*) + p^2\alpha_3\alpha_3^*$. Hence

$$n^2 + 2n\lambda\sigma = p^2\alpha_3\alpha_3^* + \lambda\sigma p(\alpha_3 + \alpha_3^*). \tag{6}$$

We now compute $\lambda\sigma p\alpha_3$ as follows: $\delta^\phi\delta^*\sigma = \delta^\phi(n + \lambda)\sigma = (n + \lambda)\delta^\phi\sigma = (n + \lambda)(\delta\sigma)^\phi = (n + \lambda)^2\sigma$. (Here we repeatedly use $(\sum a_x x)\sigma = (\sum a_x)\sigma$.) But from (4) we have $\delta^\phi\delta^*\sigma = \lambda\sigma^2 + p\alpha_3\sigma = \lambda v\sigma + p\alpha_3\sigma$. Equating these two expressions for $\delta^\phi\delta^*\sigma$ gives $(n + \lambda)^2\sigma = \lambda v\sigma + p\alpha_3\sigma$. But, by Exercise 3.2, $\lambda(v - 1) = k(k - 1)$, which, with $n = k - \lambda$, gives $p\alpha_3\sigma = n\sigma$. Using the fact that * is an automorphism of FG we also have $p\alpha_3^*\sigma = (p\alpha_3\sigma)^* = (n\sigma)^* = n\sigma$. Thus (6) gives

$$n^2 = p^2\alpha_3\alpha_3^*. \tag{7}$$

Now it is easy to see that if as many as two elements of G were actually present in α_3 with non-zero coefficients, then (since all the coefficients are non-negative) the expression $\alpha_3\alpha_3^*$ could not equal a multiple of the identity element, but would have to have a non-identity group element in it (namely the product of the two elements with non-zero coefficient in α_3). Thus, since $\alpha_3\alpha_3^* = (n/p)^2$, it follows that $\alpha_3 = (n/p)g$ for some g in G.

If $\beta = (1/n) - \{\lambda\sigma/[n(\lambda v + n)]\}$, we see that

$$(n+\lambda\sigma)\beta = (n+\lambda\sigma)\frac{1}{n} - \frac{\lambda\sigma}{n(\lambda v+n)} = 1 + \left(\frac{\lambda}{n} - \frac{\lambda}{\lambda v+n} - \frac{\lambda^2 v}{n(\lambda v+n)}\right)\sigma = 1.$$

But this means that $\delta\beta$ is an inverse for δ^*. Thus, since $\delta^\phi\delta^* = ng + \lambda\sigma$,

$$\delta^\phi = \delta^\phi\delta^*\delta\beta = (\lambda\sigma + ng)\delta\beta = \delta(\lambda\sigma + ng)\frac{1}{n} - \frac{\lambda\sigma}{n(\lambda v+n)}$$

$$= \sigma\left[g + \left(\frac{\lambda}{n} - \frac{\lambda}{(\lambda v+n)} - \frac{\lambda^2 v}{n(\lambda v+n)}\right)\sigma\right] = \delta g.$$

But in terms of the group, G, $\delta^\phi = \delta g$ is equivalent to $D^\phi = Dg$; i.e. μ_p is a multiplier. □

Although the condition $p > \lambda$ is crucial to our proof, it is widely conjectured that this is an unnecessary restriction. Certainly in all known examples μ_p is a multiplier for every prime p with $(p, v) = 1$ and $p \mid n$. It is also worth noting that if μ_p and μ_q are multipliers then, clearly, μ_{pq}, which is equal to $\mu_p\mu_q$, is also a multiplier. (If μ_s is a multiplier then we shall often refer to the integer s as the multiplier.) We also call μ_s a *numerical multiplier*.

Problem 2.1. Remove the restriction $p > \lambda$ from Theorem 2.15.

If G is abelian then $b_1^{-1}Db_2 = Db_1^{-1}b_2$ and so if ϕ is a multiplier we define the element $b \in N$ associated with it by $(P^c)^b = P^{c\phi}$ and $(x^c)^b = x^{(b_1^{-1}b_2)c\phi}$. This enables us to give an immediate proof of

Lemma 2.16. If G is an abelian Singer group for $\mathscr{D}$ and if ϕ is a multiplier with associated automorphism b of $\mathscr{D}$ then b fixes at least one block.

Proof. Clearly, $P^b = P$ and so by Lemma 1.40 there is a block fixed by b. □

Corollary 2.17. If G is an abelian Singer group for $\mathscr{D}$ and if ϕ is a multiplier of G then there is a difference set of G which is fixed by ϕ; i.e. there is a choice of D such that $D^\phi = D$.

Proof. As we saw in the proof of Theorem 2.9, the blocks of $\mathscr{D}$ are the sets Dc as c varies over G. But, by Lemma 2.7, Dc is also a difference set for G and so any block fixed by b gives a difference set which is fixed by ϕ. □

From the proof of the last theorem and its corollary it is clear that saying a multiplier fixes a difference set is equivalent to saying that the associated automorphism of the design fixes a block. Having shown that each multiplier fixes a difference set we will now show that, if $(v, k) = 1$, there is a difference set which is fixed by every multiplier.

Theorem 2.18. Let G be an abelian Singer group for a symmetric 2-(v, k, λ) $\mathscr{D}$ with $(v, k) = 1$. Then there is a difference set which is fixed by all multipliers of G.

Proof. Let D_0 be any difference set for G and let $b = \prod_{d \in D_0} d$. Then, since $(v, k) = 1$, there is an element c in G with $c^k = b$ and so $\prod_{d \in D_0} (dc^{-1}) = (\prod_{d \in D_0} d)c^{-k} = 1$. Thus we can choose $D = D_0 c^{-1}$ and then $\prod_{d \in D} d = 1$. Now let ϕ be a multiplier of G so that, since G is abelian, $D^\phi = Da$ for some $a \in G$ (Lemma 2.11). Then $\prod_{d \in D} (d^\phi) = (\prod_{d \in D} d)^\phi = 1$. But $\prod_{d \in D} d^\phi = \prod_{d \in D} (da) = (\prod_{d \in D} d)a^k = a^k$. Thus $a^k = 1$ and, since $(v, k) = 1$, this implies $a = 1$, and proves the theorem. □

When constructing a difference set for a cyclic group G of order v it is usual to write G additively as the integers modulo v. An integer t is then a multiplier of a difference set D if $tD = \{tx \mid x \in D\}$ is equal to $D + y$ for some $y \in G$. (This explains the term 'multiplier'.) We shall use a multiplier p to construct a difference set D with $pD = D$. (If G has a difference set at all then it has one with this property, by Corollary 2.17.)

Example 2.4. $v = 7$, $k = 3$, $\lambda = 1$.

The only prime satisfying Theorem 2.15 is 2. Suppose $a \in D$ then, since $2D = D$, $2a \in D$ and $4a \in D$ (note $8a = a$). Choosing $a = 1$ we get $\{1, 2, 4\}$ as the required difference set.

Example 2.5. $v=37$, $k=9$, $\lambda=2$.

This time 7 is the only prime satisfying Theorem 2.15. If $a \in D$ then, remembering that all arithmetic is modulo 37, we must work out the powers of 7 to determine the rest of D. Simple calculations give $7^2=12, 7^3=10, 7^4=33, 7^5=9, 7^6=26, 7^7=34, 7^8=16, 7^9=1$. Thus, putting $a=1$, we get $\{1,7,9,10,12, 16, 25, 33, 34\}$.

Exercise 2.14. Find a 1-difference set for
(a) the cyclic group of order 13, and
(b) the cyclic group of order 31.

In both the examples and the above exercise there is only one possible multiplier given by the theorem. We now consider a situation where there are more. We already know that the smallest value for which there might exist a symmetric design of order n with $\lambda=1$, but for which none is known is $n=10$. We now use Corollary 2.17 to show that if such a design exists then it cannot have a cyclic Singer group.

Example 2.6. There is no cyclic difference set with parameters $v=111$, $k=11$, $\lambda=1$.

Proof. By Theorem 2.15, 5 and 2 are both multipliers for the cyclic group of order 111. Suppose that D is a difference set with $2D=D$ and $5D=D$ (D must exist, by Theorem 2.18). If $a \in D$ then a, $2a$, $4a$, $5a$ are all in D. But $2a-a=5a-4a$ which, since $\lambda=1$, implies $2a \equiv 5a \pmod{111}$ and $a \equiv 4a \pmod{111}$. But this is only possible if $a \equiv 0, 37, 74 \pmod{111}$, and so we cannot get 11 elements in D. □

Exercise 2.15. Use Theorem 2.15 to construct a cyclic difference set for $(19, 9, 4)$. Compare your solution with Exercise 2.11.

We now illustrate the limitations of the Multiplier Theorem by giving some examples for which it gives no information.

Example 2.7. The set $D=\{0,1,2,4,5,8,10\}$ is a 3-difference set for the additive group of the integers modulo 15. In this case $n=4$ and there is no prime p satisfying $p \mid n$ and $p>\lambda$. (Note $n=4$ and $\lambda=3$.) However, in this example 2 *is* a numerical multiplier. (Check this!)

Example 2.8. The difference set of Exercise 2.10 is such that the Multiplier Theorem again gives no information, for $n=4$, $v=16$, and there is no prime dividing n and relatively prime to v. But, as we shall see, the multiplier group of this Singer group is very large. First, we note that, since the group is the additive group of the vector space of dimension 4 over the field $GF(2)$, the automorphism group of G is the group $GL(4,2)$ of all 4×4 non-singular matrices over $GF(2)$. If P is any permutation matrix in $GL(4,2)$ then P fixes the elements $(0,0,0,0)$ and $(1,1,1,1)$ and permutes the four elements $(1,0,0,0)$, $(0,1,0,0)$, $(0,0,1,0)$ and $(0,0,0,1)$ in some order. So P fixes D. Thus the multiplier group M of G includes the 24 permutation matrices. The following matrix

$$Q=\begin{bmatrix}1&0&0&0\\0&1&0&0\\0&0&1&0\\1&1&1&1\end{bmatrix}$$

interchanges the elements $(0,0,0,1)$ and $(1,1,1,1)$ and fixes the other elements of D. So M includes the subgroup H generated by all the permutation matrices and Q; since H is transitive on the five elements of D other than $(0,0,0,0)$, and the subgroup of H fixing $(1,1,1,1)$ is the symmetric group on the other four elements, H is isomorphic to the symmetric group S_5 (acting on the five elements of D other than $(0,0,0,0)$).

Consider the point $(0,0,0,0)$ of $\mathscr{D}$; it is on six blocks $D+\mathbf{v}$, and these are the six blocks with $\mathbf{v}\in D$. The group H is 5-transitive on the five elements $\mathbf{v}\in D$ ($\mathbf{v}\neq(0,0,0,0)$), so H fixes the block D and is 5-transitive on the other five blocks on $(0,0,0,0)$. Let R be the matrix:

$$R=\begin{bmatrix}1&0&0&0\\1&1&0&0\\1&0&1&0\\1&0&0&1\end{bmatrix}.$$

Then R sends D onto the set $\{(0,0,0,0), (1,0,0,0), (1,1,0,0), (1,0,1,0), (1,0,0,1), (0,1,1,1)\}$, which is $D+(1,0,0,0)$. So R is in M, and then M is transitive on the six blocks on $(0,0,0,0)$. So M contains the symmetric group

S_6 acting on the six blocks that contain $(0,0,0,0)$. If there are any more elements in M, then they can be assumed to fix all six blocks on the point $(0,0,0,0)$.

Suppose that b is an automorphism of $\mathscr{D}$ fixing all six blocks on the point $(0,0,0,0)$ and let X be any other point. Then X is joined to $(0,0,0,0)$ by two blocks and these blocks meet only in $(0,0,0,0)$ and X. Thus, since $(0,0,0,0)$ is fixed, b must fix X. But, since X was arbitrary, this means b fixes all points of $\mathscr{D}$, i.e. b is the identity. Hence M is isomorphic to S_6 and has order 720.

This last example illustrates that a Singer group can have many non-numerical multipliers. However, this cannot happen when the group is cyclic.

Theorem 2.19. If G is a cyclic Singer group then all its multipliers are numerical.

Proof. Let $G=\langle g\rangle$; i.e. G consists of the v powers g^i $(i=0,1,\ldots,v-1)$ of g. Then if ϕ is any automorphism of G it follows that $\phi: g\rightarrow g^r$ for some integer r with $1\leqslant r\leqslant v-1$. Thus $\phi: g^i\rightarrow g^{ri}=(g^i)^r$ and ϕ is numerical. □

Exercise 2.16. Use Theorem 2.19 to show there cannot exist a Singer group for a symmetric design for $(61, 16, 4)$.

As the reader will have discovered in the exercises, when we are discussing a *cyclic* Singer group we can assume that it is the group C_v of integers modulo v under addition; all multipliers are numerical, and (if $(v,k)=1$) we choose a difference set D such that $tD=D$ for all multipliers t. The multipliers t can be generated by our Theorem 2.15, i.e. find primes p which are (i) prime to v, (ii) greater than λ, and (iii) divide n, and consider the multiplicative group modulo v generated by all such p. This was essentially the first multiplier theorem, but, because of the reasonable conjecture that $p>\lambda$ is *not* necessary in order that p should be a multiplier, many extensions of Theorem 2.15 have been established, without, however, providing yet that p need not be greater than λ. We mention one of these results, since it is both powerful and convenient to use.

Result 2.4. Let G be an abelian Singer group with parameters (v, k, λ) and let n_1 be a square-free divisor of $n=k-\lambda$, with $n_1>\lambda$. Suppose t is a positive integer, t

relatively prime to v, such that for every prime p dividing n_1 there is an integer $a = a(p)$ for which $p^a \equiv t \pmod{v}$. Then μ_t is a multiplier of G.

A proof of Result 2.4 for the case that G is cyclic can be found in [1], and that proof, together with most of the others for cyclic G, extends in a straightforward manner to the case that G is abelian.

2.6 Arithmetical relations and Hadamard 2-designs

If $\mathscr{D}$ is a non-trivial symmetric design for (v, k, λ) then $2 \leqslant k \leqslant v-2$ and so, by Corollary 1.12 and Lemma 1.13, $\mathscr{C}(\mathscr{D})$ is a symmetric design for $(v, v-k, v-2k+\lambda)$, with the same order as $\mathscr{D}$. Since v and n are two parameters in common between a symmetric design and its complement, a relationship between them has some interest.

Theorem 2.10. If $\mathscr{D}$ is a non-trivial symmetric design for (v, k, λ) with order n, then $4n-1 \leqslant v \leqslant n^2+n+1$.

Proof. Since v and n are the same for $\mathscr{D}$ and $\mathscr{C}(\mathscr{D})$, we can assume $2k \leqslant v$. However, $\lambda(v-1) = k(k-1)$ implies that k cannot divide v, so we have $2k < v$, and, since $\lambda \geqslant 1$, this yields

$$2n < v. \tag{1}$$

Another consequence of $v \neq 2k$ is that the arithmetic mean of k and $v-k$ is strictly greater than their geometric mean, i.e.

$$v > 2\sqrt{[k(v-k)]}. \tag{2}$$

But $k = n+\lambda$, so that $k(v-k) = (n+\lambda)(v-k) = nv + \lambda v - nk - \lambda k = nv + \lambda v - k^2$. However, putting $v = [k(k-1)+\lambda]/\lambda$, $\lambda v - k^2 = \lambda - k = -n$ which gives $k(v-k) = n(v-1)$. Substituting for $k(v-k)$ in (2) and squaring gives $v^2 > 4nv - 4n$. So $(v-2n)^2 > 4n^2 - 4n$ or $(v-2n)^2 \geqslant (2n-1)^2$. Clearly, $2n > 1$ and so, from (1), both $v-2n$ and $2n-1$ are positive integers. Hence $v - 2n \geqslant 2n-1$ or $v \geqslant 4n-1$, as required.

If a and b are any two positive integers then $a+b-1 \leqslant ab$. Since $\lambda \geqslant 1$ and $v > 2k$ we can apply this simple inequality with $a = \lambda$ and $b = v-2k+\lambda$ to get $\lambda(v-2k+\lambda) \geqslant v-2k+2\lambda-1$. But similar calculations to those above give $\lambda(v-2k+\lambda) = n(n-1)$ so that $n(n-1) \geqslant v-2n-1$ or $v \leqslant n^2+n+1$. □

Exercise 2.17. If $\mathscr{D}$ is a symmetric 2-(v, k, λ) of order n, show that $n(n-1)=\lambda(v-2k-\lambda)$.

The two extreme cases of Theorem 2.20 can actually occur and are of special interest. If $v=n^2+n+1$ then, by Exercise 2.17, $n(n-1)=\lambda(n^2-n+1-\lambda)$. This simplifies to $\lambda^2-\lambda(n^2-n+1)+n^2-n=0$ or $(\lambda-1)(\lambda-(n^2-n))=0$. If $\lambda=1$ then $\mathscr{D}$ is a 2-$(n^2+n+1, n+1, 1)$ design. If, on the other hand, $\lambda=n^2-n$ then $k=n^2$ and $\mathscr{D}$ is a 2-$(n^2+n+1, n^2, n-1)$. But in this case $\mathscr{C}(\mathscr{D})$ is a 2-$(n^2+n+1, n+1, 1)$, i.e. $\mathscr{C}(\mathscr{D})$ has $\lambda=1$. Thus if $v=n^2+n+1$, either $\mathscr{D}$ or its complement has $\lambda=1$.

If $v=4n-1$ a similar calculation gives $(k-2\lambda)(k-(2\lambda+1))=0$. If $k=2\lambda+1$ then $\mathscr{D}$ is a 2-$(4\lambda+3, 2\lambda+1, \lambda)$. Any design with these parameters, for any λ, is called a *Hadamard* 2-design. If $k=2\lambda$, then $\mathscr{D}$ is a 2-$(4\lambda-1, 2\lambda, \lambda)$ so that $\mathscr{C}(\mathscr{D})$ is a 2-$(4\lambda-1, 2\lambda-1, \lambda-1)$ which again is a Hadamard 2-design. (Note: $2\lambda-1=2(\lambda-1)+1$ and $4\lambda-1=4(\lambda-1)+3$.) Thus if $v=4n+1$, $\mathscr{D}$ is either a Hadamard 2-design or the complement of one. So the two 'extreme' cases of symmetric 2-designs with given order are designs with $\lambda=1$ and Hadamard 2-designs. There exist many infinite families of Hadamard 2-designs and we will discuss some of them in Chapter 3. In Chapter 1 we showed that, for any prime power q, the projective plane $\mathscr{P}(2, q)$ is a 2-$(q^2+q+1, q+1, 1)$ design, i.e. a symmetric design with $\lambda=1$. (See Exercises 1.28 and 1.30.) We will now show any symmetric design with $\lambda=1$ is a projective plane and that every finite projective plane is a symmetric design with $\lambda=1$.

Theorem 2.21. The class of finite projective planes is the same as the class of symmetric designs with $\lambda=1$ and $k>2$.

Proof. If $\mathscr{D}$ is a symmetric design with $\lambda=1$, then $k=n+1$ and $\mathscr{D}$ is a 2-$(n^2+n+1, n+1, 1)$. Since $\lambda=1$ any two points are on a unique common block, so we shall call the blocks *lines*. Since $k>2$ each line contains at least three points and, since $\mathscr{D}$ is symmetric, any two distinct lines meet in a common point (see Theorem 1.26). Finally, since $n^2+n+1>n+1$, given any line l, there is a point A not on l. Thus, if B and C are on l, A, B and C are three points which are not on a line. We have now established that $\mathscr{D}$ satisfies the axioms (P1)–(P5) for a projective plane given in Chapter 1.

Let $\mathscr{P}$ be a finite projective plane, let y be a line of $\mathscr{P}$ and let A be any point not on y. Then A is joined to every point of y by a unique line and, conversely, every line through A meets y in a unique point. So, if we let k denote the number of points on y, every point not incident with y is on exactly k lines. A similar argument shows that any line z not passing through A must contain exactly k points. Remembering that A can be any point not on y, we have shown that either every line contains exactly k points or there exist two lines y and z such that every point of $\mathscr{P}$ is on at least one of them. If every point of $\mathscr{P}$ is on y or z then, since y and z intersect in a unique point, if X, Y and Z are three points not on a common line, which exist by axiom (P3), one of them, Y say, must be on y but not z and another, Z say, must be on z but not y. By axiom (P2) the line joining Y and Z must contain a third point which, clearly, cannot be on y or z. This shows that it is impossible for every point of $\mathscr{P}$ to be on one of two lines and establishes that every line of $\mathscr{P}$ has k points. Axiom (P3) also guarantees the existence of three lines not on a common point and a similar argument now shows that each point is on k lines. Since any two points are on a unique line $\mathscr{P}$ is a 2-structure with $\lambda=1$. If P is any point of $\mathscr{P}$ then, since every other point is on exactly one of the lines through P, the number of points of $\mathscr{P}$ is $k(k-1)+1$ and $\mathscr{P}$ is a 2-$(k(k-1)+1, k, 1)$ or, writing $n+1$ for k, a 2-$(n^2+n+1, n+1, 1)$ with $r=n+1$. Straightforward use of Corollary 1.4 shows that $\mathscr{P}$ is symmetric. □

We will discuss projective planes in Chapter 3.

2.7 The Dembowski–Wagner Theorem

The 'classical' designs are the symmetric designs $\mathscr{P}_n(q)$ and their residual designs $\mathscr{A}_n(q)$. It is important therefore to be able to give purely combinatorial characterisations of these two families of designs. In this section we will prove a number of characterisations of $\mathscr{P}_n(q)$ but, before we can do this, we must first introduce the concept of a line and a plane in an arbitrary 2-design.

If $\mathscr{D}$ is a 2-design, then, for any pair of distinct points X, Y of $\mathscr{D}$, the *line* XY is defined as the intersection of all the blocks containing X and Y. (Here, of course, we are regarding a block as a point set.) Thus if $\lambda=1$ a line is merely a block.

Lemma 2.22. If $\mathscr{D}$ is a 2-design then any pair of distinct points lie on exactly one line.

Proof. Clearly, since $\mathscr{D}$ is a 2-design, every pair of points lie on at least one common block and, hence, on at least one line. But if X', Y' are distinct points on the line XY then the λ blocks containing X and Y must also be the λ blocks containing X' and Y', i.e. $XY = X'Y'$. □

Given any line l and any block x then either l lies in x or x intersects l in 0 or 1 points, and all three possibilities may occur in the same 2-design.

Exercise 2.18. Find the lines of example $\mathscr{E}$ and find a line which

(a) intersects e_1 in one point;
(b) does not intersect e_1;
(c) is contained in e_1.

Exercise 2.19. Show that any line of $\mathscr{P}_n(q)$ contains $q+1$ points and that every block intersects every line.

Despite Exercise 2.19 it is not true in general that all the lines of a 2-design will contain the same number of points. What is true is the following:

Lemma 2.23. If $\mathscr{D}$ is a 2-(v, k, λ) then, for any given line l of $\mathscr{D}$, $2 \leqslant |l| \leqslant (b-\lambda)/(r-\lambda)$ and $|l| = (b-\lambda)/(r-\lambda)$ if and only if every block intersects l.

Proof. Clearly, $2 \leqslant |l|$. In order to establish the upper bound we count the number of blocks which must meet l. For each point of l there are r blocks through it. Hence, since there are λ blocks containing l, for each point of l there are $r-\lambda$ blocks through it which intersect l in exactly that point. Consequently there are $|l|(r-\lambda)$ blocks which intersect l in exactly one point. This gives $b \geqslant \lambda + |l|(r-\lambda)$ or $|l| \leqslant (b-\lambda)/(r-\lambda)$ as required. Clearly, equality holds precisely when each block intersects l. □

If a set of points of a 2-design $\mathscr{D}$ lie on a line then we say that they are *collinear*. Any three non-collinear points form a *triangle*, and a *plane* is defined as the intersection of all the blocks through three points in a triangle. If there is no block through a given three non-collinear points then the plane defined by them consists of all the points of $\mathscr{D}$. (Note: if $\lambda = 1$ then lines are blocks so there cannot be a block containing three non-collinear points.) Unless $\mathscr{D}$ is a 3-

design there is no reason why three non-collinear points should determine a unique plane. In fact, it is even possible for one plane to be a proper subset of another.

Exercise 2.21. Show that every plane of $\mathscr{P}_n(q)$, $n \geqslant 3$, is contained in a constant number of blocks. Show also that any three non-collinear points lie in a unique plane.

Exercises 2.19, 2.20 and 2.21 show that the lines and planes of $\mathscr{P}_n(q)$ behave more nicely than we have any right to expect in an arbitrary 2-design. This is the basis of the Dembowski–Wagner Theorem. Before proving this result we set one more exercise.

Exercise 2.22. If $\mathscr{D}$ is a 2-design with $\lambda > 1$ in which the number of blocks through a plane is constant, μ say, show

(a) any three non-collinear points determine a unique plane;

(b) if l and m are any two intersecting lines then the number of blocks containing l and m is μ;

(c) if x is a block and u is a plane which is not contained in x then $x \cap u$ must be either empty, a point or a line.

Whenever we know that three non-collinear points X, Y, Z determine a unique plane we will refer to that plane as the plane XYZ.

Theorem 2.24 (Dembowski–Wagner). If $\mathscr{D}$ is a non-trivial symmetric 2-(v, k, λ) then the following are equivalent:

(a) $\mathscr{D} \cong \mathscr{P}_n(q)$ for some $n \geqslant 2$ and some q, or $\mathscr{D}$ is a projective plane;

(b) every line meets every block;

(c) each line has $(b-\lambda)/(r-\lambda)$ points on it;

(d) every plane is contained in the same number of blocks.

Proof. It is easy to check that (a) implies (b), (c) and (d). Most of these properties were either established in Chapter 1 or have been given as exercises. However, the reader should recheck them all. Also Lemma 2.23 implies that (b) and (c) are equivalent. We will prove the theorem by showing that (d) is equivalent to (b) and (c) and then showing that (b), (c) and (d) imply (a).

Suppose $\mathscr{D}$ satisfies (d). Let P be a point of $\mathscr{D}$ and let l be a line containing P such that $m = |l| \geqslant |l'|$ for all lines l' with P on l'. Then, by Lemma 2.23,

$$m \leqslant (b-\lambda)/(r-\lambda) = (v-\lambda)/(k-\lambda) \tag{1}$$

(since $\mathscr{D}$ is symmetric). We now define an incidence structure $\mathscr{D}(P)$ whose points are the lines containing P, whose blocks are the blocks of $\mathscr{D}$ containing P and whose incidence is set theoretic inclusion. Denoting the parameters of $\mathscr{D}(P)$ by $v(P)$, $k(P)$, etc., we have $b(P)=r$, $r(P)=\lambda$ and if $\lambda(P)$ is the number of blocks through two points of $\mathscr{D}(P)$ then, by Exercise 2.22, $\lambda(P)$ is the number of blocks through a plane of $\mathscr{D}$ and this is a constant (by hypothesis). Hence $\mathscr{D}(P)$ is a proper 2-structure with more than one block and so, by Corollary 1.23, $b(P)\geqslant v(P)$, i.e. $k=r\geqslant v(P)$. However, since every point of $\mathscr{D}$ is on a unique line through P, we have

$$v-1=\sum_{l'\in P}(|l'|-1)\leqslant\sum_{l'\in P}(m-1)=v(P)(m-1)$$
$$\leqslant v(P)\left(\frac{v-\lambda}{k-\lambda}-1\right)=v(P)\frac{v-k}{k-\lambda}. \tag{2}$$

But, since $v(P)\leqslant k$, $v(P)((v-k)/(k-\lambda))\leqslant k(v-k)/(k-\lambda)=v-1$ (by Corollary 1.4). Thus each inequality in (2) must be an equality and, in particular, $|l'|=m=(v-\lambda)/(k-\lambda)$ for all l' on P. Since P was an arbitrary point this establishes (c).

Now suppose that $\mathscr{D}$ satisfies (b) and (c).

Let A, B, C be any three non-collinear points and let ζ be the number of blocks through the plane which they define. Then we must show that ζ is independent of the choice of A, B, C. Clearly, any block which contains B and C must contain all points on the line BC. Thus the number of blocks through A which do not contain the line BC is precisely the number of blocks which contain A but do not contain the plane ABC. This is equal to $k-\zeta$. Let X be any point on the line BC. There are λ blocks containing A and X. Since only ζ of these contain the line BC there must be $\lambda-\zeta$ blocks through A which meet BC only in the point X. This is true for any point X of BC and so, since by assumption there are $(v-\lambda)/(k-\lambda)$ points on BC and since, also by assumption, every block meets BC, the number of blocks through A which do not contain BC is $(\lambda-\zeta)(v-\lambda)/(k-\lambda)$. Thus $k-\zeta=(\lambda-\zeta)(v-\lambda)/(k-\lambda)$ which can be solved to express ζ in terms of k, v and λ. Thus ζ is independent of the choice of A, B, C and we have established the equivalence of (b), (c) and (d).

Now suppose that $\mathscr{D}$ satisfies (b), (c) and (d). We need to show that $\mathscr{D}$ satisfies the axioms (P1), (P2), (P3), (P4), that is, slightly rephrased:

(P1) there is exactly one line on two distinct points;

(P2) each line contains at least three points;

(P3) there exist three non-collinear points;

(P4) whenever W, X, Y, Z are distinct points with $WX\cap YZ\neq\phi$, then $WY\cap XZ\neq\phi$.

Axiom (P1) follows from Lemma 2.22 while (P2) and (P3) follow from (c) and the fact that $\mathscr{D}$ is non-trivial. Thus we only have to prove (P4).

If A, B, C are any three non-collinear points then, by Exercise 2.22, they determine a unique plane ABC. Exercise 2.22 also says that if a block does not contain ABC it must meet it in 0 points, 1 point or a line. Now suppose we are given W, X, Y, Z as in (P4). If any three are collinear then $WY \cap XZ \neq \phi$ so we may assume that no three of them are collinear. If $A = WX \cap YZ$ and if $\mathscr{E}$ is the plane AWY, then, by Exercise 2.22, $\mathscr{E} = AXZ$ so that XZ is contained in $\mathscr{E}$. Thus for any block x such that XZ is in x but $\mathscr{E}$ is not in x we have $x \cap \mathscr{E} = XZ$. Since, by (b), all blocks and lines meet, $\phi \neq WY \cap x \subseteq x \cap \mathscr{E} = XZ$; i.e. $WY \cap XZ \neq \phi$, as required. □

There are many ways in which the Dembowski–Wagner Theorem can be generalised. For instance, for properties (b) and (c) it is not necessary to assume that $\mathscr{D}$ is symmetric. In other words, it can be shown that a non-trivial 2-design in which every line meets every block (or equivalently every line has $(b-\lambda)/(r-\lambda)$ points) has to be symmetric. However, the assumption that $\mathscr{D}$ be symmetric is definitely necessary for property (d).

Exercise 2.23. Show that in $\mathscr{A}_n(q)$ $(n \geqslant 3)$, every plane is contained in the same number of blocks.

It is also possible to replace (b) by the much weaker assumption that there exists a block which meets every line.

We end this chapter by noting that if $\lambda > 1$ then the Dembowski–Wagner Theorem has an interesting corollary:

Corollary 2.25. If $\mathscr{D}$ is a symmetric 2-design with $\lambda > 1$ which admits an automorphism group transitive on its planes then $\mathscr{D} \cong \mathscr{P}(q)$, for some $n > 2$, and some q. □

Proof. $\mathscr{D}$ satisfies (d) and is not a projective plane. □

References

Reference [1] contains more information about multipliers and cyclic Singer groups, [3] gives a proof that for $\lambda = 2$ a quasiresidual design is uniquely embeddable in a symmetric design and [5] contains a discussion of various more general conditions for a quasiresidual design to be embeddable. The existence of Example 2.1 is established in

[2] and [3] is a reference book for the standard number theory results.

[1] Baumert, L. D. *Cyclic Difference Sets.* Lecture Notes in Mathematics 182, Berlin–Heidelberg–New York, Springer-Verlag, 1971.
[2] Bhattacharya, K. N. 'On a new symmetrical balanced incomplete block design'. *Bull. Calcutta Math. Soc.*, **36** (1944), 91–6.
[3] Hall, M. & Connor, W. S. 'An embedding theorem for balanced incomplete block designs'. *Canad. J. Math.*, **6** (1954), 35–41.
[4] Leveque, W. J. *Topics in Number Theory.* Reading, Mass., Addison Wesley, 1956.
[5] Singhi, N. M. & Shrikhande, S. S. 'Embedding of quasi-residual designs'. *Geom. Dedicata*, **2** (1974), 509–17.

3 Some families of symmetric designs

3.1 Introduction

In this chapter we continue the study of symmetric designs but in a somewhat more specific way than in Chapter 2. Section 3.2 contains a detailed discussion of the relation between projective and affine planes and develops some of the theory of non-desarguesian planes. (This latter development is primarily concerned with translation planes, quasifields and semifields. It has a different algebraic flavour than the rest of the book and, although the results are important, the proofs may be skipped if necessary.) Affine planes lead naturally to a discussion of latin squares in Section 3.3 followed by nets which are a very important class of 1-designs; in Section 3.4 one of the applications of nets discussed is the construction of a new infinite family of symmetric designs. Section 3.5 deals with Hadamard designs and Hadamard matrices and contains a construction of the Paley designs. Section 3.6 has a fairly detailed discussion of biplanes (symmetric designs with $\lambda=2$). In Section 3.7 we study the special class of graphs called 'strongly regular' and develop their elementary theory (e.g. eigenvalues and multiplicities), as well as giving a number of infinite families. Such graphs enable us to construct some new symmetric designs, and in addition they will be used again later in the book (see Chapters 7 and 8). The connections between strongly regular graphs and design theory are among the most important examples of the fruitful relationship between graphs and designs.

3.2 Projective and affine planes

As we have already seen, a projective plane of order n, which is another terminology for a symmetric 2-$(n^2+n+1, n+1, 1)$, exists for any prime power n. We also know, from Corollary 2.4, that there does not exist a projective plane of order n if $n \equiv 1$ or 2 (mod 4) and n is not the sum of two integral squares. In fact, for the other values of n we have no additional knowledge: no projective planes of non-prime-power order are known but no non-existence theorems beyond Corollary 2.4 have been proved. The smallest undecided value is $n=10$. There are, however, many projective planes which are not isomorphic to the designs $\mathscr{P}_2(q)$. Indeed, for every prime power q which is not a prime or equal to 4 or 8, there exists at least one projective plane of order q not isomorphic to $\mathscr{P}_2(q)$. For $q \leqslant 8$ it is known that $\mathscr{P}_2(q)$ is the only projective plane of order q, but for any prime $p>8$ it is again not known if $\mathscr{P}_2(p)$ is the only plane of order p. Before we discuss some methods for constructing finite projective planes we need to establish the close relation between affine and projective planes. The reader should refer back to Section 1.3 for the definition of an affine plane.

Theorem 3.1. If $\mathscr{P}$ is a projective plane and y is a line of $\mathscr{P}$ then $\mathscr{P}^y$ is an affine plane. Conversely, if $\mathscr{A}$ is any affine plane there exists a unique (up to isomorphism) projective plane $\mathscr{P}$ such that $\mathscr{A}$ is isomorphic to $\mathscr{P}^y$, for some line y of $\mathscr{P}$.

Proof. Suppose $\mathscr{P}$ is a projective plane and y is a line of $\mathscr{P}$, and let $\mathscr{A}=\mathscr{P}^y$. (So the points of $\mathscr{A}$ are the points of $\mathscr{P}$ not on y and the lines of $\mathscr{A}$ are the lines of $\mathscr{P}$ other than y.) Clearly, any two points of $\mathscr{A}$ are on the line which joined them in $\mathscr{P}$, so axiom (A1) is satisfied. Let z be any line of $\mathscr{A}$ and P a point of $\mathscr{A}$ not on z. If X is the point of $\mathscr{P}$ where y and z meet then, of course, X is not a point of $\mathscr{A}$. Every line of $\mathscr{P}$ through P meets z in a point of $\mathscr{P}$ and each of these, except the line PX, meetz z in $\mathscr{A}$. Thus there is a unique line of $\mathscr{A}$ through P not meeting z and axiom (A2) holds. The verification of axiom (A3) is straightforward and we leave it for the reader.

Now suppose that $\mathscr{A}$ is an affine plane. If y and z are lines of $\mathscr{A}$ we define $y \parallel z$ (in words: y is *parallel* to z) if $y=z$ or if y and z have no point in common. First we show that $\parallel$ is an equivalence relation, i.e. that:

(a) $y \parallel y$;
(b) $y \parallel z$ implies $z \parallel y$;
(c) $x \parallel y$ and $y \parallel z$ implies $x \parallel z$.

Now (a) and (b) are trivial, so we consider (c). If any two of x, y, z are equal, we are done, so we suppose they are all mutually distinct. If x is not parallel to z then there is a point P on x and z. But then x and z are both lines on P, neither meeting y, and, since $x \neq z$, this violates axiom (A2).

Now we construct $\mathscr{P}$. The points of $\mathscr{P}$ are the points of $\mathscr{A}$, and also for each parallel class $\{y\}$ of lines, $\{y\}$ is a point of $\mathscr{P}$. The lines of $\mathscr{P}$ are the lines of $\mathscr{A}$, plus one new line which we denote by l_∞. The incidence rules in $\mathscr{P}$ are: a point P of $\mathscr{A}$ and a line y of $\mathscr{A}$ are incident in $\mathscr{P}$ if and only if they are incident in $\mathscr{A}$; a point $\{y\}$ of $\mathscr{P}$ and a line z of $\mathscr{A}$ are incident in $\mathscr{P}$ if and only if $z \in \{y\}$; finally every point $\{y\}$ is on l_∞. We consider two points of $\mathscr{P}$. If the points are P, Q, both in $\mathscr{A}$ $(P \neq Q)$, then they are clearly on only the line y of $\mathscr{P}$; where y is the line they are on in $\mathscr{A}$. If the points are P and $\{y\}$, then P is on a unique line $z \in \{y\}$ (for P is either on y or on a unique line $z \parallel y$), and $\{y\}$ is on z. If the points are $\{y\}$ and $\{z\}$, y not parallel to z, then these points are clearly on the unique line l_∞. Now we consider two lines of $\mathscr{P}$. If the lines are y and z, both in $\mathscr{A}$, $y \neq z$, then either y and z meet in a point of $\mathscr{A}$, whence y is not parallel to z, and so there is no point $\{w\}$ on both lines; or if y and z do not meet in a point of $\mathscr{A}$, then $y \parallel z$, so both contain the point $\{y\}$. A line y, of $\mathscr{A}$, and the line l_∞ meet in the point $\{y\}$.

Certainly $\mathscr{P}$ contains a set of three points, not on a common line, since even $\mathscr{A}$ does. We must still show that every line of $\mathscr{P}$ contains at least three points. If a line y of $\mathscr{A}$ contains at least two points in $\mathscr{A}$, then it contains at least three points in $\mathscr{P}$, for y also is incident with $\{y\}$ in $\mathscr{P}$. Let A, B, C be three points of $\mathscr{A}$, not on a common line, and let a, b, c be the three lines of $\mathscr{A}$ containing, respectively, B and C, A and C, and A and B. If y is a line of $\mathscr{A}$ on none of A, B or C, then it cannot be parallel to more than one of a, b, c, so three of these lines must meet y in two distinct points. If y is on A, say, but not equal to b or c, then on B there is a line parallel to b, which must meet y in a point X, and on C there is a line parallel to c which must meet y in a point Y, and $X \neq Y$. (The reader should draw a picture.) Finally the three lines a, b, c are mutually not parallel, so $\{a\}$, $\{b\}$ and $\{c\}$ are three distinct points on l_∞.

It is not difficult to prove that $\mathscr{P}$ is unique; see Exercise 3.1. □

Exercise 3.1. Prove that if $\mathscr{P}_1$ and $\mathscr{P}_2$ are projective planes with lines y_1, y_2 respectively, such that $\mathscr{P}_1^{y_1} \cong \mathscr{P}_2^{y_2}$, then $\mathscr{P}_1 \cong \mathscr{P}_2$. □

Exercise 3.2. Show that if x and y are lines of a projective plane $\mathscr{P}$, then $\mathscr{P}^x \cong \mathscr{P}^y$ if and only if there is an automorphism α of $\mathscr{P}$ with $x^\alpha = y$. Deduce that the number of non-isomorphic affine planes that arise from $\mathscr{P}$ is equal to the number of (point or line) orbits of Aut $\mathscr{P}$ (see Exercise 1.41).

Corollary 3.2. The class of finite affine planes is equal to the class of 2-designs with parameters $(n^2, n, 1)$, with $n > 1$.

Proof. If $\mathscr{P}$ is a finite projective plane of order n then $\mathscr{P}$ is a symmetric 2-$(n^2+n+1, n+1, 1)$ with $n>1$ and so, by Theorem 2.1, for any line y, $\mathscr{P}^y$ is a 2-$(n^2, n, 1)$ design, with $n>1$. Conversely, if $\mathscr{A}$ is a 2-$(n^2, n, 1)$ design with $n>1$ then, by Corollary 1.4(c), $\mathscr{A}$ has $r=n+1$ lines on every point. So if P is a point and y a line of $\mathscr{A}$, P not on y, P is joined to the n points of y by n different lines of $\mathscr{A}$. This leaves exactly one line through P not meeting y. □

Note that the order of a 2-$(n^2, n, 1)$ is $r-\lambda=(n+1)-1=n$. Thus the values of n for which there exist projective planes of order n are precisely those for which there exist affine planes.

Exercise 3.3. Show that an affine plane of order n has $n+1$ parallel classes with n lines in each class and hence has a total of n^2+n lines.

In order to construct projective planes, it suffices to construct affine planes, and many of the known projective planes are constructed in this fashion.

We will now give a sketch of how to construct, for some prime powers q, a projective plane of order q which is not isomorphic to $\mathscr{P}_2(q)$. The discussion is 'algebraic' and we make no attempt to justify most of our claims, but leave them as optional exercises for the interested reader (for more detail see [5]). Our main objective is to give some idea of the numerous examples that exist, and an indication of how many of them are constructed.

Suppose Q is a set with at least the two distinct elements 0 and 1, and suppose Q has two binary operations, $+$ (*addition*), and $\cdot$ (*multiplication*, where we also write xy for $x\cdot y$), all satisfying:

(Q1) $(Q, +)$ is an abelian group, with 0 as identity;

(Q2) if Q^* is the set of elements x of Q, $x\neq 0$, then $(Q^*, \cdot)$ is a loop, with 1 as identity (i.e. if $a, b\in Q^*$, then $ax=b$ and $ya=b$ are satisfied for unique elements $x, y\in Q^*$, while $1\cdot x=x\cdot 1=x$ for all $x\in Q^*$);

(Q3) $a(b+c)=ab+ac$ for all $a, b, c\in Q$;

(Q4) if $a, b, c\in Q$ with $a\neq b$, then there is a unique $x\in Q$ such that $ax=bx+c$;

(Q5) $0\cdot x=0$ for all $x\in Q$.

Then $(Q, +, \cdot)$, or merely Q, is a *left quasifield*, or merely a *quasifield*. (A right quasifield, which we do *not* call a quasifield, has the appropriate changes made in (Q3), (Q4) and (Q5).)

Lemma 3.3. Suppose $(Q, +, \cdot)$ satisfies:

(Q1)′ $(Q, +)$ is a group, with 0 as identity;

(Q3) $a(b+c)=ab+ac$ for all $a, b, c \in Q$.

Then Q also satisfies:

(i) if $-x$ is the additive inverse of x (so $x+(-x)=0$), then $x(-y)=-(xy)$ for all $x, y \in Q$;

(ii) $x \cdot 0=0$ for all $x \in Q$.

Proof. Let $a, b \in Q$ and consider $ab=a(b+0)=ab+a \cdot 0$. Since $ab+x=ab$ has the unique solution $x=0$, it follows that $a \cdot 0=0$. Now $y+(-y)=0$, so $x(y+(-y))=xy+x(-y)=x \cdot 0=0$. Thus $x(-y)=-(xy)$. □

We can write $x-y$ for $x+(-y)$, and do so freely.

Lemma 3.4. If Q is a *finite* system with two binary operations $+$ and $\cdot$, and at least the two distinct elements 0 and 1, satisfying

(Q1)′ $(Q, +)$ is a group with identity 0,

as well as (Q2), (Q3) and (Q5), then Q is a quasifield.

Proof. For each $a \in Q^*$, let L_a be the mapping of Q into Q given by:

$$xL_a=ax.$$

Then clearly, from (Q2), L_a is a bijection of Q^* onto Q^*, and, since $0L_a=0$ from Lemma 3.3, L_a is a bijection of Q. But from (Q3)

$$(x+y)L_a=xL_a+yL_a,$$

so L_a is an automorphism of $(Q, +)$. From (Q2), the set $\{L_a\}$, for all $a \in Q^*$, has the property: if $x, y \in Q^*$, then there is a (unique) L_a such that $xL_a=y$. So the automorphism group of $(Q, +)$ is transitive on the elements $\neq 0$, and hence every element of Q^* has the same additive order in $(Q, +)$. Since every finite group has an element of prime order it follows that every element $\neq 0$ of Q has additive order p, a prime. So $(Q, +)$ is a p-group. Now in any p-group, there is a non-identity element which commutes with the entire group, so there is an element $c \in Q^*$ such that $x+c=c+x$ for all $x \in Q$. But then cL_a commutes with all of Q as well, and hence the set of elements ac, i.e. *all* elements of Q, commute with all of Q. So $(Q, +)$ is abelian (and is a p-group).

To solve $ax=bx+c$, we note that Lemma 3.3 applies to our Q. Consider the mapping $T: x \to ax-bx$. If T is not a bijection, then there are elements $x, y \in Q$, $x \neq y$, such that $ax-bx=ay-by$, or $a(x-y)=b(x-y)$. But if $a \neq b$, this implies $x=y$, a contradiction. So T is a bijection, so there is a unique element $x \in Q$ such that $xT=c$, i.e. $ax=bx+c$. □

Corollary 3.5. A finite quasifield has a prime-power number of elements. □

Now let Q be a quasifield, and let $\mathscr{A}=\mathscr{A}(Q)$ be the structure (of points and lines) defined as follows:

The points of $\mathscr{A}$ are the ordered pairs (x, y) for all $x, y \in Q$; the lines of $\mathscr{A}$ are the ordered pairs $[m, k]$ and the 'singletons' $[k]$, for all $m, k \in Q$; (x, y) is on $[m, k]$ if $mx+y=k$; (x, y) is on $[k]$ if $x=k$.

Theorem 3.6. $\mathscr{A}(Q)$ is an affine plane of order $|Q|$.

Proof. Let (x, y) and (u, v) be distinct points of $\mathscr{A}$. Then $[m, k]$ is on the two points if and only if $mx+y=mu+v=k$, while $[k]$ is on both points if and only if $k=x=u$.

If $x\neq u$, then $mx+y=mu+v$ is equivalent to $m(x-u)=v-y$, which has a unique solution for $m\in Q$, so k $(=mx+y)$ is uniquely determined as well.

If $x=u$, then $mx+y=mx+v$ is not possible, since $y\neq v$, so the two points are on the unique line $[x]$.

Hence two distinct points are on a unique line of $\mathscr{A}$.

Now we assert: the only lines that fail to meet a line $[k]$ are the lines $[m]$, with $m\neq k$, while the only lines that fail to meet a line $[m, k]$ are the lines $[m, s]$, with $s\neq k$. (See Exercise 3.4.) So to prove the 'parallel' postulate, it suffices to show: if $x\neq k$, then there is a unique line $[s]$ on (x, y), while if $mx+y\neq k$, then there is a unique line $[m\ s]$ on (x, y). But both statements are easy: the line $[s]$ is $[x]$, while the line $[m, s]$ is $[m, mx+y]$.

Finally $(0, 0)$, $(1, 0)$ and $(0, 1)$ determine the 'triangle' of lines $[0, 0]$, $[1, 1]$ and $[0]$. □

Exercise 3.4. Prove the assertion about parallel classes in the proof of Theorem 3.6.

A *division ring*, or a *semifield*, is a set D with at least the two distinct elements 0 and 1, and two binary operations $+$ and $\cdot$, satisfying:

(D1) $(D, +)$ is a group with identity element 0;

(D2) $(D^*, \cdot)$ is a loop with identity 1;

(D3) $a(b+c)=ab+ac$, for all $a, b, c\in D$;

(D4) $(a+b)c=ac+bc$, for all $a, b, c\in D$.

Exercise 3.5.

(a) By considering $(a+b)(c+d)$, show that a semifield has commutative addition.

(b) Show that, in a semifield, $(-y)x = -(yx)$ for all x, y.

Exercise 3.6. Prove that a semifield is a quasifield.

Now we construct some examples.

Example 3.1. Let K be a finite field $GF(q)$ and σ an automorphism of $K, \sigma \neq 1$, (i.e. q is not a prime). Let D be a 2-dimensional vector space over K, with elements represented as $x+\lambda y$, where $x, y \in K$ and λ is an indeterminate. So addition in D is given by

$$(x+\lambda y)+(u+\lambda v)=(x+u)+\lambda(y+v).$$

Let $a, b \in K$ be fixed elements, chosen so that $x^2 = a+bx$ has no solution in K (this is always possible since there are always irreducible quadratics $x^2 - bx - a$ over any finite field). Then define a multiplication in D by:

$$(x+\lambda y)(u+\lambda v)=(xu+a\cdot y^\sigma v)+\lambda(yu+x^\sigma v+by^\sigma v).$$

Exercise 3.7.* Show that D, as in Example 3.1, is a semifield, and is not a field.

Example 3.2. Let K be a finite field $GF(q)$, and $f(x)=x^2+ax+b$ an irreducible quadratic over K. Let Q be a 2-dimensional vector space over K, with elements $x+\lambda y$, as in Example 3.1, and with addition as in that example. Define multiplication in Q by:

$$x(u+\lambda v)=xu+\lambda\cdot xv$$

and, if $y \neq 0$,

$$(x+\lambda y)(u+\lambda v)=xu-y^{-1}vf(x)+\lambda(yu-(a+x)v).$$

Exercise 3.8.* Show that Q, as in Example 3.2, is a quasifield and is never a semifield except when $K=GF(2)$, in which case it is even a field.

Example 3.3. Let K be a field $GF(q^2)$, where q is an odd prime power. Then the mapping $\sigma : x \to x^q$ is an automorphism of K of order 2. Furthermore, half the non-zero elements are square and these elements form a (multiplicative) subgroup of the multiplicative group of K. For each $x \in K$, define σ_x as follows:

$$\sigma_x : y \to y \qquad \text{if } x \text{ is a square;}$$

$$\sigma_x : y \to y^\sigma = y^q \quad \text{if } x \text{ is not a square.}$$

Let K_0 be the same set of elements as K, with the same addition, but with a new multiplication, written $\circ$, defined as follows:

$$x \circ y = xy^{\sigma_x}.$$

Exercise 3.9.* Show that K_0 is a quasifield, has associative, but not commutative, multiplication and is never a semifield.

If Q is a quasifield, then let $\mathscr{A} = \mathscr{A}(Q)$ be the affine plane 'coordinatised' by Q, as above, and let $\mathscr{P} = \mathscr{P}(Q)$ be the projective plane constructed as in Theorem 3.1, so that $\mathscr{P}^{l_\infty} \cong \mathscr{A}$. We say that $\mathscr{B}$ is a *translation* affine plane, or merely that $\mathscr{B}$ is a translation plane, if it is isomorphic to $\mathscr{A}(Q)$, for some quasifield Q; similarly, we say that $\mathscr{S}$ is a *translation* projective plane if it is isomorphic to $\mathscr{P}(Q)$ for some quasifield. In fact, we shall use 'translation plane' indifferently to refer to both $\mathscr{A}(Q)$ and $\mathscr{P}(Q)$, but in the case of $\mathscr{P}(Q)$ it is necessary to know which line y plays the role of l_∞. (Since affine and projective planes are so closely related, by Theorem 3.1, this apparent ambiguity need cause no confusion.) If the quasifield Q happens to be the finite field $GF(q)$, we now have two ways to construct a projective plane from Q. But:

Theorem 3.7. If K is the finite field $GF(q)$, then $\mathscr{P}_2(q) \cong \mathscr{P}(K)$.

Proof. The points of the plane $\mathscr{P}_2(q)$ are the 1-dimensional subspace of the 3-dimensional vector space V over K; choosing a basis for V, we may represent any such point by $\langle(x, y, z)\rangle$, where $(x, y, z) \neq (0, 0, 0)$. One of the 2-dimensional subspaces of V is the space L of all vectors $\{(x, y, 0)\}$, so L contains all the points $\langle(x, y, 0)\rangle$. Every point of $\mathscr{P}_2(q)$ not on L can then be represented as $\langle(x, y, 1)\rangle$, since if $z \neq 0$, $\langle(x, y, z)\rangle = \langle(xz^{-1}, yz^{-1}, 1)\rangle$.

Now a 2-dimensional subspace of V can be identified by a linear equation

$$ax + by + cz = 0, \quad \text{where not all of } a, b, c \text{ are zero,} \tag{1}$$

i.e. the subspace contains all the points $\langle(x, y, z)\rangle$ which satisfy (1). Replacing a, b, c in (1) by any multiple ka, kb, kc, where $k \neq 0$, does not change the subspace, i.e. the line. The line L above is the one with $a = b = 0$, $c = 1$. We shall let $\mathscr{P}=\mathscr{P}_2(q)$, and show that $\mathscr{P}^L$ is isomorphic to $\mathscr{A}(K)$; by Theorem 3.1, this will finish the proof.

The points of $\mathscr{P}$ not on L have the form $\langle(x, y, 1)\rangle$; we let ϕ map the points of $\mathscr{P}^L$ to the points of $\mathscr{A}=\mathscr{A}(K)$ as follows:

$$\phi : \langle(x, y, 1)\rangle \rightarrow (x, y).$$

The lines of $\mathscr{P}$ not equal to L can be identified by triples $[a, b, c]$ where not both a and b are zero, and $\langle(x, y, 1)\rangle$ is on $[a, b, c]$ if $ax+by+c=0$, as in (1). If $b=0$, then $a\neq 0$, so we may assume $a=1$. So we define ϕ by:

$$\phi : [1, 0, c] \rightarrow [-c].$$

Finally, if $b\neq 0$, we may assume $b=1$, so ϕ is given by:

$$\phi : [a, 1, c] \rightarrow [-a, -c].$$

It is now easy to see that ϕ is the desired isomorphism.

Exercise 3.10. Show that ϕ in the proof of Theorem 3.7 is an isomorphism. □

Now we investigate the reason for calling an affine plane $\mathscr{A}(Q)$, where Q is a quasifield, a 'translation plane'. If $\mathscr{P}$ is a projective plane and y is a line of $\mathscr{P}$, then an automorphism α of $\mathscr{P}$ which fixes every point of y, and either all other points (so $\alpha=1$), or no other points, is called a *translation*, or an *elation*, with axis y.

Exercise 3.11.* If $\alpha\neq 1$ is a translation of the projective plane $\mathscr{P}$, with axis y, show that there is a unique point X, on y, such that every line on X is fixed by α. (X is called the *centre* of α.)

Exercise 3.12. If $\mathscr{P}$ is a projective plane and y is a line of $\mathscr{P}$, then show that the set $T(y)$ of all translations of $\mathscr{P}$ with axis y is a subgroup of Aut $\mathscr{P}$.

It is in fact true that if $T(y)$ contains non-identity elements with different centres, then $T(y)$ is abelian, and every non-identity element of $T(y)$ has the same order. So if $T(y)$ is also finite (e.g. if $\mathscr{P}$ is finite) $T(y)$ is an elementary abelian p-group, for some prime p. (For a more detailed discussion of this theorem, and related matters, see [5].)

Theorem 3.8. If $\mathscr{P}$ is a projective plane, y a line of $\mathscr{P}$, and $\mathscr{A}=\mathscr{P}^y$, then Aut $\mathscr{A}=$ (Aut $\mathscr{P})_y$.

Proof. Certainly if $\alpha \in \operatorname{Aut} \mathscr{P}$ fixes y, then α is an automorphism of $\mathscr{P}^y$. Conversely, if $\alpha \in \operatorname{Aut} \mathscr{A}$, then α must preserve the parallel classes of $\mathscr{A}$ (why?), so α induces an automorphism of $\mathscr{P}$, fixing y (which plays the role of l_∞ in Theorem 3.1). □

Corollary 3.9. If $\mathscr{P}$ is a projective plane and $\alpha \in \operatorname{Aut} \mathscr{P}$, then α is a translation of $\mathscr{P}$ with axis y if and only if α is an automorphism of $\mathscr{A} = \mathscr{P}^y$ which maps any line z of $\mathscr{A}$ onto a line z^α which is parallel to z. □

Theorem 3.10. If Q is a quasifield, then the group $T(l_\infty)$ is transitive on the points of $\mathscr{P}(Q)$ which are not on l_∞, i.e. on the points of $\mathscr{A}(Q)$.

Proof. Let $a, b \in Q$, and define a mapping $\tau = \tau(a, b)$ on the points and lines of $\mathscr{A}(Q)$ as follows:

$$\tau : (x, y) \rightarrow (x+a, y+b)$$
$$[m, k] \rightarrow [m, k-ma+b]$$
$$[k] \rightarrow [k+a].$$

Then τ is an automorphism of $\mathscr{A}(Q)$; for instance, (x, y) is on $[m, k]$ if and only if $y = mx+k$, while $(x+a, y+b)$ is on $[m, k-ma+b]$ if and only if $y+b = m(x+a)+k-ma+b$, which is equivalent to $y = mx+k$. Since the parallel classes of $\mathscr{A}(Q)$ are, first the set of lines $[k]$, for $k \in Q$ and, secondly, a set $[m, k]$ with fixed m and all $k \in Q$, τ is a translation of $\mathscr{A}(Q)$, hence of $\mathscr{P}(Q)$ (and with axis l_∞). Clearly, the set of all τ is transitive on the points not on l_∞. □

It is, in fact, true that the projective plane $\mathscr{P}$ has a line y such that $T(y)$ is transitive on the points of $\mathscr{P}^y$ if and only if $\mathscr{P} \cong \mathscr{P}(Q)$, where Q is a quasifield, and the line y is the 'line at infinity' l_∞. It is also true that $\mathscr{P}$ has *two* such lines if and only if Q is a field (in the finite case). Again, see [5] for more detail. So a projective plane is a translation plane if and only if it has a special line y such that $T(y)$ is transitive on the points of $\mathscr{P}^y$, which is the explanation of the term. (There are also many finite projective planes which are not translation planes.)

The automorphism group of the plane $\mathscr{P}_2(q)$ then contains translations with *every* axis. The full group Aut $(\mathscr{P}_2(q))$ is known:

Theorem 3.11. Aut $(\mathscr{P}_2(q)) \cong P\Gamma L(3, q)$, the group of all non-singular semi-linear transformations of the 3-dimensional vector space V over $GF(q)$, modulo the group of all scalar multipliers. (A scalar multiplier of V is a mapping $\mathbf{v} \to k\mathbf{v}$, where $\mathbf{v} \in V$ and k is a fixed non-zero element of $GF(q)$.)

Proof. See [5]. □

3.3 Latin squares

Let $\mathscr{A}$ be an affine plane of order n, i.e. a 2-$(n^2, n, 1)$ design. Then $\mathscr{A}$ has $n+1$ parallel classes, which we label $J_\infty, J_0, J_1, \ldots, J_{n-1}$. If $\mathscr{R} = \{0, 1, 2, \ldots, n-1\}$ then, for each individual parallel class J_x, we label the lines of J_x as $0, 1, \ldots, n-1$. Thus each line is uniquely labelled by an ordered pair (J_x, y), where $x = \infty$ or $x \in \mathscr{R}$ and $y \in \mathscr{R}$. Each of the n^2 points of $\mathscr{A}$ is on exactly one line of each parallel class so we 'coordinatise' the point X by (x,y) if X is on (J_∞, x) and (J_0, y). So the line joining (x, y_1) and (x, y_2), $y_1 \neq y_2$, is (J_∞, x) and the line joining (x_1, y), (x_2, y) $(x_1 \neq x_2)$ is (J_0, y). If we now consider the line (J_x, z) $(z \neq \infty, 0)$, then this line contains n points (x, y) of $\mathscr{A}$ and these n points must have n distinct values for x and n distinct values for y. This observation leads us to define a latin square. A *latin square* of order n is an n by n matrix with entries from $\mathscr{R}$ (we fix on this $\mathscr{R}$ for convenience; clearly, the definition extends to other sets) such that each entry occurs exactly once in each row and in each column. We can construct latin squares from our affine plane $\mathscr{A}$ of order n as follows: for each $s \in \mathscr{R}$ $(s \neq 0)$, let L_s be the n by n matrix with rows and columns indexed by $\mathscr{R}$, whose entry in the position labelled (i, j) is z, where z is the line of J_s which contains the point (i, j).

Lemma 3.12. L_s is a latin square for any $s \in \mathscr{R}$ $(s \neq 0)$.

Proof. Since each line of J_s meets the line (J_0, i) exactly once, row i contains each entry from $\mathscr{R}$. Similarly, since each line of J_s also meets the line (J_∞, j) once, column j also contains each entry. □

The set of latin squares $L_1, L_2, \ldots, L_{n-1}$ has another interesting property.

Lemma 3.13. If $s, r \in \mathscr{R}$ $(s \neq r)$, neither equal to 0, then L_s and L_r have the property: for each ordered pair $x, y \in \mathscr{R}$, there is exactly one ordered pair $i, j \in \mathscr{R}$ such that L_s has x in position (i, j) and L_r has y in position (i, j).

Proof. Line x of J_s meets line y of J_r in exactly one point P of $\mathscr{A}$. If P is named (i, j), then this is the ordered pair needed. □

Two latin squares with the property of Lemma 3.13 are called *orthogonal*. Another way to see it is to imagine L_r superimposed on L_s, obtaining an n by n matrix whose entries are ordered pairs from $\mathscr{R}$; then orthogonality is the condition that these n^2 ordered pairs should take on n^2 different values.

Example 3.4. The following pair of matrices is a pair of orthogonal latin squares of order 3.

$$\begin{bmatrix} 0 & 1 & 2 \\ 1 & 2 & 0 \\ 2 & 0 & 1 \end{bmatrix} \quad \begin{bmatrix} 0 & 1 & 2 \\ 2 & 0 & 1 \\ 1 & 2 & 0 \end{bmatrix}.$$

Example 3.5. The following three matrices are latin squares and any two are orthogonal.

$$\begin{bmatrix} 0 & 1 & 2 & 3 \\ 1 & 0 & 3 & 2 \\ 2 & 3 & 0 & 1 \\ 3 & 2 & 1 & 0 \end{bmatrix} \quad \begin{bmatrix} 0 & 1 & 2 & 3 \\ 2 & 3 & 0 & 1 \\ 3 & 2 & 1 & 0 \\ 1 & 0 & 3 & 2 \end{bmatrix} \quad \begin{bmatrix} 0 & 1 & 2 & 3 \\ 3 & 2 & 1 & 0 \\ 1 & 0 & 3 & 2 \\ 2 & 3 & 0 & 1 \end{bmatrix}.$$

If $M_1, M_2, \ldots, M_m$ are a set of latin squares of order n, we say that they form a set of *mutually orthogonal* latin squares (or, a 'set of MOLS') if any two are orthogonal.

Theorem 3.14. Let $\mathscr{R} = \{0, 1, \ldots, n-1\}$. If $M_1, M_2, \ldots, M_{n-1}$ is a set of $n-1$ MOLS with entries from $\mathscr{R}$, then there is an affine plane $\mathscr{A}$ of order n such that the M_i play the role of the squares L_i in the construction above.

Proof. We let the points of $\mathscr{A}$ be the ordered pairs (x, y), where $x, y \in \mathscr{R}$. We define lines of $\mathscr{A}$ as follows:

(a) for each $z \in \mathscr{R}$, $[z]$ is a line;
(b) for each $u, v \in \mathscr{R}$, $[u, v]$ is a line.

Incidence is given by:

(c) (x, y) is on $[z]$ if $x = z$;
(d) (x, y) is on $[0, v]$ if $y = v$;
(e) (x, y) is on $[u, v]$, $u \neq 0$, if the entry in position (x, y) of M_u is v.

Now it is easy to see that a line $[z]$ contains n points; the n points (z, y) as y ranges over $\mathscr{R}$. Similarly $[0, v]$ contains the n points (x, v), for all $x \in \mathscr{R}$. The line $[u, v]$ contains those points (x, y) such that v occurs in position (x, y) of M_u, but since v occurs once in each row and column of M_u it occurs n times in M_u. So all lines have n points, and hence $\mathscr{A}$ is a uniform (and obviously reduced) structure.

Let (x, y) and (u, v) be distinct points. If $x = u$ but $y \neq v$, then the two points occur on the line $[x]$ and no other. If $x \neq u$ but $y = v$, then the points are on $[0, y]$, and on no other. So we assume $x \neq u$ and $y \neq v$. We want to find a unique square M_a such that M_a has the same entry in position (x, y) and (u, v). For $a \in \mathscr{R}$ $(a \neq 0)$, if the entry in position (x, y) of M_a is c there is a unique position in row u of M_a where c occurs. Let the position be $(u, a\theta)$, so θ is a mapping from $\mathscr{R}^*$ into $\mathscr{R}^*$, ($\mathscr{R}^*$ is the set of $j \in \mathscr{R}, j \neq 0$). We want to find a unique $a \in \mathscr{R}^*$ such that $a\theta = v$, i.e. we want to show that θ is a bijection on $\mathscr{R}^*$. If not, then there are elements $a, b \in \mathscr{R}^*$ such that $a \neq b$ but $a\theta = b\theta$. This means that M_a has c in position (x, y) and $(u, a\theta)$, while M_b has the same entry, d say, in position (x, y) and $(u, b\theta) = (u, a\theta)$. But this violates the orthogonality of M_a and M_b since the ordered pair (c, d) would occur twice in the 'superimposition' of M_a and M_b. So θ is a bijection and we are done: $\mathscr{A}$ is a 2-$(n^2, n, 1)$, i.e. an affine plane of order n.

It is easy to see that these M_i are the squares L_i of the construction, using the appropriate labelling of parallel classes and lines in $\mathscr{A}$. □

If we want to represent an affine plane $\mathscr{A}$ by a set of MOLS, we may think of the *positions* in the squares as representing the points. One parallel class J_∞ has the columns for its lines and a second parallel class J_0 has the rows. We then think of each square as representing a parallel class and a line of that class contains all those positions in the square which contain the same entry. As an illustration, if we look at Example 3.4, which represents an affine plane of order 3, the lines of the parallel class of the first square are, as point sets, $\{(0, 0), (1, 2), (2, 1)\}$, $\{(0, 1), (1, 0), (2, 2)\}$ and $\{(0, 2), (1, 1), (2, 0)\}$.

Since the existence of an affine plane of order n is equivalent to the existence of a set of $n-1$ MOLS of size n, there are obviously restrictions on the values of

n for which one can find such a set (namely the restrictions of the BRC Theorem).

It is true that, for any n, one can construct an n by n latin square but it is easy to find squares for which there is no orthogonal mate. (An *orthogonal mate* is a latin square which is orthogonal to the given square.)

Exercise 3.13. If G is any finite group show that the Cayley table of G is a latin square (thus proving that n by n latin squares exist for all n).

Exercise 3.14. Show that there is no latin square orthogonal to

$$\begin{bmatrix} 0 & 1 & 2 & 3 \\ 1 & 2 & 3 & 0 \\ 2 & 3 & 0 & 1 \\ 3 & 0 & 1 & 2 \end{bmatrix}.$$

There are various ways to 'normalise' a set of MOLS. Firstly, any permutation of the symbols of $\mathscr{R}$, within just one square of the set, certainly leaves that square latin, and, in fact, leaves it orthogonal to all the others. Secondly, any permutation of the rows of *all* the squares (i.e. the set) leaves the squares all latin and mutually orthogonal; similarly for columns. Hence it is not difficult to see:

Lemma 3.15. A set of MOLS on the set $\mathscr{R}=\{0,1,\ldots,n-1\}$ can be assumed to have the property that every top row is $0, 1, \ldots, n-1$, in that order, and the left column of *one* square also has $0, 1, \ldots, n-1$, in that order, from top to bottom. □

We shall say that a set of MOLS is in *normal form* if it has the form of Lemma 3.15. Notice that the geometric effect of permuting the symbols of one square is to rename the lines of that parallel class, permuting the rows (columns) is geometrically equivalent to relabelling the lines of J_0 (J_∞).

Exercise 3.15. By using Lemma 3.15 and considering the elements in position $(1, 0)$, show that if $M_1, M_2, \ldots, M_m$ is a set of MOLS of order n then $m \leqslant n-1$.

Since, for any prime power q, there is an affine plane of order q we know there exist $q-1$ MOLS of order q. In fact, if Q is any finite quasifield we can construct a set of $q-1$ MOLS associated with $\mathscr{A}(Q)$ as follows: for any $m \in Q$

($m \neq 0$), the parallel class J_m is the set of lines $[m, k]$ and if we label $[m, k]$ as (J_m, k) then k is in position (x, y) on L_m if and only if (x, y) is on $[m, k]$, i.e. if and only if $mx + y = k$. So position (x, y) of L_m has $mx + y$ in it.

An interesting question now arises: if n is not a prime power then what is the maximum number of MOLS of order n? If n is a number for which there is known to be no affine plane of order n (see the BRC Theorem), then this number is certainly not $n-1$. It is, in fact, known that if $n \neq 1, 2$ or 6 then there are always at least two MOLS of order n. (How is Exercise 3.14 consistent with this fact?). The case $n = 6$ appears to be quite exceptional (clearly, there cannot be two MOLS of order 2, by Exercise 3.15): Euler made a famous conjecture that there are never two MOLS of order n for any $n \equiv 2 \pmod 4$. His grounds for the conjecture seem to have been his difficulties in finding two MOLS of order 6. The fact that there is no pair of MOLS of order 6 was proved by Tarry in a series of papers published about 1900. (However, the reader will find a shorter proof in [4].) The falsity of the Euler conjecture was established in [1] where two MOLS of order 22 were found. However, it was not until 1960 that it was finally shown that if $n \equiv 2 \pmod 4$ and $n \neq 2$ or 6 then there is a pair of MOLS of order n. (See [2].) If $n \not\equiv 2 \pmod 4$ then it is easy to construct fairly large sets of mutually orthogonal latin squares.

Theorem 3.16. Let $n = q_1 q_2 \ldots q_t$, where $q_1, q_2, \ldots, q_t$ are powers of distinct primes. Then there is a set of $q_i - 1$ MOLS of order n, where q_i is the smallest of the numbers $q_1, q_2, \ldots, q_t$.

The interested reader will find a proof of this in [6].

Exercise 3.16. Show that the only latin square of order 4, in normal form, which has an orthogonal mate is:

$$\begin{bmatrix} 0 & 1 & 2 & 3 \\ 1 & 0 & 3 & 2 \\ 2 & 3 & 0 & 1 \\ 3 & 2 & 1 & 0 \end{bmatrix},$$

and that this square has two orthogonal mates, which are orthogonal to each other. (Compare Exercise 3.14.)

Exercise 3.17. Using Exercise 3.16, conclude that there is up to isomorphism only one affine or projective plane of order 4.

Exercise 3.18. As in the two preceding exercises, show that the affine and projective planes of order 3 are unique.

Exercise 3.19.* Show that the affine and projective planes of order 5 are unique. (This problem is starred only because it is 'fussy' and somewhat tedious; it is not deep.)

These results are of considerable importance, but the use of MOLS to find all planes of a given order n becomes more difficult as n grows. The proof of the uniqueness of the affine and projective planes of order 8, for example, uses latin squares and a computer, and the problem for $n=9$ appears too big even today (where the answer cannot be that the planes are unique since there are four known non-isomorphic projective planes and a much larger number of affine planes).

3.4 Nets

If $M_1, M_2, \ldots, M_s$ is a set of s MOLS with entries from $\mathscr{R}=\{0, 1, \ldots, n-1\}$ then there are a number of interesting questions one might ask. For instance, if $s<n-1$ can we find a latin square which is orthogonal to each of them? Can we extend them to a set $M_1, \ldots, M_s, M_{s+1}, \ldots, M_{n-1}$ of MOLS and thus construct an affine plane? If we can use them to construct an affine plane, then how large must s be for $M_1, M_2, \ldots, M_s$ to uniquely determine the plane? We will not even begin to discuss these problems (at least not in the way they are phrased above), and refer the interested reader to [3]. Instead, we will consider another problem. Is there any geometric or design significance to this set of s MOLS of order n if $s<n-1$? Examination of the proof of Theorem 3.14 shows that there is. We use them to define a structure $\mathscr{N}$ as follows: the points of $\mathscr{N}$ are the n^2 ordered pairs (x, y) with $x, y \in \mathscr{R}$. For each $i=1, 2, \ldots, s$ and each $k \in \mathscr{R}$ we define the line $[i, k]$ to be the set of points (x, y) such that M_i has k in position (x, y). In addition $[0, k]$ is the line of all points (x, k), while $[k]$ contains all points (k, y). The geometric properties of $\mathscr{N}$ are shown in the following theorem:

Theorem 3.17. The structure $\mathscr{N}$ is uniform and regular with n^2 points on a line and $s+2$ lines on a point. It has $n(s+2)$ lines which can be partitioned into

$s+2$ 'parallel classes' $J_\infty, J_0, J_1, \ldots, J_s$, all with the following properties:

(a) two distinct points of $\mathcal{N}$ are on at most one common line;

(b) if P is a point and J_x is a parallel class then there is a unique line of J_x which contains P.

Proof. Obvious. □

Motivated by Theorem 3.17 we define a *net* $\mathcal{N}$ of order n to be a uniform structure of block size n whose blocks can be partitioned into t classes $C_1, C_2, \ldots, C_t$ such that two points of $\mathcal{N}$ are on at most one block and each point is on exactly one block of each class. We shall call the blocks of a net *lines.*

Exercise 3.20. Show that in a net $\mathcal{N}$ of order n each line class has exactly n lines and deduce that $\mathcal{N}$ has n^2 points.

The line classes $C_1, C_2, \ldots, C_t$ of the net $\mathcal{N}$ are called the *parallel classes* and we call t the *class number*. Thus an affine plane of order n is a net of order n and class number $n+1$. Clearly, the class number cannot exceed $n+1$ and we define the *defect* of a net of order n and class number t to be $n+1-t$. Thus the defect is the number of parallel classes which must be added to the net to construct an affine plane. Of course, it may not be possible to add any extra parallel classes to the net. To ask when a net can be 'extended' to an affine plane is merely a different formulation of one of our earlier questions about s MOLS.

As well as being of interest themselves, certain nets can be used to construct other structures. For example, suppose $\mathcal{N}$ is a net of even order n with class number $n/2$. We construct $\mathcal{H} = \mathcal{H}(\mathcal{N})$ as follows. Each point P of $\mathcal{N}$ defines a point (P) and a block $[P]$ of $\mathcal{H}$. The point (P) is on the block $[Q]$ if $P \neq Q$ but P and Q are joined by a line of $\mathcal{N}$.

Theorem 3.18. If $\mathcal{N}$ is a net of even order n and class number $n/2$ then $\mathcal{H}(\mathcal{N})$ is a symmetric $(n^2, n(n-1)/2, n(n-2)/4)$ design.

Proof. Since $\mathcal{N}$ has n^2 points and each point of $\mathcal{N}$ determines a point and a block of $\mathcal{H}$, $\mathcal{H}$ has n^2 points and blocks. Thus it is square. If $[Q]$ is a block then (P) is on $[Q]$ if and only if P and Q are on a common line of $\mathcal{N}$. But there are $n/2$

lines of $\mathscr{N}$ through Q and each contains $n-1$ points distinct from Q. Thus there are $(n(n-1))/2$ points on each block. If (A) and (B) are two distinct points of $\mathscr{H}$ then in order to count the number of blocks containing them both we must consider separately the cases where A and B are joined or not joined in $\mathscr{N}$.

(a) If there is a line y joining A and B then (A) and (B) are on each of the $n-2$ blocks $[X]$ where X is a point of y ($X \neq A$ or B). But (A) and (B) will also lie on the block $[Y]$ if Y is a point of $\mathscr{N}$ which is not on y but which is joined to both A and B in $\mathscr{N}$. If z is any line through A ($z \neq y$), then there is a unique line, z' say, through B in the parallel class of z. If x is any line through B ($x \neq y$ or z'), then x will intersect z in a point Y and (A) and (B) will be on $[Y]$ in $\mathscr{H}$. There are $(n/2)-1$ possibilities for z but only $(n/2)-2$ choices for x. Thus there are $((n/2)-1)((n/2)-2)$ possibilities for Y. Hence (A) and (B) are on $n-2+((n/2)-1)((n/2)-2)=(n(n-2))/4$ blocks of $\mathscr{H}$.

(b) If there is no line of $\mathscr{N}$ through A and B then any line z through A will intersect every line through B except the one which is parallel to z. Thus (A) and (B) are on $(n/2)((n/2)-1)=(n(n-2))/4$ blocks.

Thus $\mathscr{H}$ is a square uniform 2-structure for $(n^2, (n(n-1))/2, (n(n-2))/4)$ which, by Theorem 1.26, is a symmetric design. □

This theorem has many interesting applications. For instance, putting $n=4$ it tells us how to construct a symmetric 2-(16, 6, 2) from a net of order 4 and class number 2. But this net is easily obtainable. It is merely a set of eight lines in two parallel classes, as in Figure 3.1.

For $n=6$ we need a net of order 6 and class number 3, which is equivalent to a single latin square of order 6. Thus, for instance, the Cayley table of either of the two groups of order 6 allows us to construct a symmetric design for (36, 15, 6). (See Exercise 3.13.)

Exercise 3.21. Show that if $\mathscr{S}$ is a symmetric 2-(36, 15, 6) design constructed from the Cayley table of a group G of order 6, then $\mathscr{S}$ has a Singer group

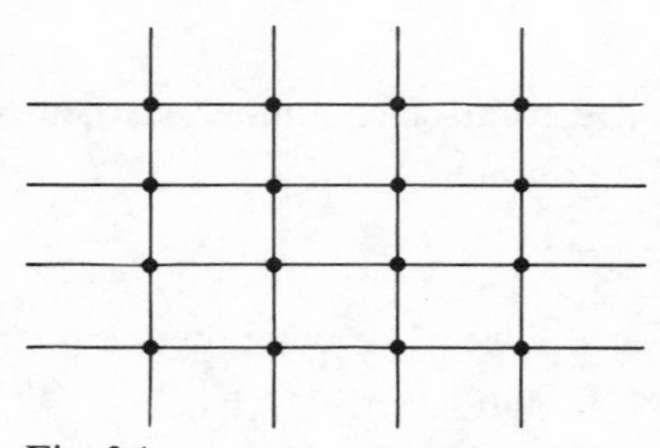

Fig. 3.1

isomorphic to $G \times G$. (So, taking G as S_3, this gives an example of a non-abelian Singer group, while $G = C_6$ gives an abelian, non-cyclic example.)

There are many interesting unsolved problems about these 'half-net designs' $\mathscr{H}(\mathscr{N})$. It is known, for instance, that there are 17 'different' latin squares of order 6. Yet it is not known how many of the 17 resulting half-net designs are isomorphic. (The two in Exercise 3.21 are indeed non-isomorphic, which is by no means obvious.)

Problem 3.1. Find the numbers of non-isomorphic symmetric designs for 2-(36, 15, 6) which arise as half-net designs. How many of these have Singer groups?

For larger values of n (e.g. $n = 2^t$ or $n = 12$) there exist half-net designs about which very little is known.

3.5 Hadamard matrices and Hadamard 2-designs

A square n by n matrix H is called a *Hadamard matrix of order n* if all the entries of H are ± 1 and $HH^{\mathrm{T}} = nI$.

For brevity we shall often call a Hadamard matrix an $\mathscr{H}$-matrix.

Lemma 3.19. If H is an $\mathscr{H}$-matrix then H^{T} is also an $\mathscr{H}$-matrix.

Proof. Clearly, every entry of H^{T} is ± 1 so we have to show that $H^{\mathrm{T}}(H^{\mathrm{T}})^{\mathrm{T}} = H^{\mathrm{T}}H = nI$. But, since $HH^{\mathrm{T}} = nI$, H is non-singular and $H^{\mathrm{T}}H = H^{-1}(HH^{\mathrm{T}})H = H^{-1}(nI)H = nH^{-1}H = nI$. □

If A is any n by n matrix and H is a Hadamard matrix of order n then $(AH)(AH)^{\mathrm{T}} = AHH^{\mathrm{T}}A^{\mathrm{T}} = nAA^{\mathrm{T}}$. Thus AH is a Hadamard matrix if and only if every entry of AH is ± 1 and $AA^{\mathrm{T}} = I$. Similarly HA is a Hadamard matrix if and only if every entry of HA is ± 1 and $AA^{\mathrm{T}} = I$. Clearly, if A and B are n by n permutation matrices then AH, HB and hence AHB are all $\mathscr{H}$-matrices. However, we shall now define a larger class of matrices for which the same statement is true.

A *generalised permutation matrix* is a square matrix whose entries are all 0 or ± 1 such that in each row and in each column there is exactly one non-zero entry. Thus a generalised permutation matrix in which all the entries are 0 or $+1$ is an ordinary permutation matrix.

Theorem 3.20. If H is an $\mathcal{H}$-matrix of order n and if A and B are n by n generalised permutation matrices then AHB is an $\mathcal{H}$-matrix.

Proof. Clearly, $AA^{\mathrm{T}} = I = BB^{\mathrm{T}}$ so that $(AHB)(AHB)^{\mathrm{T}} = AHBB^{\mathrm{T}}H^{\mathrm{T}}A^{\mathrm{T}} = nI$. Since AH is merely H with its rows permuted and some of its rows multiplied by -1, every entry of AH is ± 1. Similarly $AHB(=(AH)B)$ is AH with its columns permuted and some of them multiplied by -1. Hence each entry of AHB is ± 1 and AHB is an $\mathcal{H}$-matrix. □

If H_1 and H_2 are two $\mathcal{H}$-matrices of order n such that there exist generalised permutation matrices A and B with $H_2 = AH_1B$, then we shall say that H_1 and H_2 are *$\mathcal{H}$-equivalent*. This defines an equivalence relation on the Hadamard matrices of a given order. Before looking for a 'best' representative from each equivalence class it is perhaps worth noting that, although any two $\mathcal{H}$-equivalent Hadamard matrices are certainly equivalent as matrices over the field of rationals, if two Hadamard matrices are equivalent as matrices over the rationals they need not be $\mathcal{H}$-equivalent.

Lemma 3.21. Any $\mathcal{H}$-matrix is $\mathcal{H}$-equivalent to an $\mathcal{H}$-matrix with every entry in the first row and first column equal to $+1$.

Proof. If we multiply the ith row of H by -1 then that is equivalent to premultiplying by the generalised permutation diagonal matrix which has every diagonal entry equal to $+1$ except for the (i, i)-entry which is -1. Hence, by Lemma 3.19, multiplying any row of H by -1 gives a new $\mathcal{H}$-matrix which is $\mathcal{H}$-equivalent to H. Similarly multiplying any column by -1 gives a new $\mathcal{H}$-equivalent matrix. Thus if we multiply by -1 precisely those rows and columns of H whose first entries are -1, we obtain an $\mathcal{H}$-equivalent matrix of the desired form. □

If a Hadamard matrix has every entry in the first row and column equal to $+1$

then we say it is *normalised.* If every entry in the first row (column) is $+1$ then it is *row* (*column*) *normalised.*

It is easy to see that if

$$H=\begin{pmatrix}+1 & +1\\ +1 & -1\end{pmatrix}$$

then H is a normalised Hadamard matrix of order 2. (In fact, it is the only one.) However, if the reader tries to construct a Hadamard matrix of order 3 he will soon discover that it is impossible.

In fact, the possible values for the order of a Hadamard matrix are restricted by the following theorem:

Theorem 3.22. If H is an $\mathscr{H}$-matrix of order n then $n=1$ or 2 or $n\equiv 0 \pmod 4$.

Proof. Clearly, $n=1$ is a trivial possibility, so we shall now assume $n>1$. Since by Lemma 3.21, H is $\mathscr{H}$-equivalent to a matrix with its first row having each entry $+1$, we may assume that the first row of H has each entry $+1$. The (i, j)-entry of HH^{T}, which by the definition of a Hadamard matrix is 0 if $i\neq j$, is the scalar product of the ith and jth rows of H. Thus the scalar product of the first and second rows of H must be zero. But, since each entry in the first row is $+1$, this scalar product is merely the sum of the entries in the second row. Hence n must be even and, if $n=2m$, the second row has m of its entries $+1$ and m entries equal to -1.

By Theorem 3.20 any permutation of the columns of H gives another $\mathscr{H}$-matrix which is $\mathscr{H}$-equivalent to H and still, clearly, has each entry in its first row equal to $+1$. Thus we may assume that the first m entries of the second row are $+1$ and the second m entries are -1.

If $n\neq 2$ we now consider the third row of H. Let s be the number of $+1$s in the first m columns of the third row and let t be the number of $+1$s in the second m columns of the third row. Then, clearly, the number of -1s in the first (second) m columns of the third row is $m-s$ $(m-t)$. Thus the scalar product of the first and third rows is $2s+2t-2m$, and that of the second and third rows is $2s-2t$. However, as H is Hadamard, both these scalar products are zero so that $s=t$ and $m=2s$. Hence, since $n=2m$, $n=4s$, i.e. $n\equiv 0 \pmod 4$. □

In the following exercise we list a few simple properties of a normalised Hadamard design. The problem is virtually solved in the above proof.

Exercise 3.22. If H is a normalised $\mathscr{H}$-matrix of order n ($n \geqslant 2$), show:

(a) each row, other than the first, has exactly half its entries equal to $+1$;

(b) each column, other than the first, has exactly half its entries equal to $+1$;

(c) if $H = (h_{ij})$ then, for any fixed l and k with $l \neq k$, $l \neq 1 \neq k$, the n ordered pairs (h_{li}, h_{ki}) $(i = 1, \ldots, n)$ are such that $n/4$ of them are $(+1, +1)$, $n/4$ are $(+1, -1)$, $n/4$ are $(-1, +1)$ and $n/4$ are $(-1, -1)$.

It is widely conjectured that Hadamard designs exist for all orders $n \equiv 0 \pmod 4$. Although this has not yet been proved, Hadamard matrices have been constructed for all orders $n \equiv 0 \pmod 4$ with $n \leqslant 200$. We give here a simple construction which shows, given Hadamard matrices of orders a and b, how to construct a Hadamard matrix of order ab.

Theorem 3.23. If A is a Hadamard matrix of order m and $\{B_i\}$ $(i = 1, 2, \ldots, m)$ are Hadamard matrices of order s, then $A \otimes \{B_i\} = (a_{ij}B_i)$ is a Hadamard matrix of order ms.

Proof. For any n let I_n be the n by n identity matrix and let 0_n be the n by n matrix with each entry zero.

Since all the entries of A and each B_i are ± 1 it follows that all the entries of $A \otimes \{B_i\}$ are ± 1. Thus we have only to show that $(A \otimes \{B_i\})(A \otimes \{B_i\})^{\mathrm{T}} = msI_{ms}$.

Clearly,

$$(A \otimes \{B_i\})(A \otimes \{B_i\})^{\mathrm{T}} = \begin{pmatrix} a_{11}B_1 & \cdots & a_{1m}B_1 \\ \vdots & & \vdots \\ a_{m1}B_m & \cdots & a_{mm}B_m \end{pmatrix} \begin{pmatrix} a_{11}B_1^{\mathrm{T}} & \cdots & a_{m1}B_m^{\mathrm{T}} \\ \vdots & & \vdots \\ a_{1m}B_1^{\mathrm{T}} & \cdots & a_{mm}B_m^{\mathrm{T}} \end{pmatrix}$$

and the matrix in the (i, j)-position is

$$\sum_{k=1}^{m} a_{ik}B_i a_{jk}B_j^{\mathrm{T}} = \left(\sum_{k=1}^{m} a_{ik}a_{jk}\right)B_iB_j^{\mathrm{T}}.$$

If $i \neq j$ then $\sum_{k=1}^{m} a_{ik}a_{jk}$, which is the scalar product of the ith and jth rows of A, is zero and the matrix in the (i, j)-position is 0_s. Similarly the entry in the (i, i)-position is

$$\left(\sum_{k=1}^{m} a_{ik}^2\right)B_iB_i^{\mathrm{T}}.$$

But since A is Hadamard, $\sum_{k=1}^{m} a_{ik}^2 = m$ and, since B_i is Hadamard, $B_iB_i^{\mathrm{T}} = sI_s$ and so the matrix in the (i, i)-position is msI_s. There are m such diagonal entries and the theorem is proved. □

Corollary 3.24. If there exist $\mathscr{H}$-matrices of orders a and b, then there exists an $\mathscr{H}$-matrix of order ab; in particular, if there is an $\mathscr{H}$-matrix of order a, then there is one of order $2a$. □

Some of the other methods used for the construction of $\mathscr{H}$-matrices are very complicated, and we do not include them here (except for the 'Paley' construction later in this section). The interested reader can consult [8]. However, for completeness, we include without proof a statement of the major results.

Theorem 3.25. An $\mathscr{H}$-matrix of order n exists if n satisfies any one of the following conditions:

(a) $n=ab$ where $\mathscr{H}$-matrices of orders a and b exist;
(b) $n=p^s+1$ where p is a prime and $p^s\equiv 3 \pmod 4$;
(c) $n=m(p^s+1)$ where $m>1$, an $\mathscr{H}$-matrix of order m exists and p is an odd prime;
(d) $n=mp^s(p^s+1)$ where $m>1$, an $\mathscr{H}$-matrix of order m exists and p is a prime, $p^s\equiv 1 \pmod 4$;
(e) $n=mtp^s(p^s+1)$ where $m>1, t>1$, $\mathscr{H}$-matrices of orders m and t exist and p is an odd prime;
(f) $n=m(m-1)$ where $m=2^s t$ and t is a product of numbers of the form $p^s+1\equiv 0 \pmod 4$ where p is a prime;
(g) $n=m(m+3)$ with m and $m+4$ both of the form $2^s t$ where t is as defined in (f);
(h) $n=mv(v+3)$ with $m>2$, an $\mathscr{H}$-matrix of order m exists and $v\equiv 2 \pmod 4$ such that v and $v+4$ are both of the form p^s+1 where p is a prime;
(i) $n=mtv(v+3)$ where $m>1, t>1$, $\mathscr{H}$-matrices of orders m and t exist and v and $v+4$ are both of the form p^s+1 where p is a prime;
(j) $n=m^2$ where $m-1$ and $m+1$ are both primes;
(k) $n=(m-2)^2$ where $m-3=q^s\equiv 1 \pmod 4$ for q a prime and $m=2^v t$ where t is as defined in (f);
(l) $n=(m-1)^3+1$ where $m=2^s t$ and t is as defined in (f);
(m) $n=v\{(m-1)^2+1\}$ where an $\mathscr{H}$-matrix of order v exists and $m=2^s t$ such that t is as defined in (f).

There are many integers which are multiples of 4 which are not covered by any one of the conditions in Theorem 3.25, e.g. $n=116$ or $n=188$, and these orders have been studied individually.

Despite their many interesting matrix properties, our main reason for discussing $\mathscr{H}$-matrices here is that they enable us to construct Hadamard 2-designs. We recall that a Hadamard 2-design is a symmetric design for $(4\lambda+3, 2\lambda+1, \lambda)$.

Theorem 3.26. Let H be a normalised $\mathscr{H}$-matrix of order $n>4$. If A is the $(n-1)$ by $(n-1)$ matrix obtained by deleting the first row and column of H and replacing each -1 by 0, then A is the incidence matrix of a symmetric 2-$(n-1, (n-2)/2, (n-4)/4)$ design. Conversely, the incidence matrix of any 2-$(4\lambda+3, 2\lambda+1, \lambda)$ design becomes an $\mathcal{H}$-matrix of order $4(\lambda+1)$ by the reverse procedure.

Proof. Let $A=(a_{ij})$ and let $\mathcal{D}$ be the incidence structure with $n-1$ points $P_1, P_2, \ldots, P_{n-1}$ and $n-1$ blocks $y_1, y_2, \ldots, y_{n-1}$ such that P_i is on y_j if and only if $a_{ij}=1$. By Exercise 3.18 every column of H has $n/2$ entries of $+1$ and $n/2$ of -1. Thus, since H is normalised, A has $(n/2)-1$ non-zero entries in each column and so each block of $\mathcal{D}$ contains $(n-2)/2$ points.

If P_i and P_j are any two distinct points then the number of blocks containing them both is the scalar product of the ith and jth rows of A. But, by (c) of Exercise 3.18, the ith and jth rows have $n/4$ entries of $+1$ in common columns. Thus, since the first column of H is deleted to give A, the scalar product of the ith and jth rows of A is $(n/4)-1$. Since this is independent of the choice of P_i and P_j, $\mathcal{D}$ is a 2-$(n-1, (n-2)/2, (n-4)/4)$ design. A straightforward application of Corollary 1.4 shows that $\mathcal{D}$ is symmetric.

Now suppose that $\mathcal{D}$ is a 2-$(4\lambda+3, 2\lambda+1, \lambda)$ design and let A be an incidence matrix for $\mathcal{D}$. Then, since $\mathcal{D}$ is symmetric, each row and column of A has $2\lambda+1$ entries of 1 and $2\lambda+2$ zero entries. Since any two points are on λ common blocks any two rows of A have exactly λ columns in which they both have the entry 1. Thus there are $2(\lambda+1)$ columns in which only one of the rows has a 1 and $4\lambda+3-2(\lambda+1)-\lambda=\lambda+1$ columns in which they both have the entry 0.

Let H be the n by n matrix obtained by adding a first row and column in which each entry is $+1$ and replacing each 0 of A by -1. Since each row of A has $2\lambda+1$ non-zero entries, each row of H other than the first has $2\lambda+2$ entries of $+1$ and $2\lambda+2$ of -1, so that the scalar product of any row of H with the first is zero. For any other two rows of H there are $\lambda+1$ columns in which they both have $+1$, $\lambda+1$ columns in which they both have -1, and $2(\lambda+1)$ columns in which the two rows have entries of opposite sign. Thus the inner product of any two distinct rows of H is zero and H is Hadamard. □

We have already met some small examples of Hadamard 2-designs. For $\lambda=1$ a Hadamard 2-design has parameters 2-(7, 3, 1) and so, as we have seen, is isomorphic to $\mathcal{P}_2(2)$. For $\lambda=2$ we have parameters 2-(11, 5, 2) and we constructed a design with these parameters in Exercise 2.9. In each of these instances we could use the design, which we constructed by other methods, to construct a Hadamard matrix of the appropriate size. Similarly Exercise 2.11 enables us to construct a Hadamard matrix of order 20. We now want to construct an infinite family of Hadamard 2-designs called the *Paley designs.* Before doing this, however, we note that $\mathscr{P}_n(2)$ has parameters 2-$(2^{n+1}-1, 2^n-1, 2^{n-1}-1)$. Thus $\mathscr{P}_n(2)$ has parameters 2-$(4\lambda+3, 2\lambda+1, \lambda)$ with $\lambda = 2^{n-1}-1$ and so we have already met one infinite family of Hadamard 2-designs.

If q is any prime power let $\mathscr{S}$ be the set of elements of $K = GF(q)$ and let H be the permutation group of $\mathscr{S}$ consisting of all mappings $\pi_{a,b} : x \rightarrow ax+b$, where $a, b \in GF(q)$ $(a \neq 0)$. Since the subgroup $S = \{\pi_{1,b}\}$ is transitive on K and $H_0 = \{\pi_{a,0}\}$ is transitive on $K \backslash \{0\}$, H is 2-transitive on K. Also $|H| = q(q-1)$.

In order to construct the Paley designs we need to consider the subgroup G of H consisting of these $\pi_{a,b}$ for which a is a non-zero square. We recall, from elementary field theory, that:

(a) if q is even then every element of $K^*(=K\backslash\{0\})$ is a square;
(b) if q is odd then $(q-1)/2$ elements of K^* are squares;
(c) -1 is a square if and only if q is even (in which case $-1=1$), or $q \equiv 1 \pmod 4$.

Thus if q is even $G = H$, while if q is odd $|G| = (q(q-1))/2$. If q is odd then, since $|G| < q(q-1)$, G cannot be 2-transitive on K. If, however, $q \equiv 3 \pmod 4$ G does have a similar but weaker property. A permutation group is called *t-homogeneous* if it is transitive on the (unordered) subsets of t distinct elements. (Note that this is not the same as being t-transitive. The latter implies transitivity on the *ordered* subsets of t distinct elements.)

Lemma 3.27. If $q \equiv 3 \pmod 4$ then G is 2-homogeneous on $\mathscr{S}$.

Proof. We know that $|G| = (q(q-1))/2$ and that the number of unordered subsets of two elements is also $(q(q-1))/2$. Let $\mathscr{T} = \{0, 1\}$. If $G_{\mathscr{T}}$ is the subgroup stabilising $\mathscr{T}$ as a set then the number of images of $\mathscr{T}$ under G is $|G|/|G_{\mathscr{T}}|$. Thus the lemma will be true if and only if $|G_{\mathscr{T}}| = 1$. If $\pi_{a,b} \in G_{\mathscr{T}}$ then either $0\pi_{a,b} = 0$ and $1\pi_{a,b} = 1$ or $0\pi_{a,b} = 1$ and $1\pi_{a,b} = 0$. If $0\pi_{a,b} = 0$ and $1\pi_{a,b} = 1$ then $a \cdot 0 + b = 0$ and $a \cdot 1 + b = 1$, which implies $a = 1$, $b = 0$; i.e. $\pi_{a,b}$ is the identity. If $1\pi_{a,b} = 0$ and $0\pi_{a,b} = 1$ then $a + b = 0$ and $a \cdot 0 + b = 1$. Thus $b = 1$ and $a = -1$. But if $q \equiv 3 \pmod 4$ then -1 is not a square and $\pi_{-1,1} \notin G$. Thus if $q \equiv 3 \pmod 4$, $|G_{\mathscr{T}}| = 1$ and the lemma is proved. (Note, by the way, that if $q \equiv 1 \pmod 4$ then G is definitely not 2-homogeneous.) □

We will now use G and K to construct the Paley designs. For any prime power $q \equiv 3 \pmod 4$ we define a structure $\mathscr{L}(q)$ as follows: The points of $\mathscr{L}(q)$ are the elements of K and the blocks are the sets $D + x$, where x is any element of K and

D is the set of the $(q-1)/2$ non-zero squares. It is clear that $\mathscr{L}(q)$ has q points, $(q-1)/2$ points on each block and q blocks. However, it may not be reduced, for we do not, as yet, know that $D+y$ and $D+x$ are different if $x \neq y$.

Lemma 3.28. G is an automorphism group of $\mathscr{L}(q)$.

Proof. The point x is on the block $D+y$ if and only if $x=d^2+y$ for a non-zero d. If $\pi_{a^2,b}$ is in G, then $x\pi_{a^2,b}=(d^2+y)a^2+b=(ad)^2+a^2y+b\in D+(a^2y+b)$. Hence $\pi_{a^2,b}$ is an automorphism of $\mathscr{L}(q)$. □

Theorem 3.29. $\mathscr{L}(q)$ is a Hadamard 2-design with parameters 2-$(q, (q-1)/2, (q-3)/4)$.

Proof. Since G is an automorphism group of $\mathscr{L}(q)$ which is 2-homogeneous on the points, $\mathscr{L}(q)$ must be a 2-structure. (The automorphism mapping one pair of points to another maps the blocks on one pair to the blocks on the other.) Thus $\mathscr{L}(q)$ is a uniform square 2-structure for $(q, (q-1)/2, \lambda)$ where, as yet, we do not know λ. Thus, by Theorem 1.26, $\mathscr{L}(q)$ is a symmetric design, and knowing $v=(k(k-1)+\lambda)/\lambda$ gives $\lambda=(q-3)/4$. □

The designs $\mathscr{L}(q)$ enable us to construct Hadamard matrices of orders 8, 12, 20, 24, 28, 32, 44, 48, 60, ... It is worth noting that, although the parameters occasionally agree, the only time a Paley design $\mathscr{L}(q)$ is isomorphic to a member of our other infinite family, i.e. $\mathscr{P}_n(2)$, is when $\lambda=1$.

Exercise 3.23. Show that a Hadamard 2-design exists for all values of $\lambda \leqslant 32$, using the construction methods of this section, except possibly for $\lambda=8, 12, 18, 22, 24, 28, 30$. Using Theorem 3.25 show that only $\lambda=22, 28$ are left open. (See also Exercise 3.51 for the case $\lambda=8$.)

Exercise 3.24.* Show that $\mathscr{L}(31)$ is not isomorphic to $\mathscr{P}_4(2)$.

Exercise 3.25.* Let H be a Hadamard matrix of order $n>2$. Define $\mathscr{D}=\mathscr{D}(H)$ as follows: for each $i=1, 2, \ldots, n$, $(i, +1)$ and $(i, -1)$ are points, and $[i, +1]$

and $[i, -1]$ are blocks; incidence is given by (i, a) is on $[j, b]$ if $h_{ij} = ab$ (here h_{ij} is the (i, j)-entry in H).

(a) Show that $\mathscr{D}$ has the properties:

- (i) any three distinct points (blocks) of $\mathscr{D}$ are incident with either 0 or $n/4$ common blocks (points);
- (ii) given a point P (block y) of $\mathscr{D}$, there is a unique point P' (block y') such that P and P' are on no common block (y and y' do not meet);
- (iii) any two distinct points (blocks) of $\mathscr{D}$ are incident with either 0 or $n/2$ common blocks (points).

Show:

(b) If H_1 and H_2 are $\mathscr{H}$-equivalent Hadamard matrices, then $\mathscr{D}(H_1) \cong \mathscr{D}(H_2)$.

(c)* If $\mathscr{D}$ is a structure and there is a constant $n > 0$ such that $\mathscr{D}$ satisfies (i) and (ii), then $\mathscr{D} \cong \mathscr{D}(H)$ for some Hadamard matrix H.

(The design $\mathscr{D}(H)$ is called a *generalised Hadamard design.*)

Exercise 3.26.* Let H_n be the Hadamard matrix of order $2n$ obtained from H by letting $H_n = H \otimes H_{n-1}$, $H_1 = H$, where

$$H = \begin{bmatrix} 1 & 1 \\ 1 & -1 \end{bmatrix}.$$

Show that the Hadamard 2-design constructed from H_n is isomorphic to the Hadamard 2-design $\mathscr{P}_{n-1}(2)$.

3.6 Biplanes

It was noticed some years ago that $\lambda = 1$ is the only value for which infinitely many symmetric 2-(v, k, λ) designs are known to exist. This led people to wonder if, in fact, there are only a finite number of symmetric 2-designs for every other λ. A natural problem is to try to settle the question for the next value of λ, i.e. $\lambda = 2$. A symmetric design with $\lambda = 2$ is called a *biplane.* We have already met some examples of biplanes. We note that (a) the trivial 2-(4, 3, 2) is a biplane with $k = 3$, (b) the complement of the projective plane of order 2 is a 2-(7, 4, 2), i.e. a biplane with $k = 4$, (c) the Hadamard 2-design with $\lambda = 2$ is a 2-(11, 5, 2), i.e. a biplane with $k = 5$ (see Exercise 2.9), (d) the half-net design $\mathscr{H}(\mathscr{N})$ of order 4 and class number 2 is a 2-(16, 6, 2), i.e. a biplane with $k = 6$ (see Theorem 3.18), (e) Example 2.5 is a 2-(37, 9, 2), i.e. a biplane with $k = 9$. Since the BRC Theorem implies that there cannot exist biplanes with $k = 7$, 8 or 10 we already know that there exist biplanes for all permissible values up to $k = 10$.

Exercise 3.27. Show that there are no biplanes with $k=7$, 8 or 10.

We begin our discussion of biplanes by establishing some fundamental properties. Since, for a symmetric 2-(v, k, λ), $v=(k(k-1)+\lambda)/\lambda$, we note that, for a biplane, $v=1+(k(k-1))/2$. (So if $k \equiv 2$ or 3 (mod 4), v is even and, by the BRC Theorem, $k-2$ must be a square.) If y is a block of a biplane $\mathscr{B}$ with v points then there are $v-k=[(k-1)(k-2)]/2$ points not on y. Each block $z \neq y$ meets y in exactly two points and, conversely, every pair of points on y are on a unique block other than y. Thus we have established:

Lemma 3.30. There is a one-to-one correspondence between the unordered pairs of points on a block y of a biplane $\mathscr{B}$ and the blocks of $\mathscr{B}$ other than y. □

Now suppose that we have a biplane $\mathscr{B}$ with a block y and let P be any point of $\mathscr{B}$ not on y. We define a graph $\langle P \rangle$ on the set of points of y as follows: the vertices of $\langle P \rangle$ are the points of y and two points X, Y are adjacent in $\langle P \rangle$ if and only if the unique block $z \neq y$ containing them also contains P.

Lemma 3.31. The graph $\langle P \rangle$ is divalent (that is, each vertex has two edges on it), and hence is a union of polygons, each with at least three edges.

Proof. If X is a point of y, then X is joined to P by two blocks, w and z; each of w and z meets y in one more point, and these two points are distinct. □

Lemma 3.32. If P and Q are distinct points, neither on y, then $\langle P \rangle$ and $\langle Q \rangle$ have exactly two edges in common.

Proof. Since P and Q are joined by two blocks, these blocks give two common edges for $\langle P \rangle$ and $\langle Q \rangle$. □

These last few lemmas give:

Theorem 3.33. If $\mathscr{B}$ is a biplane with block size k, and y is a block of $\mathscr{B}$, then the $(k-1)(k-2)/2$ graphs $\langle P \rangle$, for P not on y, have the properties:

(1) each $\langle P \rangle$ is divalent;

(2) two different graphs $\langle P \rangle$ and $\langle Q \rangle$ have exactly two edges in common. □

Theorem 3.33 has a very interesting converse:

Theorem 3.34. Let $\mathscr{Y}$ be a set of k vertices, and let $G_1, G_2, \ldots, G_{(k-1)(k-2)/2}$ be a set of divalent graphs with vertex set $\mathscr{Y}$ such that, for $i \neq j$, G_i and G_j have exactly two edges in common. Then the structure $\mathscr{B}$ given below is a biplane with block size k:

the points of $\mathscr{B}$ are (a) the vertices of $\mathscr{Y}$, and (b) the symbols (i), for $i = 1, 2, \ldots, (k-1)(k-2)/2$;

the blocks of $\mathscr{B}$ are (a) the symbol y, and (b) the unordered pairs $[X, Y]$ of distinct vertices $X, Y \in \mathscr{Y}$;

the incidence rules are: a vertex X of $\mathscr{Y}$ is on y and on every block $[X, Y]$ for $Y \neq X$; a point (j) is on $[X, Y]$ if XY is an edge of G_j.

Proof. Let $v = k(k-1)/2 + 1$. Then the number of points of $\mathscr{B}$ is $k + (k-1)(k-2)/2 = v$, while the number of blocks is $1 + k(k-1)/2 = v$. So $\mathscr{B}$ is square. A point $P \in \mathscr{Y}$ is on y and on the $k-1$ blocks $[P, Q]$ for $Q \neq P$, while a point (j) is on $[P, Q]$ if PQ is an edge of G_j. But a divalent graph on k vertices has k edges, so (j) is on k blocks (see Exercise 3.28). Hence $\mathscr{B}$ is regular. Finally, if P, Q are distinct vertices of $\mathscr{Y}$, then the points P and Q are on two blocks, y and $[P, Q]$, while if (i) and (j) are two distinct points, then since G_i and G_j have exactly two common edges (i) and (j) are on exactly two common blocks. If P is a vertex of $\mathscr{Y}$, and G_i is a graph, then P is on exactly two edges PQ and PR of G_i, so P and (i) are on just two blocks. Thus $\mathscr{B}$ is a square regular 2-structure for $\lambda = 2$. Hence, by Lemma 1.29, $\mathscr{B}$ is uniform. So it is a symmetric design, i.e. a biplane. □

Exercise 3.28. Let G be a divalent graph on k vertices. Show that G has k edges.

The graphs $\langle P \rangle$ of Theorem 3.33, or the graphs G_i of Theorem 3.34 (notice that they are the 'same', i.e. $\langle (i) \rangle$ is G_i in the $\mathscr{B}$ of Theorem 3.34), are called *Hussain graphs*: a *complete set* of Hussain graphs is a set of $(k-1)(k-2)/2$ divalent graphs, on a set of k vertices, such that any two meet in exactly two edges.

Lemma 3.35. In a complete set of Hussain graphs, $G_1, G_2, \ldots, G_{(k-1)(k-2)/2}$, the two edges in common with G_i and G_j, where $i \neq j$, do not have a common vertex.

Proof. This is merely the observation that two blocks on distinct points P and Q which are not on y cannot meet again in y (or anywhere else, for that matter). □

In view of Theorem 3.34, a single block of $\mathscr{B}$, together with its complete set of Hussain graphs, completely determines $\mathscr{B}$. So let $\{G_i\}$ and $\{H_i\}$ be two complete sets of Hussain graphs on vertex sets $\mathscr{Y}$ and $\mathscr{U}$, each with k vertices, and let $\mathscr{B}$ and $\mathscr{C}$ be the resulting biplanes, where y and u are, respectively, the blocks $\mathscr{Y}$ and $\mathscr{U}$. Then there is an isomorphism α from $\mathscr{B}$ to $\mathscr{C}$ such that $y^\alpha = u^\alpha$ if and only if there is a bijection α^* from $\mathscr{Y}$ to $\mathscr{U}$ which maps the set $\{G_i\}$ onto the set $\{H_i\}$. So:

Theorem 3.36. Two biplanes $\mathscr{B}_1$ and $\mathscr{B}_2$ are isomorphic if and only if for any block $y_1 \in \mathscr{B}_1$ there is a block $y_2 \in \mathscr{B}_2$ such that the complete set of Hussain graphs determined by y_1 is isomorphic to the complete set of Hussain graphs determined by y_2. In particular, there is an automorphism α of the biplane $\mathscr{B}$ sending the block y onto the block z if and only if the complete set of Hussain graphs determined by y is isomorphic to the complete set of Hussain graphs determined by z. □

A special case of Theorem 3.36 is:

Theorem 3.37. If $\mathscr{B}$ is a biplane and y is a block of $\mathscr{B}$, then an automorphism α fixing y is determined by a permutation of the points of y which maps the complete set of Hussain graphs determined by y onto itself. □

Theorem 3.38. If an automorphism α of the biplane $\mathscr{B}$ fixes all the points on a block, or all the blocks on a point, then $\alpha = 1$.

Proof. This is a corollary of Lemma 3.30. □

We now apply some of these ideas for small values of k.

Example 3.6. For $k = 4$, we need three graphs on four points. But any one graph must be a single 4-gon, and it is easy to see that the three graphs must be as in Figure 3.2. Since this set of graphs is unique, Theorem 3.36 implies that the automorphism group of $\mathscr{B}$ is transitive on blocks (and hence on points). In addition, *any* permutation of the four points preserves the set of graphs. So Aut $\mathscr{B}$ has order $7 \cdot 4! = 7 \cdot 24 = 168$. (It is in fact the simple group $PGL(3,2)$, which we could have seen another way: any automorphism of $\mathscr{B}$ is an automorphism of $\mathscr{C}(\mathscr{B})$, and $\mathscr{C}(\mathscr{B})$ is the (unique) projective plane of order 2, whose group is well known to be $PGL(3,2)$.) In addition, since it is the unique biplane with $k=4$, $\mathscr{B}$ is self-dual.

Example 3.7. For $k=5$, we need six graphs on five vertices. But a divalent graph on five points must be a pentagon. Using Lemma 3.35, it is easy to see that the six graphs are those of Figure 3.3.

Hence $\mathscr{B}$ is again unique and, by Theorem 3.36, its automorphism group is transitive on blocks (and points). The automorphism group fixing a block is the alternating group A_5 and Aut $\mathscr{B}$ has order $11 \cdot 60 = 660$.

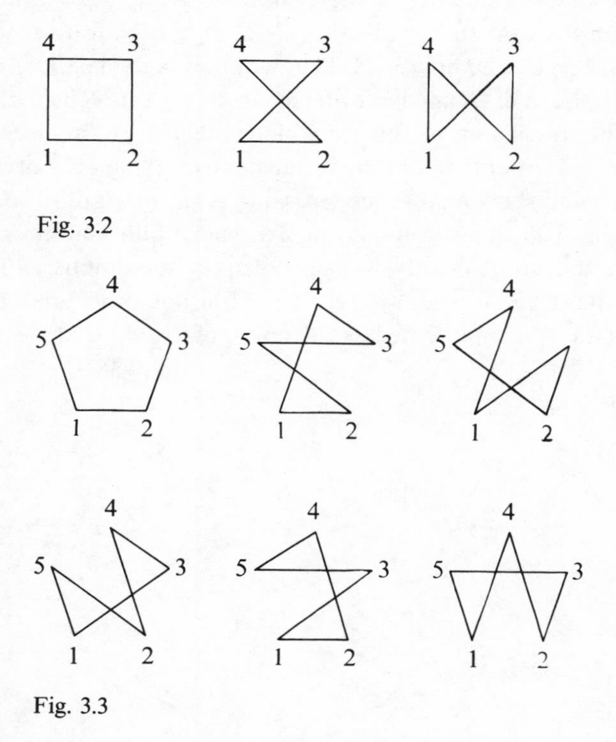

Fig. 3.2

Fig. 3.3

Exercise 3.29. Show that the automorphism group fixing a block in Example 3.7 is A_5.

Exercise 3.30.* Show that Aut $\mathscr{B}$, for Example 3.7, is simple. (It is in fact abstractly isomorphic to $PSL(2, 11)$.)

Example 3.8. For $k=6$, we need $v-k=10$ graphs, and here we have a choice: any single graph can be a 6-gon or two 3-gons. First, how many pairs of triangles are there on six points? We can choose the first one in $\binom{6}{3}$ ways, and then the other one is determined. But each pair of triangles is counted twice this way, so that the total number is $\frac{1}{2}\binom{6}{3}=10$. Now any two graphs, each consisting of a pair of triangles, *must* meet in two edges (see Figure 3.4). So one possibility is to use the 10 triangle pairs to construct a biplane, which we denote by $\mathscr{B}_2$. Note that if we let y be the block containing these 10 triangles then we do not know anything yet about the Hussain graphs on the other blocks. However, since any permutation of six letters is also a permutation of the 10 pairs of triangles, $(\text{Aut } \mathscr{B}_2)_y$ is the symmetric group S_6. Thus, since $(\text{Aut } \mathscr{B}_2)_y$ is 2-transitive on the points of y, $(\text{Aut } \mathscr{B}_2)_y$ is transitive on the remaining blocks of $\mathscr{B}_2$ (see Lemma 3.30). Thus we know (a) that either Aut $\mathscr{B}_2$ is transitive on the blocks of $\mathscr{B}_2$ and has order $6!\cdot 16$, or (b) Aut $\mathscr{B}_2$ has order $6!$ and fixes y, and the Hussain graphs on every block other than y are identical. We will now sketch a proof that Aut $\mathscr{B}_2$ is transitive on the blocks of $\mathscr{B}_2$ by choosing a block z and showing that the Hussain graphs on z are 10 triangle pairs. (By Theorem 3.36 this will imply the existence of an automorphism mapping y onto z and, consequently, will show Aut $\mathscr{B}_2$ does not fix y.) To do this we represent the 16 points of $\mathscr{B}_2$ by $1, 2, 3, 4, 5, 6$ (the points of y) and $P_1, P_2, \ldots, P_{10}$. Each point P_i is represented as the union of its two triangles on y. Thus

$$P_1 = 123,\ 456$$
$$P_2 = 124,\ 356$$

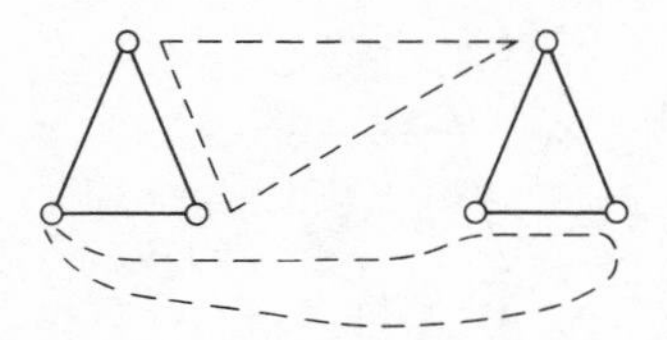

Fig. 3.4

$P_3 = 125, 346$

$P_4 = 126, 345$

$P_5 = 134, 256$

$P_6 = 135, 246$

$P_7 = 136, 245$

$P_8 = 145, 236$

$P_9 = 146, 235$

$P_{10} = 156, 234.$

The blocks of $\mathscr{B}_2$ are given by:

Blocks	*Points on a block*					
y	1	2	3	4	5	6
[12]	1	2	P_1	P_2	P_3	P_4
[13]	1	3	P_1	P_5	P_6	P_7
[14]	1	4	P_2	P_5	P_8	P_9
[15]	1	5	P_3	P_6	P_8	P_{10}
[16]	1	6	P_4	P_7	P_9	P_{10}
[23]	2	3	P_1	P_8	P_9	P_{10}
[24]	2	4	P_2	P_6	P_7	P_{10}
[25]	2	5	P_3	P_5	P_7	P_9
[26]	2	6	P_4	P_5	P_6	P_8
[34]	3	4	P_3	P_4	P_5	P_{10}
[35]	3	5	P_2	P_4	P_6	P_9
[36]	3	6	P_2	P_3	P_7	P_8
[45]	4	5	P_1	P_4	P_7	P_8
[46]	4	6	P_1	P_3	P_6	P_9
[56]	5	6	P_1	P_2	P_5	P_{10}.

The next step is to determine the Hussain graphs for the block $z = [12]$, say. The set $\mathscr{S}_1$ will consist of the six points 1, 2, P_1, P_2, P_3, P_4. The graphs for points not on [12] are then determined. For instance, consider the point 3. The blocks on 3 are: y, [13], [23], [34], [35], [36]. These meet [12] in, respectively: $\{1, 2\}, \{1, P_1\}, \{2, P_1\}, \{P_3, P_4\}, \{P_2, P_4\}, \{P_2, P_3\}$. So the graph is a union of two triangles, i.e. $(1, P_1, 2)$ and (P_2, P_3, P_4). Repeating this for the other nine points not on [12], we find that every graph is a union of two triangles. So the set of Hussain graphs for [12] is isomorphic to the set for the block y, and so *all* the sets are isomorphic. Thus Aut ($\mathscr{B}_2$) is transitive on points and on blocks, and $|\text{Aut}(\mathscr{B}_2)| = 16 \cdot 720$.

Aut $\mathscr{B}_2$ is not simple and in fact has a normal Singer group of order 16 (see Exercise 3.31).

Exercise 3.31. Let $\mathscr{B}$ be the biplane (16, 6, 2) given by the Singer group $V_4(2)$ of Exercise 2.10. Show that $\mathscr{B}$ is isomorphic to $\mathscr{B}_2$, and that its Singer group is normal in Aut $\mathscr{B}$.

Example 3.9. Another alternative for $k=6$ is to take the graphs of Figure 3.5 for $P_1, \ldots, P_{10}$.

The following are automorphisms of the resulting biplane $\mathscr{B}_3$:

$$\alpha=(123456),\quad \beta=(26)(35),\quad \gamma=(23)(56),\quad \delta=(36).$$

The permutation α shows that (Aut $\mathscr{B}_3)_y$ is transitive on the points of y. By considering the triangles containing 1 it is straightforward to argue that the subgroup fixing 1 also fixes 4. Similarly 2 and 5 are 'fixed together' and also 3 and 6.

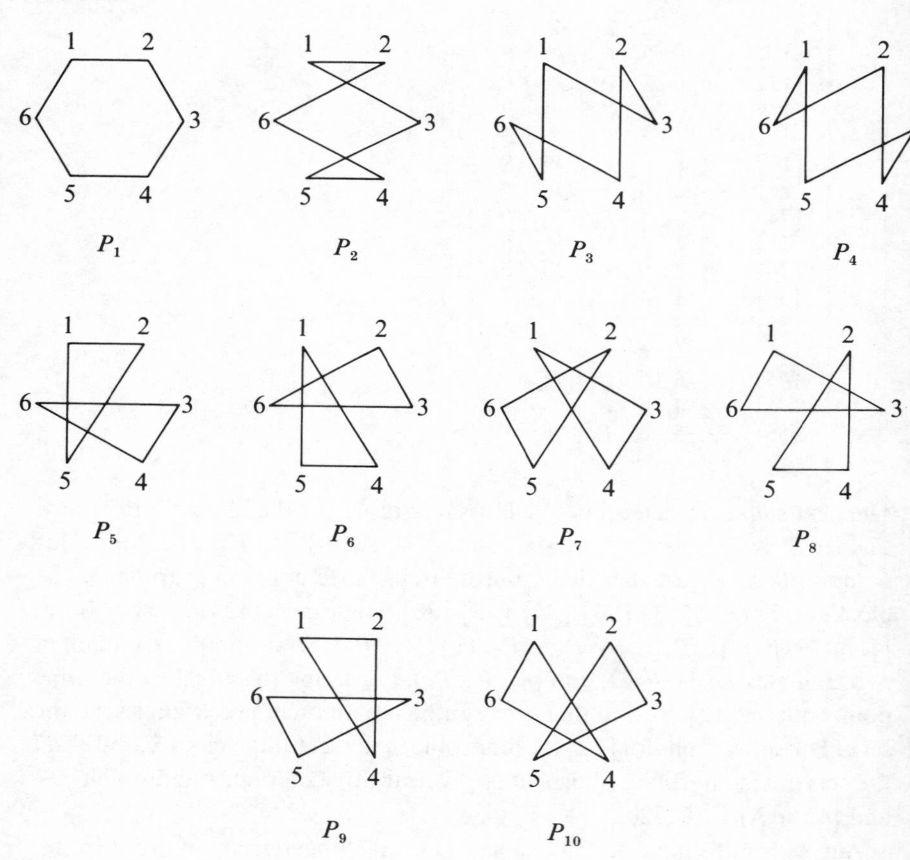

Fig. 3.5

The group $\langle \beta, \gamma \rangle$ has order 4, and is transitive on the four points 2, 3, 5, 6; the subgroup fixing 1 and 2 also fixes 4 and 5, and is $\langle (36) \rangle$, of order 2. So $(\text{Aut }\mathscr{B}_3)_y$ has order $6\cdot 4\cdot 2=48$.

Other properties of Aut $\mathscr{B}_3$ are:

(a) $(\text{Aut }\mathscr{B}_3)_y$ has two orbits of 2-sets on y; namely $\mathscr{C}_1=\{(1,4), (2,5), (3,6)\}$ and $\mathscr{C}_2=\{(i,j)\,|\,(i,j)\notin\mathscr{C}_1\}$ and three orbits of points of $\mathscr{B}_3$; namely $\{1,2,3,4,5,6\}$, $\{P$ not on $y\,|\,\langle P\rangle$ is a hexagon$\}$, $\{P$ not on $y\,|\,\langle P\rangle$ is a triangle pair$\}$;

(b) $(\text{Aut }\mathscr{B}_3)_y$ has two orbits of blocks $\neq y$;

(c) the set of Hussain graphs for the block [14] is isomorphic to the given one on y;

(d) the set of Hussain graphs for the block [12] is isomorphic to the given one on y.

Therefore every block has a set of Hussain graphs isomorphic to the given one, and

(e) Aut $\mathscr{B}_3$ is transitive and the subgroup fixing a block has three orbits of blocks of size 1, 3 and 12; $|\text{Aut }\mathscr{B}_3|=16\cdot 48$.

Exercise 3.32.* Prove all the statements made above about $\mathscr{B}_3$.

Example 3.10. A third example for a set of Hussain graphs is given in Figure 3.6.

The following are automorphisms:

$$\alpha=(12)(45),\quad \beta=(15)(24),\quad \gamma=(36),\quad \delta=(245).$$

We call this biplane $\mathscr{B}_4$. This set of Hussain graphs is, up to isomorphism, the only other one besides those already listed (see Exercise 3.33), and hence Aut $\mathscr{B}_4$ is transitive. $(\text{Aut }\mathscr{B}_4)_y$ is not transitive on y but has two orbits of points, namely $\{3,6\}$ and $\{1,2,4,5\}$; $|(\text{Aut }\mathscr{B}_4)_y|=4\cdot 2\cdot 3=24$, so $|\text{Aut }\mathscr{B}_4|=16\cdot 24$.

$(\text{Aut }\mathscr{B}_4)_y$ has three orbits of 2-sets: $\{(3,6)\}$, $\{(1,2), (1,4), (1,5), (2,4), (2,5), (4,5)\}$ and $\{(1,3), (2,3), (4,3), (5,3), (1,6), (2,6), (4,6), (5,6)\}$. So $(\text{Aut }\mathscr{B}_4)_y$ has four orbits of blocks with sizes 1, 1, 6, 8.

Exercise 3.33.* Prove that the three sets of Hussain graphs given above are the only ones possible, up to isomorphism, for $k=6$.

Exercise 3.34. Prove all the statements made above about $\mathscr{B}_4$.

Exercise 3.35.* Using the Singer group for $\mathscr{B}$, the biplane with parameters $(37, 9, 2)$ given in Example 2.5, find a set of Hussain graphs for $k=9$. Show that the full automorphism group, Aut $\mathscr{B}$, of $\mathscr{B}$ has order $37 \cdot 9$ and that the Singer group is normal in Aut $\mathscr{B}$.

Besides the biplanes we have constructed, only a few others are known at the present time. For $k=3, 4, 5$, there is (up to isomorphism) only one biplane; for $k=6$, there are exactly the three we have found. For $k=9$, there are, besides the one we have seen (e.g. see Example 2.5), exactly three others (see Salwach & Mezzaroba [7]). For $k=11$, we shall construct one in Chapter 8, and there are at least three others (but the total number is not yet known). For $k=13$, two biplanes (duals of each other), are known, and they complete the list of known biplanes. A computer search for biplanes with abelian Singer groups has shown that if $k>9$ there can be none with $k \leqslant 5000$.

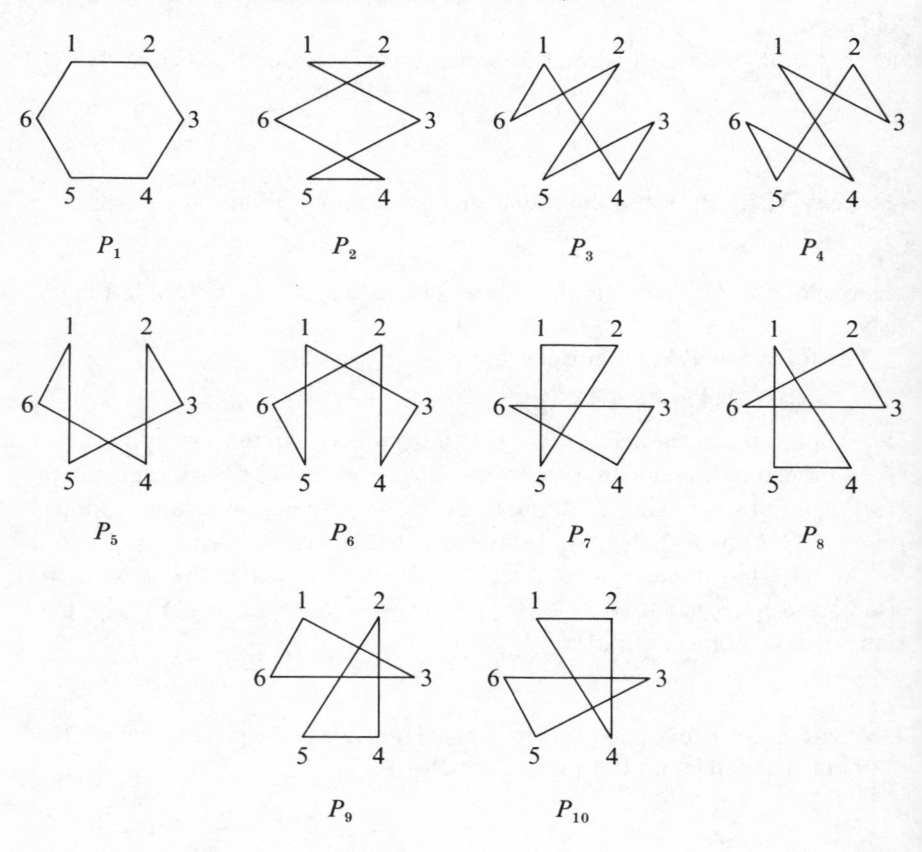

Fig. 3.6

3.7 Strongly regular graphs

In this section we carry the subject of graph theory, introduced in Chapter 1, a little further. The special class of graphs that we consider are extremely interesting in themselves but, while we shall indicate some of their intrinsic qualities, we are chiefly interested in their connections to design theory.

Let Γ be a regular graph of degree k, with v vertices. If there are integers λ, μ, such that:

(SR1) if P, Q are adjacent vertices, then there are exactly λ vertices X adjacent to both P and Q,

(SR2) if P, Q are non-adjacent (distinct) vertices, then there are exactly μ vertices X adjacent to both P and Q,

then Γ is a *strongly regular* ('SR') graph. We say that Γ has *parameters* (v, k, λ, μ). If Γ and its complement Γ^c are both connected, then we say that Γ is a *non-trivial* SR graph; Γ is *trivial* otherwise.

Exercise 3.36. Let $\mathscr{A}$ be an affine plane of order n, and Γ its block-adjacency graph. Show that Γ is SR and find its parameters. Also show that Γ^c is not connected, so that Γ is trivial. (In fact, Γ^c is the 'disjoint union' of $n+1$ complete graphs K_n.)

If Γ is an SR graph with parameters (v, k, λ, μ), then if we choose a vertex P we can divide the remaining vertices into two sets: Γ_1, the set of vertices joined to P, and Γ_2, the set of vertices not joined to P. Each vertex in Γ_1 is joined to P and to λ vertices in Γ_1, hence to $k-\lambda-1$ vertices in Γ_2. Each vertex in Γ_2 is joined to μ vertices in Γ_1, and so to $k-\mu$ vertices in Γ_2. This information can be conveniently pictured in Figure 3.7.

Theorem 3.39. If Γ is an SR graph with parameters (v, k, λ, μ), then $k(k-\lambda-1) = \mu(v-k-1)$.

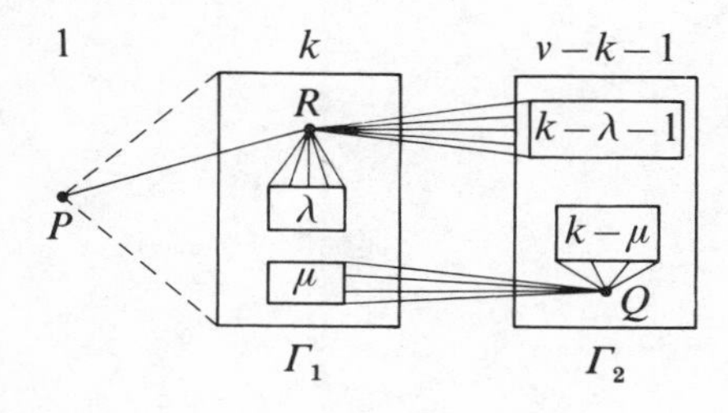

Fig. 3.7

Proof. Choose a vertex P and consider the diagram above. Count pairs (i.e. edges) (R, Q), where R is in Γ_1, Q is in Γ_2, and R, Q are joined. This immediately gives the result. □

Theorem 3.40. If Γ is an SR graph with parameters (v, k, λ, μ), then Γ^c is an SR graph with parameters $(v, v-k-1, v-2k+\mu-2, v-2k+\lambda)$.

Exercise 3.37. Prove Theorem 3.40. □

Exercise 3.38. Let Γ be SR, with parameters (v, k, λ, μ). Show that Γ is trivial if any one of the following conditions hold:

(a) Γ is bipartite; (d) $\lambda \geqslant k-1$;

(b) $v=2k$; (e) $v=k+1$;

(c) $\mu=0$; (f) $v+\lambda \leqslant 2k$.

Exercise 3.39. Show that if Γ is SR of degree $k=1$, then Γ is trivial.

Exercise 3.40. Show that if Γ is SR of degree $k=2$, then Γ is trivial unless Γ has parameters (5, 2, 0, 1).

Exercise 3.41. Let Γ be the graph with vertices $1, 2, \ldots, n$, and edges (1, 2), (2, 3), ..., $(n-1, n)$, $(n, 1)$. Show that if $n>5$ then Γ is not SR; if $n=5$, Γ is non-trivial SR; if $1<n<5$, then Γ is trivial SR.

Exercise 3.42. Demonstrate that the *Petersen graph*, pictured in Figure 3.8, is non-trivial SR, and find its parameters.

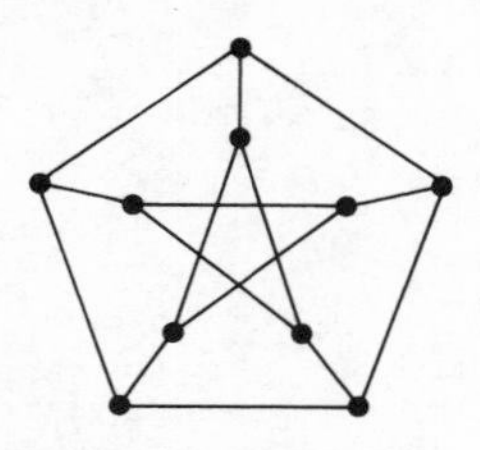

Fig. 3.8

Example 3.11. Let $\mathcal{N}$ be a net of order n, with m parallel classes. Let $\Gamma = \Gamma_{\mathscr{P}}(\mathcal{N})$ be the point-adjacency graph of $\mathcal{N}$. If $1 < m < n$, then Γ is a non-trivial SR graph with parameters $(n^2, m(n-1), m^2 - 3m + n, m(n-1))$.

Example 3.12. Let $\mathscr{B}_n$, $n > 4$, be the set of integers $\{1, 2, \ldots, n\}$, and define $T(n)$, the *triangular graph*, as follows: the vertices of $T(n)$ are the 2-sets (i, j) from $\mathscr{B}_n$, and (i, j) is joined to (k, l) if the set i, j, k, l consists of exactly three distinct elements of $\mathscr{B}_n$. Then $T(n)$ is a non-trivial SR graph with parameters $(\binom{n}{2}, 2(n-2), n-2, 4)$.

Example 3.13. Let K be the field $GF(q)$, where $q \equiv 1 \pmod 4$. Define a graph $\Gamma(q)$ whose vertices are the elements of K, and where x is joined to y if $x - y$ is a non-zero square. Then $\Gamma(q)$ is a non-trivial SR graph with parameters $(q, (q-1)/2, (q-5)/4, (q-1)/4)$.

Exercise 3.43. Show that the graphs of Examples 3.11 and 3.12 are indeed non-trivial SR with parameters as given.

Exercise 3.44.* Show that $\Gamma(q)$, as in Example 3.13, is non-trivial SR, with the given parameters. (Hint: use the fact that the group $x \to a^2x + b$, where $a, b \in K$ $(a \neq 0)$ is an automorphism group of $\Gamma(q)$. Also see the proof of Theorem 3.29.)

Exercise 3.45. Show that the complement of $T(5)$ is isomorphic to the Petersen graph.

Lemma 3.41

(a) A $(0, 1)$-matrix A is the adjacency matrix for a graph if and only if $A = A^{\mathrm{T}}$ and $\operatorname{tr}(A) = 0$ (i.e. all entries on the main diagonal are 0).

(b) The graph Γ is regular if and only if $A = A(\Gamma)$ satisfies $JA = kA$ (and then the degree is k).

(c) The graph Γ is regular of degree k if and only if $A = A(\Gamma)$ satisfies $\mathbf{j}A = k\mathbf{j}$, which is equivalent to saying that k is an eigenvalue, with eigenvector $\mathbf{j}$.

The proof is trivial, and we omit it. $\square$

Theorem 3.42. Let Γ be a graph with v vertices. Then Γ is SR if and only if there are real constants a, b, c such that $A = A(\Gamma)$ satisfies:

$$A^2 = aI + bA + cJ.$$

In that case, Γ is SR with parameters $(v, a+c, b+c, c)$.

Proof. As $A = A^{\mathrm{T}}$, we have $A^2 = AA^{\mathrm{T}}$, so the entry in position (i, j) of A^2 is the inner product of row i and row j of A, and counts the number of vertices X joined to both P_i and P_j. If $i = j$, this is $a + c$, so every vertex is joined to $a + c$ others, and hence Γ is regular. If $i \neq j$, then this is $b + c$ if P_i is joined to P_j (i.e. if A has $+1$ in position (i, j)), and is c if P_i is not joined to P_j. But this is the definition of an SR graph. □

Corollary 3.43. If Γ is SR, with parameters (v, k, λ, μ) and $A = A(\Gamma)$ is an adjacency matrix for Γ, then

$$A^2 = (k-\mu)I + (\lambda - \mu)A + \mu J,$$

$$\mathbf{j}A = k\mathbf{j} \quad \text{and} \quad JA = kJ. \ \square$$

We shall use the fact from linear algebra that if M is a real (square) symmetric matrix of order n, then M has n (not necessarily distinct) real eigenvalues. We want to consider the eigenvectors of A, one of which we already know is $\mathbf{j}$ (with eigenvalue k). Since the trace of A, tr (A), is zero, the sum of the eigenvalues of A (with their multiplicities, of course) is zero, and we can exploit this fact.

Theorem 3.44. An adjacency matrix A for a non-trivial SR graph Γ with parameters (v, k, λ, μ) has three distinct real eigenvalues k, r_1, r_2, with respective multiplicities $1, m_1, m_2$, satisfying

$$m_1 + m_2 = v - 1,$$

$$r_1 m_1 + r_2 m_2 = -k.$$

Furthermore r_1, r_2 are the zeros of $f(x) = x^2 - (\lambda - \mu)x - (k - \mu)$, and both are integral unless Γ has parameters $(4\mu + 1, 2\mu, \mu - 1, \mu)$.

Proof. First, let us assume $v > 2k$, and prove the theorem for such SR graphs; we deal with $v \leqslant 2k$ at the end. Let $\mathbf{v}$ be an eigenvector for A, with eigenvalue r, and $\mathbf{v} \notin \langle \mathbf{j} \rangle$. Then $\mathbf{v}A = r\mathbf{v}$ implies $\mathbf{v}A^2 = r^2\mathbf{v}$. So

$$\mathbf{v}A^2 = (k-\mu)\mathbf{v} + (\lambda - \mu)\mathbf{v}A + \mu\mathbf{v}J$$

which gives

$$f(r)\cdot \mathbf{v}=[r^2-(\lambda-\mu)r-(k-\mu)]\mathbf{v}=\mu|\mathbf{v}|\mathbf{j},$$

where $|\mathbf{v}|$ is the sum of the coordinates of $\mathbf{v}$. Since $\mathbf{v}\notin\langle\mathbf{j}\rangle$, this implies

$$f(r)=\mu|\mathbf{v}|=0.$$

Hence r is a zero of $f(x)$. So A has at most three eigenvalues: k, r_1, r_2, where r_1, r_2 are the zeros of $f(x)$. If k is a zero of $f(x)$, then

$$k^2-(\lambda-\mu)k-k+\mu=0.$$

But from Theorem 3.39

$$k^2-k\lambda-k+\mu k=\mu(v-1),$$

so

$$\mu(v-1)+\mu=\mu v=0,$$

which is impossible. So k is distinct from r_1 and r_2.

Since A has v eigenvalues, if we let m_i be the multiplicity of r_i, for $i=1, 2$, then $1+m_1+m_2=v$, and, since the sum of the eigenvalues is the trace of A, we have $k+r_1m_1+r_2m_2=0$. This gives the first two equations of the theorem.

Now $f(x)=(x-r_1)(x-r_2)=x^2-(r_1+r_2)x+r_1r_2$. Hence

$$r_1+r_2=\lambda-\mu.$$

But if we solve $f(x)=0$, we find that it has zeros

$$\tfrac{1}{2}(\lambda-\mu\pm\sqrt{D}),$$

where $D=(\lambda-\mu)^2+4(k-\mu)$. From $m_1r_1+m_2r_2=-k$, this gives

$$(m_1+m_2)(\lambda-\mu)+(m_1-m_2)\sqrt{D}=-2k.$$

Thus if $m_1\neq m_2$, we can solve for $\sqrt{D}$, and we see that $\sqrt{D}$ is a rational number. Thus r_1, r_2 are rational. But if a monic quadratic with integral coefficients has two rational zeros r_1, r_2, then r_1, r_2 must be integers. Hence r_1, r_2 are integral unless $m_1=m_2$.

Suppose $m=m_1=m_2$. Then $1+2m=v$ and $2m(\lambda-\mu)=-2k$, from the first two equations of the theorem. This gives $(v-1)(\mu-\lambda)=2k$.

Now $v>2k$, so $v-1\geqslant 2k$, hence $(v-1)(\mu-\lambda)=2k\leqslant v-1$. So $\mu-\lambda\leqslant 1$. But $\mu-\lambda\leqslant 0$ implies $k\leqslant 0$, and so $\mu-\lambda=1$, from which $v=2k+1$. Now, using Theorem 3.39, it follows that Γ has parameters $(4\mu+1, 2\mu, \mu-1, \mu)$.

It remains to show that $r_1\neq r_2$, and that both zeros r_1, r_2 actually occur as eigenvalues, which is equivalent to showing that neither m_1 nor m_2 is zero. From the form for the zeros of $f(x)$ it is clear that $r_1=r_2$ if and only if $D=(\lambda-\mu)^2+4(k-\mu)=0$. But $k\geqslant\mu$, so $D=0$ implies $\lambda=\mu=k$, which is impossible since $\lambda\leqslant k-1$. Now if $m_2=0$, then $m_1=v-1$ and $r_1m_1=-k$, so $(-r_1)(v-1)=k$, and $-r_1>0$. This is impossible since $(-r_1)(v-1)\geqslant v-1$, but $k<v-1$.

Finally, if Γ does not satisfy $v>2k$, then Γ^c has parameters (v, k', λ', μ'). Since

$v \neq 2k$ (see Exercise 3.38), we have $v > 2k'$ in Γ^c, so Theorem 3.44 certainly applies to Γ^c. The rest of the proof is left as an exercise.

Exercise 3.46. Let Γ be a non-trivial SR graph with parameters (v, k, λ, μ), and Γ^c its complement. If r_1, r_2 are the eigenvalues for Γ, as in Theorem 3.44, show that $-r_1-1$, $-r_2-1$ are the corresponding eigenvalues for Γ^c. Hence show that if $v < 2k$, then the conclusions of Theorem 3.44, applied to Γ^c, imply that Theorem 3.44 holds for Γ. $\square$

Exercise 3.47. Show that if q is not a square, then the eigenvalues r_1, r_2 for $\Gamma(q)$ (see Example 3.13) are not integral.

There is one easy connection between SR graphs and designs that we can develop:

Theorem 3.45. Let Γ be an SR graph with parameters

(a) (v, k, λ, λ)

or

(b) $(v, k, \lambda, \lambda+2)$.

Then if $A = A(\Gamma)$ is an adjacency matrix for Γ, the matrix B is an incidence matrix for a symmetric design $\mathscr{S}$, where:

(a) $B = A$, and $\mathscr{S}$ has parameters (v, k, λ)

or

(b) $B = A + I$, and $\mathscr{S}$ has parameters $(v, k+1, \lambda+2)$.

Proof. In case (a), $A^2 = AA^{\mathrm{T}} = (k-\lambda)I + \lambda J$ and, in case (b), $B^2 = BB^{\mathrm{T}} = (A+I)^2 = A^2 + 2A + I = (k-\lambda-2)I - 2A + (\lambda+2)J + 2A + I = (k-\lambda-1)I + (\lambda+2)J$. $\square$

Exercise 3.48. Let $\mathscr{N}$ be a net of order n, with m parallel classes. For what values of n and m will the SR graph $\Gamma_{\mathscr{P}}(\mathscr{N})$ lead to a symmetric design as in Theorem 3.45? In particular, show that symmetric designs with parameters $(16, 6, 2)$, $(36, 15, 6)$, and $(64, 36, 20)$ can be constructed in this way. (Compare with Theorem 3.18.)

The most important use of Theorem 3.44 is as a non-existence result, where it is very powerful. For example, consider a possible SR graph Γ with parameters $(21, 5, 1, 1)$. Then $f(x) = x^2 - 4$, so the eigenvalues r_1, r_2 of Theorem 3.44 are

± 2. Then there are integers m_1, m_2 such that

$$m_1+m_2=20,$$
$$2m_1-2m_2=-5.$$

But the second of these equations is impossible, since $2m_1-2m_2$ is even. So no such Γ exists. Or suppose there is an SR graph Γ with parameters (15, 7, 0, 6). Then the quadratic $f(x)=x^2+6x+1$ does not have integral zeros, and so Γ cannot exist.

Another use of SR graphs in design theory is given by:

Theorem 3.46. Let $\mathscr{S}$ be a 2-design for (v, k, λ), and suppose that there are integers μ_1, μ_2, where $\mu_1 \neq \mu_2$, such that any two distinct blocks of $\mathscr{S}$ always have μ_1 or μ_2 points in common. Let Γ be the graph whose vertices are the blocks of $\mathscr{S}$, two vertices being joined if the corresponding blocks meet in μ_2 points. Then Γ is SR.

Proof. Let A be an incidence matrix for $\mathscr{S}$, and suppose $\mu_2 > \mu_1$ (otherwise, we consider the complementary graph Γ_1^c, whose edges in Γ_1 are defined by two blocks meeting in μ_1 points). Then an adjacency matrix M for Γ is given by

$$M=\frac{1}{\mu_2-\mu_1}[A^{\mathrm{T}}A-\mu_1 J-(k-\mu_1)I].$$

Now $AA^{\mathrm{T}}=nI+\lambda J$, where $n=r-\lambda$, so $(A^{\mathrm{T}}A)^2=A^{\mathrm{T}}(nI+\lambda J)A=nA^{\mathrm{T}}A+\lambda k^2 J$. Then

$$(\mu_2-\mu_1)^2M^2=(A^{\mathrm{T}}A)^2+\mu_1^2bJ+(k-\mu_1)^2I-\mu_1(A^{\mathrm{T}}AJ+JA^{\mathrm{T}}A)$$
$$-2(k-\mu_1)A^{\mathrm{T}}A+2\mu_1(k-\mu_1)J.$$

Using $A^{\mathrm{T}}AJ=rkJ=JA^{\mathrm{T}}A$, $(A^{\mathrm{T}}A)^2=nA^{\mathrm{T}}A+\lambda k^2J$, and $A^{\mathrm{T}}A=(\mu_2-\mu_1)M+\mu_1 J+(k-\mu_1)I$, it is easy to see that this yields

$$M^2=\alpha I+\beta M+\gamma J$$

for real constants α, β, γ. So Γ is SR, by Theorem 3.42. □

Exercise 3.49. Evaluate the constants α, β, γ in the proof of Theorem 3.46, and hence find conditions under which Γ is non-trivial.

Exercise 3.50. Let $\mathscr{S}$ be a 2-design for $(v, k, 1)$. Then show that the conditions of Theorem 3.46 always hold and that Γ is merely the block-adjacency graph of

$\mathcal{S}$, and is SR with parameters

$$(b,\ k(r-1),\ r-1+k^2-2k,\ k^2).$$

Exercise 3.51. Let $\mathcal{S}$ be the 2-design of points and lines in $\mathcal{P}(3,2)$. Using Theorem 3.46, show that $\mathcal{S}$ leads to an SR graph with parameters $(35,18,9,9)$, and that hence there exists a symmetric design for $(35,17,8)$ (i.e. a Hadamard 2-design, see Exercise 3.23).

Exercise 3.52. Suppose $\mathcal{S}$ is a symmetric design for (v,k,λ), with polarity σ. If σ has no absolute points, then show that there is an incidence matrix A for $\mathcal{S}$ which is also the adjacency matrix of an SR graph Γ. Then, by considering the polynomial $f(x)$ for Γ, and the multiplicities m_1, m_2 (see Theorem 3.44), give another proof of Corollary 1.37: $n=k-\lambda$ is a square, and $\sqrt{n}$ divides λ.

References

[3], [5] and [8] are standard references for latin squares, projective planes and Hadamard designs respectively. Proofs of our specific results on latin squares can be found in [1], [2], [4] and [6], while [7] contains a discussion of biplanes with $k=9$.

[1] Bose, R. C. & Shrikhande, S. S. 'On the falsity of Euler's conjecture about the non-existence of two orthogonal latin squares of order $4t+2$'. *Proc. Nat. Acad. Sci., USA*, **45** (1959), 734–7.

[2] Bose, R. C., Shrikhande, S. S. & Parker, E. T. 'Further results on the construction of mutually orthogonal latin squares and the falsity of Euler's conjecture'. *Canad. J. Math.*, **12** (1960), 189–203.

[3] Denes, J. & Keedwell, A. D. *Latin Squares and Their Applications.* English Universities Press, 1974.

[4] Fisher, R. A. & Yates, F. 'The 6 × 6 latin squares'. *Proc. Camb. Phil. Proc.*, **30** (1934), 492–507.

[5] Hughes, D. R. & Piper, F. C. *Projective Planes.* Berlin–Heidelberg–New York, Springer, 1973.

[6] MacNeish, H. F. 'Euler squares'. *Ann. Math.*, **23** (1922), 221–7.

[7] Salwach, C. J. & Mezzaroba, J. A. 'The four biplanes with $k=9$'. *Journal of Combinatorial Theory* (*A*), **24** (1978), 141–5.

[8] Wallace, J., Street, A. P. & Wallace, W. *Combinatorics: Room Squares, Sum-free Sets and Hadamard Matrices.* Lecture Notes in Mathematics, Berlin–Heidelberg–New York, Springer-Verlag, vol. 292, 1972.

4 3-Designs and related topics

4.1 Introduction

There are many known examples of 2-designs and, even with the restriction that they are symmetric, we have already constructed several infinite families. But t-designs with $t > 2$ are much rarer. Although (as we proved in Theorem 1.5) t-structures exist for arbitrarily high values of t, there are no known non-trivial 7-designs and, in fact, the first non-trivial 6-designs were not discovered until 1982 (see [6]). In this chapter we look at some of the important t-designs with $t \geqslant 3$ and at various methods of constructing such designs.

Any t-design must be the extension of a $(t-1)$-design and in Section 4.2 we prove Cameron's Theorem. This says that there are very few possibilities for starting with a symmetric design and extending it to a 3-design. This leads us to extending symmetric Hadamard 2-designs to get Hadamard 3-designs which, in turn, leads to the construction of the little Mathieu designs in Section 4.4.

Just as we used the 2-homogeneous groups $ASL(2, q)$ for $q \equiv 3 \pmod 4$ to construct the Paley designs $\mathscr{L}(q)$, in Section 4.3 we use the 3-transitive groups $PGL(2, q)$ to construct some 3-designs. We then, in Section 4.5, discuss some of Alltop's ideas for constructing $(t+1)$-designs from t-transitive groups and indicate how to use the 3-transitive groups $PGL(4, 2^n)$ to construct some 4-designs which, in fact, turn out to be 5-designs. Finally, in Section 4.6 we prove a generalisation of Fisher's Inequality for $2s$-designs with $s \geqslant 1$, which leads to the concept of tight designs. Both Sections 4.5 and 4.6 contain very little detail and any interested reader will have to consult the relevant references.

4.2 Cameron's Theorem and Hadamard 3-designs

In our brief discussion of extensions in Chapter 1 we saw, in Corollary 1.17, that a necessary condition for a t-(v, k, λ) to have an extension is $k+1 \mid b(v+1)$. As an illustration of the significance of this simple divisor property we will show:

Lemma 4.1. If $n \neq 2, 4$ or 10 then a projective plane of order n cannot have an extension.

Proof. Let $\mathscr{P}$ be a projective plane and assume that it has an extension $\mathscr{P}^*$. By Theorem 2.21 $\mathscr{P}$ is a 2-(n^2+n+1), $n+1$, 1) with n^2+n+1 blocks. Since $\mathscr{P}^*$ exists, then, by Corollary 1.17, $n+2 \mid (n^2+n+1)(n^2+n+2)$. Clearly, $n^2+n+1=(n-1)(n+2)+3$ and $n^2+n+2=(n-1)(n+2)+4$. Thus $(n^2+n+1)(n^2+n+2) \equiv 12 \pmod{(n+2)}$ and $n+2 \mid (n^2+n+1)(n^2+n+2)$ only if $n+2 \mid 12$. Since $n>1$, $n+2>3$ which implies $n+2=4, 6$ or 12, i.e. $n=2, 4$ or 10. □

In our small examples in Chapter 1, Example $\mathscr{F}$ was a 2-(7, 3, 1), i.e. a projective plane of order 2, while Example $\mathscr{K}$ was a 3-(8, 4, 1). But straightforward verification shows that $\mathscr{K}_8 \cong \mathscr{F}$ and thus we know that a finite projective plane of order 2 has an extension. In Chapter 8 we shall show that a projective plane of order 4 also has an extension (in fact, this extension is one of the most important designs known). Since we do not know whether a projective plane of order 10 exists we cannot say much about its extension. All we can say is that if a plane of order 10 exists then it might be extendable.

Corollary 1.17 has similar consequences for biplanes.

Exercise 4.1. If $k \neq 3, 5$ or 11 show that a biplane with k points on a block (i.e. a symmetric 2-$((k^2-k+2)/2, k, 2)$ with $k>2$) cannot have an extension. Show also that if $k=3$ then the extension exists and is unique.

Exercise 4.2. Show that if $n>2$ then $\mathscr{P}_n(q)$ cannot have an extension unless $q=2$.

Exercise 4.3. Show that $\mathscr{A}_n(q)$ cannot have an extension unless $n=2$. Show, further, that $\mathscr{A}_2(2)$ is extendable.

Lemma 4.1 and Exercises 4.1 and 4.2 are, essentially, special cases of the following important theorem about the extendability of symmetric designs:

Theorem 4.2 (Cameron's Theorem). If $\mathscr{S}$ is a non-trivial symmetric design with parameters (v, k, λ), and $\mathscr{S}$ has an extension $\mathscr{S}^*$, then one of the following holds:

(a) $v=4\lambda+3$, $k=2\lambda+1$, i.e. $\mathscr{S}$ is a Hadamard 2-design;
(b) $v=(\lambda+2)(\lambda^2+4\lambda+2)$, $k=\lambda^2+3\lambda+1$;
(c) $v=111$, $k=11$, $\lambda=1$, i.e. $\mathscr{S}$ is a projective plane of order 10;
(d) $v=495$, $k=39$, $\lambda=3$.

Proof. The number of blocks in $\mathscr{S}^*$ is $b^*=v(v+1)/(k+1)$, from Theorem 1.2. Writing $v=(k^2-k+\lambda)/\lambda$, this gives

$$b^*=\frac{(k^2-k+\lambda)(k^2-k+2\lambda)}{\lambda^2(k+1)},$$

and so $k+1$ divides $(k^2-k+\lambda)(k^2-k+2\lambda)$. But

$$(k^2-k+\lambda)(k^2-k+2\lambda)\equiv 2(\lambda+1)(\lambda+2) \pmod{k+1},$$

and so

$$k+1 \text{ divides } 2(\lambda+1)(\lambda+2). \tag{1}$$

Now let y be a fixed block of $\mathscr{S}^*$, and $\mathscr{S}^*(y)$ the structure of all points of $\mathscr{S}^*$ not on the block y, and of all blocks of $\mathscr{S}^*$ which do not meet y. Next we note a general fact about $\mathscr{S}^*$: if two blocks of $\mathscr{S}^*$ meet, in the point X, say, then the two blocks are blocks of the symmetric design $\mathscr{S}^*_X$, hence meet λ more times. That is, two blocks of $\mathscr{S}^*$ meet in 0 or $\lambda+1$ points. So the blocks of $\mathscr{S}^*(y)$ are those which do not meet y in $\lambda+1$ points (and, of course, y is not in $\mathscr{S}^*(y)$).

Suppose A, B are distinct points of $\mathscr{S}^*(y)$. We count flags (X, w), where X is on y and w is on A and B. Let m be the number of blocks of $\mathscr{S}^*$ on A and B which also meet y; then the number of such flags is $m(\lambda+1)$. There are λ blocks w on X, A, B, for each of $k+1$ choices of X, so the number of such flags is $\lambda(k+1)$. Thus $m=\lambda(k+1)/(\lambda+1)$. There are k blocks of $\mathscr{S}^*$ on two points of $\mathscr{S}^*$, so there are $k-m=(k-\lambda)/(\lambda+1)$ blocks of $\mathscr{S}^*(y)$ on A and B. Hence $\mathscr{S}^*(y)$ is a 2-structure for $(v-k,\ k+1,\ (k-\lambda)/(\lambda+1))$. Furthermore, $\lambda+1$ divides $\lambda(k+1)$, so it divides $k+1$ (since $\lambda+1$ is relatively prime to λ); we write $k+1=s(\lambda+1)$, where s is an integer.

If $v-k=k+1$, then $v=2k+1$, and it is easy to see that $k=2\lambda+1$ and $\mathscr{S}$ is a Hadamard 2-design; this is case (a). So we may assume that $v>2k+1$; then $\mathscr{S}^*(y)$ is a proper 2-structure and Corollary 1.23 assures us that the number of blocks of $\mathscr{S}^*(y)$ is at least as large as $v-k$. So

$$\frac{k-\lambda}{\lambda+1}\cdot\frac{(v-k)(v-k-1)}{(k+1)k}\geqslant v-k.$$

Using the expression for v at the beginning of the proof, this simplifies to $(k-\lambda)(k-\lambda-1)\geqslant\lambda(\lambda+1)(k+1)$, or

$$k^2-k(2\lambda+1)+\lambda(\lambda+1)\geqslant k\lambda(\lambda+1)+\lambda(\lambda+1),$$

and, since $k\neq 0$, we have

$$k\geqslant \lambda^2+3\lambda+1. \tag{2}$$

So

$$k+1\geqslant(\lambda+1)(\lambda+2). \tag{3}$$

Now (3) implies $s\geqslant\lambda+2$ and (1) implies $s\leqslant 2(\lambda+2)$, so

$$\lambda+2\leqslant s\leqslant 2(\lambda+2). \tag{4}$$

But (1) asserts that $s(\lambda+1)$ divides $2(\lambda+1)(\lambda+2)$, and so

$$s \text{ divides } 2(\lambda+2). \tag{5}$$

Thus (4) and (5) imply

$$s=\lambda+2 \quad \text{or} \quad s=2(\lambda+2). \tag{6}$$

Now if $s=\lambda+2$, then $k=\lambda^2+3\lambda+1$, and we have case (b) of the theorem. If $s=2(\lambda+2)$, then $k=2\lambda^2+6\lambda+3$; but $\lambda(v-1)=k(k-1)$ implies that λ divides $k(k-1)$, and thus λ divides $(2\lambda^2+6\lambda+3)(2\lambda^2+6\lambda+2)$. Hence λ divides 6.

If $\lambda=1$, then $k=11$, $v=111$, which is case (c) of the theorem.

If $\lambda=2$, then $k=23$, $v=254$; v is even but $n=21$ is not a square, so $\mathscr{S}$ cannot exist.

If $\lambda=3$, then $k=39$, $v=495$, which is case (d) of the theorem.

If $\lambda=6$, then $k=111$, $v=2036$; again, v is even, but $n=105$ is not a square, so $\mathscr{S}$ does not exist.

(For $\lambda=2$ and 6 we have used the Bruck–Ryser–Chowla Theorem.) □

(NB. Note that in case (b) of Theorem 4.2, the design $\mathscr{S}^*(y)$ is a symmetric design itself, but it always satisfies the BRC conditions, since it has order $(\lambda+1)^2$.)

As we have already observed, there is no design known with the parameters of case (c) (it would be a projective plane of order 10). The same is true for case (d). For case (b) designs with these parameters are known to exist for $\lambda=1$ and $\lambda=2$. If $\lambda=1$ then we have the projective plane of order 4 and we will discuss its extension in Chapter 8. If $\lambda=2$, we are considering biplanes with $v=56$, several of which exist: but none of the known ones have extensions. Thus it is still an open question whether there is an extendable symmetric design for case (b) and $\lambda\geqslant 2$. For case (a) the situation is totally different. We have already discussed Hadamard 2-designs in Section 3.5. We observed that it is possible they exist for all choices of λ and showed that they certainly exist for infinitely many values of λ. We shall, in fact, now show that a Hadamard 2-design always has a unique extension. Before doing so we set a simple but interesting exercise:

Exercise 4.4. Show that if $\mathscr{E}$ is an extension of a non-trivial symmetric design $\mathscr{D}$ then, unless $\mathscr{D}$ is the projective plane of order 4, $\mathscr{E}$ cannot be extended.

In order to justify our claim that all Hadamard 2-designs are extendable we first examine a more general case:

Lemma 4.3. Let $\mathscr{S}$ be a 2-structure for $(2k+1, k, \lambda)$. Then $\mathscr{S}$ always has an extension to a 3-structure $\mathscr{S}^*$ for $(2k+2, k+1, \lambda)$.

Proof. We construct $\mathscr{S}^*$ as follows: the points of $\mathscr{S}^*$ are the points of $\mathscr{S}$, plus a new point named '∞'; if y is a block of $\mathscr{S}$, then the point set $y \cup \infty$ is a block of $\mathscr{S}^*$, and the point set y' consisting of all points of $\mathscr{S}$ not on y is a block of $\mathscr{S}^*$. Incidence is natural inclusion.

Clearly, $\mathscr{S}^*$ is uniform, with $k+1$ points on every block. We need to show that there are exactly λ blocks on any 3-set of points.

(a) If P, Q are distinct points of $\mathscr{S}$, then the three points P, Q and ∞ are on the λ blocks $y \cup \infty$, where y is one of the λ blocks on P and Q in $\mathscr{S}$.

(b) Suppose P, Q and R are distinct points of $\mathscr{S}$. In $\mathscr{S}^*$ they are on $y \cup \infty$, where y is a block of $\mathscr{S}$ on all three of P, Q and R, and they are also on y', where y is a block of $\mathscr{S}$ which is on none of P, Q and R.

If we let c_0 be the number of blocks of $\mathscr{S}$ on none of P, Q and R, and let c_{PQR} be the number of blocks of $\mathscr{S}$ on all three of P, Q and R, then we want $c_0 + c_{PQR}$. But in Theorem 1.28 it is shown that $c_0 + c_{PQR} = b - 3r + 3\lambda$. (In fact, Exercise 1.49 already provides us with a solution; in view of the importance of this result, we give a proof nevertheless.) Using Theorem 1.2 to substitute for b and r in $b - 3r + 3\lambda$, we get

$$\begin{aligned} b - 3r + 3\lambda &= \lambda \frac{(2k+1)\cdot 2k}{k(k-1)} - 3\lambda \frac{2k}{k-1} + 3\lambda \\ &= \frac{\lambda}{k(k-1)} [2k(2k+1) - 6k^2 + 3k(k-1)] \\ &= \lambda. \quad \square \end{aligned}$$

Exercise 4.5. In Lemma 4.3, show that if $\mathscr{S}$ is a design, then $\mathscr{S}^*$ is also a design.

Corollary 4.4. If $\mathscr{H}$ is a Hadamard 2-design, then $\mathscr{H}$ has an extension $\mathscr{H}^*$. $\square$

But we can, in fact, prove more than Corollary 4.4: the extension $\mathscr{H}^*$ of $\mathscr{H}$ is even unique, if $\mathscr{H}$ is a Hadamard 2-design. (But Exercise 4.7, together with Theorem 4.20, will show that this is not generally true even for 2-designs for $(2k+1, k, \lambda)$.) To prove the uniqueness of $\mathscr{H}^*$, we show that in a 3-$(4\lambda+4, 2\lambda+2, \lambda)$, the complement of any block is a block; since any extension of a 2-$(4\lambda+3, \lambda+1, \lambda)$ has such parameters, this will prove that it was constructed by the 'complementation trick' of Lemma 4.3.

Theorem 4.5. A Hadamard 2-design has an extension, unique up to isomorphism, to a 3-design. (Such a 3-design, with parameters 3-$(4\lambda+4, 2\lambda+2, \lambda)$, is called a *Hadamard 3-design.*)

Proof. We suppose $\mathscr{S}$ is a 3-$(4\lambda+4, 2\lambda+2, \lambda)$; it suffices, to prove the theorem, to show that the point-complement of any block in $\mathscr{S}$ is also a block. First we note that for any point X in $\mathscr{S}$, $\mathscr{S}_X$ is a 2-$(4\lambda+3, 2\lambda+1, \lambda)$, i.e. a Hadamard 2-design which, of course, is symmetric. So if two blocks of $\mathscr{S}$ meet in a point X, they meet λ more times in $\mathscr{S}_X$. Thus we have shown that two blocks of $\mathscr{S}$ meet in 0 or $\lambda+1$ points. Furthermore, the number r of blocks on a point is the number of blocks in $\mathscr{S}_X$. In other words, $r=4\lambda+3$.

Now let w be a fixed block of $\mathscr{S}$, and let us count all flags (X, y), where X is a point on w, and $y \neq w$. There are $2\lambda+2$ choices of X and $r-1=4\lambda+2$ blocks y $(\neq w)$ on X. On the other hand, if m is the number of blocks y $(\neq w)$ which meet w, then the number of these flags is $m(\lambda+1)$. So

$$m(\lambda+1)=(2\lambda+2)(4\lambda+2),$$

and hence

$$m=8\lambda+4.$$

But from Theorem 1.2 the number of blocks in $\mathscr{S}$ is $b=8\lambda+6$. So $8\lambda+4$ blocks meet w, leaving w, and one other (unique) block z not meeting w. Thus z contains all points of $\mathscr{S}$ not in w. □

Corollary 4.6. If $\mathscr{S}$ is a Hadamard 3-design, then $(\text{Aut } \mathscr{S})_X=\text{Aut } (\mathscr{S}_X)$ for any point X in $\mathscr{S}$.

Proof. This says that any automorphism of the Hadamard 2-design $\mathscr{S}_X$ extends to an automorphism of $\mathscr{S}$. But this is clear since $\mathscr{S}$ is constructed from $\mathscr{S}_X$ by the 'complementation trick', so any automorphism of $\mathscr{S}_X$ is defined naturally and correctly on $\mathscr{S}$. □

Corollary 4.7. If $\mathscr{S}$ is a Hadamard 3-design, and X, Y are points of $\mathscr{S}$, then the Hadamard 2-designs $\mathscr{S}_X$ and $\mathscr{S}_Y$ are isomorphic if and only if there is an element $\alpha \in \operatorname{Aut} \mathscr{S}$ such that $X^\alpha = Y$. □

Exercise 4.6. Let $\mathscr{L}^*(q)$ be the Hadamard 3-design which is the extension of the Hadamard 2-design $\mathscr{L}(q)$ of Section 3.5. Show that $\mathscr{L}^*(7)$ and $\mathscr{L}^*(11)$ are the unique (up to isomorphism) Hadamard 3-designs for $\lambda = 1$ and 2 respectively, and that each of Aut $(\mathscr{L}^*(7))$ and Aut $(\mathscr{L}^*(11))$ is 3-transitive on points.

Exercise 4.7. Let $\mathscr{M}$ be a 4-(11, 5, 1). (In Theorem 4.20 we will show that such a design exists.) If x is any block of $\mathscr{M}$ show that there are 30, 20 and 15 blocks which meet x in exactly three, two and one points, respectively.

Show that $\mathscr{S} = \mathscr{C}(\mathscr{M})$ is a 4-(11, 6, 3) in which two blocks meet in four, three or two points. Hence deduce that

(a) for any point X, $\mathscr{S}_X$ is a 3-(10, 5, 3) in which the complement of a block is never a block;

(b) for any two points X, Y, $\mathscr{S}_{X,Y}$ is a 2-(9, 4, 3) with two non-isomorphic extensions.

4.3 Inversive planes and a class of 3-designs

In Exercise 4.3 we saw that the 2-design consisting of the points and hyperplanes of an affine geometry never has an extension if the (geometric) dimension exceeds 2. This is an immediate consequence of Corollary 1.17. We now look at the case where the dimension is 2 and extend the discussion to include more general affine planes. First we note that, by Corollary 3.2, an affine plane is a 2-$(n^2, n, 1)$, for some $n > 1$, with $n^2 + n$ blocks. Since $n + 1$ always divides $(n^2 + n)(n^2 + 1)$ the necessary condition for extendability given in Corollary 1.17 is always satisfied. In anticipation of our discussion we make an important definition. An *inversive plane* (also sometimes called a *Möbius plane*) $\mathscr{I}$ is a set of points with distinguished subsets of the points, called *circles* (or *blocks*) such that:

(I1) any three distinct points of $\mathscr{I}$ are in exactly one common circle;

(I2) if P, Q are points of $\mathscr{I}$ and y is a circle with $P \in y$ but $Q \notin y$ then there is a unique circle of $\mathscr{I}$ which contains both P and Q and meets y only in the point P;

(I3) $\mathscr{I}$ contains four points which are not on a common circle.

The motivation for this definition will become clear in Theorem 4.8. But first we set an exercise to give an example of an infinite inversive plane.

Exercise 4.8. Let $\mathscr{S}$ be a sphere in 3-dimensional Euclidean space over the real numbers. Let $\mathscr{I}$ be the structure whose points are those on the surface of $\mathscr{S}$ and whose circles are the intersections with the surface of $\mathscr{S}$ of those planes which meet $\mathscr{S}$ in more than one point. Show that $\mathscr{I}$ is an inversive plane.

Theorem 4.8. If $\mathscr{I}$ is an inversive plane and X is any point of $\mathscr{I}$ then $\mathscr{I}_X$ is an affine plane.

Proof. If P, Q are distinct points of $\mathscr{I}_X$ then, since P, Q, X are on a unique circle of $\mathscr{I}$, P and Q are on exactly one block of $\mathscr{I}_X$.

If y is any block of $\mathscr{I}_X$ and Q is a point of $\mathscr{I}_X$ not on y then there is exactly one block x of $\mathscr{I}$ such that Q is on x and X is the only point on both x and y. Thus in $\mathscr{I}_X$, X is the unique block which contains Q but does not intersect y.

Finally it should be clear that axiom (I3) for an inversive plane implies that $\mathscr{I}_X$ contains a set of three points on no common block and we leave this as an exercise. This completes the proof that $\mathscr{I}_X$ is an affine plane. □

Exercise 4.9. Let $\mathscr{E}$ be a real affine plane and let $\mathscr{K}$ be the structure defined as follows: the points of $\mathscr{K}$ are the points of $\mathscr{E}$ plus a new point labelled ∞, while the blocks of $\mathscr{K}$ are the circles of $\mathscr{E}$ plus the point sets $y \cup \infty$, for each line y of $\mathscr{E}$. Show that $\mathscr{K}$ is an inversive plane with $\mathscr{K}_\infty = \mathscr{E}$. Show also that $\mathscr{K}$ is isomorphic to the example $\mathscr{I}$ of Exercise 4.8.

If $\mathscr{I}$ is any finite inversive plane then, by Theorem 4.8, $\mathscr{I}$ is an extension of a finite affine plane. Thus, since a finite affine plane is a 2-$(n^2, n, 1)$ for some $n \geqslant 2$, $\mathscr{I}$ must be a 3-$(n^2+1, n+1, 1)$ (see Theorem 1.16). In fact, we can prove more.

Theorem 4.9. The finite inversive planes are exactly the class of 3-designs with parameters $(n^2+1, n+1, 1)$ with $n \geqslant 2$.

Proof. In view of the preceding discussion we need only prove that if $\mathscr{I}$ is a 3-$(n^2+1, n+1, 1)$ with $n \geqslant 2$ then $\mathscr{I}$ is an inversive plane. If we use the term circles for the blocks of $\mathscr{I}$ then, since $\mathscr{I}$ is a 3-design with $\lambda_3 = 1$, axiom (I1) is obviously satisfied. Since $n \geqslant 2$ axiom (I3) is also clearly satisfied.

To show that axiom (I2) holds, let A, B be any two points of $\mathscr{I}$ and let y be a circle which contains A but not B. Since $\mathscr{I}_A$ is an affine plane there is a unique line, x say, of $\mathscr{I}_A$ such that B is on x but x does not intersect y in $\mathscr{I}_A$. Thus x is the unique block of $\mathscr{I}$ which contains B and intersects y in the single point A. Thus $\mathscr{I}$ is an inversive plane. □

As an alternative proof that axiom (I2) is satisfied we could have counted the number of blocks through A and B and argued that exactly one of them must intersect y in B only. We set this as an exercise.

Exercise 4.10. Show that if $\mathscr{I}$ is a 3-$(n^2+1, n+1, 1)$ with $n > 1$ then $\lambda_2 = n+1$. Deduce that axiom (I2) for an inversive plane is satisfied by $\mathscr{I}$.

When we constructed the Hadamard 2-designs $\mathscr{L}(q)$ in Chapter 3 we made use of the fact that, for $q \equiv 3 \pmod 4$, the group of transformations $x \rightarrow a^2x + b$ $(a \neq 0)$ is 2-homogeneous on the elements of $GF(q)$. In order to construct an infinite family of finite inversive planes we now exploit the fact that $PGL(2, q)$ acts as a 3-transitive group of transformations on $GF(q) \cup (\infty)$.

We need some standard facts about $PGL(2, q)$.

Result 4.1

(a) The elements of $PGL(2, q)$ are the mappings $\sigma(a, b, c, d)$ for all $a, b, c, d \in GF(q)$ with $ad - bc \neq 0$, defined by

$$\sigma(a, b, c, d) : x \longrightarrow \frac{ax+b}{cx+d}.$$

(Here we use standard 'abusive' conventions such as $\infty \rightarrow a/c$ if $c \neq 0$, $\infty \rightarrow \infty$ if $c = 0$, $-d/c \rightarrow \infty$ if $c \neq 0$.)

(b) $PGL(2, q)$ has order $(q+1)q(q-1)$ and is sharply 3-transitive on the set $GF(q) \cup \infty$.

(c) $PGL(2, q^n)$ contains $PGL(2, q)$ in the natural way as a subgroup; but here $PGL(2, q)$ acts not only on the set $GF(q) \cup \infty$ (as in (a)) but on the larger set $GF(q^n) \cup \infty$.

Theorem 4.10. Let $\mathscr{I}_n(q)$ for $n > 1$ be the structure defined as follows:

the points of $\mathscr{I}_n(q)$ are the elements in $GF(q^n) \cup \infty$;
the blocks of $\mathscr{I}_n(q)$ are the subsets y^σ of $GF(q^n) \cup \infty$, where $\sigma \in PGL(2, q^n)$ and y is the set $GF(q) \cup \infty$.

Then $\mathscr{I}_n(q)$ is a 3-design for $(q^n + 1, q+1, 1)$.

Proof. Since y has $q+1$ points, every block y^σ has $q+1$ points; obviously $\mathscr{I}_n(q)$ has q^n+1 points. The group $PGL(2, q^n)$ is an automorphism group of $\mathscr{I}_n(q)$, since $x \in y^\sigma$ implies $x^{\sigma_1} \in y^{\sigma\sigma_1}$ for any $\sigma_1 \in PGL(2, q^n)$ and any $x \in GF(q^n) \cup \infty$. Thus, since $PGL(2, q^n)$ is 3-transitive on the points of $\mathscr{I}_n(q)$, the number of blocks on a 3-set of points is a constant, so $\mathscr{I}_n(q)$ is a uniform 3-structure for $(q^n+1, q+1, \lambda)$. We shall show $\lambda = 1$, which also implies that $\mathscr{I}_n(q)$ has no repeated blocks (since there is only one block on three points).

Since all the blocks are images of y, the number b of blocks in $\mathscr{I}_n(q)$ is $|PGL(2, q^n)|/|H|$, where H is the subgroup of $PGL(2, q^n)$ fixing y as a set. Clearly, $PGL(2, q)$ fixes y, so $PGL(2, q) \leqslant H$, and $|H| \geqslant (q+1)q(q-1)$. Thus

$$b = \frac{|PGL(2, q^n)|}{|H|} \leqslant \frac{(q^n+1)q^n(q^n-1)}{(q+1)q(q-1)}.$$

From Theorem 1.2,

$$b = \lambda \frac{(q^n+1)q^n(q^n-1)}{(q+1)q(q-1)},$$

so, comparing the two expressions for b, we have $\lambda \leqslant 1$. Hence $\lambda = 1$ (and so $H = PGL(2, q)$ must follow as well). □

Corollary 4.11. For every prime power q, there exists an inversive plane with $q+1$ points on a circle, namely $\mathscr{I}_2(q)$.

Proof. This is immediate from the fact that any finite field $GF(q)$ is contained in a field $GF(q^2)$. □

Exercise 4.11.* Show that $(\mathscr{I}_n(q))_\infty$ is isomorphic to $\mathscr{A}_{n,1}(q)$ (i.e. the 2-design of points and lines in the affine geometry $\mathscr{A}(n, q)$).

The inversive planes $\mathscr{I}_2(q)$, which are sometimes called the *Miquelian* inversive planes (from a configurational theorem which they satisfy), are not the only finite inversive planes known; in fact, the Miquelian examples play a role somewhat analogous to the role of the projective planes $\mathscr{P}_2(q)$ in the theory of projective planes. The theory of inversive planes is quite developed and is deep; the interested reader should consult [4] for more.

4.4 Hadamard designs, the little Mathieu designs and the little Mathieu groups

If $\mathscr{H}$ is any 2-$(4\lambda+3, 2\lambda+1, \lambda)$ Hadamard design then we shall denote its (unique) extension to a Hadamard 3-design by $\mathscr{H}^*$. As we saw in Section 4.2, for any block x of $\mathscr{H}^*$ there is a unique block which does not intersect x. Furthermore this unique block is the complement of x in $\mathscr{H}^*$. For brevity, if two blocks of $\mathscr{H}^*$ do not intersect then we shall say they are *parallel.*

For any two blocks y, z of a structure let $y * z$ denote the set of points which are in either both y and z or neither of them. We can now define a new structure $\mathscr{K} = \mathscr{K}(\mathscr{H})$ as follows: the points of $\mathscr{K}$ are the points of $\mathscr{H}$, while the blocks of $\mathscr{K}$ are of two types, namely (a) the blocks of $\mathscr{H}$, and (b) all the point sets $y * z$ for $y, z \in \mathscr{H}$. This structure has some interesting properties.

Lemma 4.12. $\mathscr{K}$ is a 2-structure for $(4\lambda+3, 2\lambda+1, 2\lambda(\lambda+1))$.

Proof. Clearly, $\mathscr{K}$ has $4\lambda+3$ points. For any two blocks y, z the symmetry of $\mathscr{H}$ implies that y and z intersect in λ points and so, since $\mathscr{H}$ has $4\lambda+3$ points and y and z each have $\lambda+1$ points not in their intersection, there are $4\lambda+3-((\lambda+1)+(\lambda+1)+\lambda)=\lambda+1$ points not in y or z. Thus $y * z$ has $2\lambda+1$ points and so, since each block of $\mathscr{H}$ also has $2\lambda+1$ points, every block of $\mathscr{K}$ has $2\lambda+1$ points. Let X, Y be distinct points of $\mathscr{K}$. Now in $\mathscr{H}$ there are λ blocks on X and Y, $\lambda+1$ blocks on X but not on Y, $\lambda+1$ blocks on Y but not on X, and $4\lambda+3-2(\lambda+1)-\lambda=\lambda+1$ blocks on neither X nor Y.

In $\mathscr{K}$, X and Y are on λ blocks of type (a), i.e. λ blocks which are also blocks of $\mathscr{H}$. They are on a block $w * z$ if and only if:

(i) X, Y are both on w and z;

(ii) X is on w and z but Y is on neither w nor z;

(iii) Y is on w and z but X is on neither w nor z;

(iv) X is on neither w nor z, and Y is on neither w nor z.

Since there are λ blocks of $\mathscr{H}$ containing X and Y there are $\binom{\lambda}{2}$ blocks $w * z$ of type (i). Similarly there are $\binom{\lambda+1}{2}$ blocks $w * z$ of each of the other three types. So X and Y are on a total of

$$\lambda+\binom{\lambda}{2}+3\binom{\lambda+1}{2}=2\lambda(\lambda+1)$$

blocks. □

As the following exercise illustrates, $\mathscr{K}$ need not be a design.

Exercise 4.12. Let $\mathscr{H}$ be the Hadamard 2-(7, 3, 1). Show that $\mathscr{K}$ has repeated blocks and hence is not a 2-design.

There are, however, occasions when $\mathscr{K}$ is a design and the existence or otherwise of repeated blocks in $\mathscr{K}$ depends upon the structure of $\mathscr{H}$ in a rather complex way. In fact, the most interesting $\mathscr{K}$ for our purpose will turn out not to have repeated blocks. No matter whether $\mathscr{K}$ is a design or not, we can apply Lemma 4.3 and conclude that $\mathscr{K}$ has an extension $\mathscr{K}^*$. We can even show how to obtain $\mathscr{K}^*$ directly from $\mathscr{H}^*$. To do this we will define a new structure $\mathscr{S}(\mathscr{H}^*)$ from $\mathscr{H}^*$ which is isomorphic to $\mathscr{K}^*$. The points of $\mathscr{S}(\mathscr{H}^*)$ are the points of $\mathscr{H}^*$. Each block of $\mathscr{H}^*$ is a block of $\mathscr{S}(\mathscr{H}^*)$ and the other blocks of $\mathscr{S}(\mathscr{H}^*)$ are the points sets $y*z$, where y and z are non-parallel blocks of $\mathscr{H}^*$. However, if $y \parallel y'$ and $z \parallel z'$ then we identify $y*z$ with $y'*z'$ and get only one block of $\mathscr{S}(\mathscr{H}^*)$. (Note that the point sets $y*z$ and $y'*z'$ are equal.)

Exercise 4.13. Show $\mathscr{K}^* \cong \mathscr{S}(\mathscr{H}^*)$.

In general, $\mathscr{K}^*$ will not be a 4-structure.

Lemma 4.13. If $\mathscr{K}^*$ is a 4-structure then $\lambda=2$.

Proof. $\mathscr{K}^*$ is a 3-structure for $(4\lambda+4,\ 2\lambda+2,\ 2\lambda(\lambda+1))$, with b blocks, where

$$b=2\lambda(\lambda+1)\,\frac{(4\lambda+4)(4\lambda+3)(4\lambda+2)}{(2\lambda+2)(2\lambda+1)2\lambda}$$
$$=4(\lambda+1)(4\lambda+3).$$

Hence if $\mathscr{K}^*$ is a 4-structure for $(4\lambda+4,\ 2\lambda+2,\ \alpha)$, we must have

$$b=\alpha\,\frac{(4\lambda+4)(4\lambda+3)(4\lambda+2)(4\lambda+1)}{(2\lambda+2)(2\lambda+1)2\lambda(2\lambda-1)},$$

so $\alpha(4\lambda+1)=2\lambda(2\lambda+1)(2\lambda-1)$. This implies that $4\lambda+1$ divides $4\lambda^3+2\lambda^2-2\lambda$, and hence divides $16\lambda^3+8\lambda^2-8\lambda=(4\lambda+1)(4\lambda^2+\lambda-2)-(\lambda-2)$. So $4\lambda+1$ divides $\lambda-2$, and this is only possible if $\lambda=2$. □

Exercise 4.14. Show that if $\mathscr{K}$ is a 3-structure, then $\lambda=2$.

Now we return to $\mathscr{H}^*$ and use it to construct yet another structure $\mathscr{K}_0$. The points of $\mathscr{K}_0$ are the parallel classes of $\mathscr{H}^*$, or, equivalently, for every block y of $\mathscr{H}$, $(y\cup\infty, y')$ is a point of $\mathscr{K}_0$. The blocks of $\mathscr{K}_0$ are the 2-sets of points of $\mathscr{H}^*$, or, equivalently, if P is a point of $\mathscr{H}$, then $[\infty, P]$ is a block of $\mathscr{K}_0$, while if P, Q are distinct points of $\mathscr{H}$, then $[P, Q]=[Q, P]$ is a block of $\mathscr{K}_0$. The point $(y\cup\infty, y')$ is on the block $[X, Y]$, where X, Y are points of $\mathscr{H}^*$ (i.e. including ∞) if X, Y are in $y\cup\infty$, or X, Y are in y'.

Lemma 4.14. $\mathscr{K}_0$ is a 2-structure for $(4\lambda+3, 2\lambda+1, 2\lambda(\lambda+1))$.

Proof. If w, w' is a parallel class of $\mathscr{H}^*$, the point (w, w') is on $[X, Y]$ if and only if w is on X, Y or w' is on X, Y. So the block size of $\mathscr{K}_0$ is the number of blocks on two points in $\mathscr{H}^*$, which is $2\lambda+1$. The number of points of $\mathscr{K}_0$ is the number of parallel classes in $\mathscr{H}^*$, and that is half the number of blocks, i.e. $4\lambda+3$.

Finally, if (w, w') and (y, y') are distinct points of $\mathscr{K}_0$, then they lie on $[X, Y]$ if X and Y lie together in one of the four subsets $w\cap y$, $w\cap y'$, $w'\cap y$, or $w'\cap y'$. There are $\lambda+1$ points in each subset, hence there are

$$4\binom{\lambda+1}{2}=2\lambda(\lambda+1)$$

blocks on two points of $\mathscr{K}_0$. □

Notice that $\mathscr{K}_0$ has the parameters of $\mathscr{K}$ but that $\mathscr{K}$ was constructed from $\mathscr{H}$, while $\mathscr{K}_0$ is constructed from $\mathscr{H}^*$. However, under certain circumstances they are, in fact, isomorphic.

Lemma 4.15. If $\mathscr{H}$ has a polarity, then $\mathscr{K}_0$ is isomorphic to $\mathscr{K}$.

Proof. Let α be a polarity of $\mathscr{H}$. We define ϕ from $\mathscr{K}_0$ to $\mathscr{K}$ as follows:

$$\phi : (y \cup \infty, y') \to y^{\alpha}$$
$$[\infty, P] \to P^{\alpha}$$
$$[P, Q] \to P^{\alpha} * Q^{\alpha},$$

where y, P, Q are in $\mathscr{H}$. It is immediate that ϕ is a one-to-one mapping of $\mathscr{K}_0$ onto $\mathscr{K}$.

If $(y \cup \infty, y')$ is on $[\infty, P]$, then P is in y, so y^{α} is on P^{α}.

If $(y \cup \infty, y')$ is on $[P, Q]$, then

(i) P, Q are in y, so y^{α} is in $P^{\alpha} \cap Q^{\alpha}$, or

(ii) P, Q are in y', so P, Q are both not in y, hence y^{α} is in neither P^{α} or Q^{α}.

In either case, y^{α} is in $P^{\alpha} * Q^{\alpha}$. This proves the lemma. □

Having constructed these various structures from $\mathscr{H}$ and $\mathscr{H}^*$ we will now look at some properties of their automorphism groups.

Theorem 4.16

(a) Aut $\mathscr{H} \leqslant$ Aut $\mathscr{K}$ and Aut $\mathscr{H}^* \leqslant$ Aut $\mathscr{K}^*$.

(b) If $\mathscr{H}$ has a polarity then Aut $\mathscr{H}^* \leqslant$ Aut $\mathscr{K}$.

(c) If $\mathscr{H}$ has a polarity and if Aut $\mathscr{H}^*$ is transitive on the points of $\mathscr{H}^*$ then Aut $\mathscr{H}^*$ occurs as a subgroup of Aut $\mathscr{K}^*$ in two different ways, i.e. as a subgroup transitive on the points of $\mathscr{K}^*$ and as a subgroup fixing a point of $\mathscr{K}^*$.

Proof. From their constructions it is clear that any automorphism of $\mathscr{H}$ or $\mathscr{H}^*$ acts as an automorphism of $\mathscr{K}$ or $\mathscr{K}^*$ respectively. Furthermore, since their point sets are identical, if Aut $\mathscr{H}^*$ is transitive on points then so is Aut $\mathscr{K}^*$. Similarly Aut $\mathscr{H}^* \leqslant$ Aut $\mathscr{K}_0$ which, if $\mathscr{H}$ has a polarity, is equivalent to Aut $\mathscr{H}^* \leqslant$ Aut $\mathscr{K}$. However, in this case, since $\mathscr{K} \cong \mathscr{K}^*_X$ for any point X of $\mathscr{K}^*$ (and here we are using the assumption that Aut $\mathscr{H}^*$ is transitive on the points of $\mathscr{H}^*$), Aut $\mathscr{H}^*$ fixes a point of $\mathscr{K}^*$. □

The only Hadamard 3-designs with automorphism groups known to act transitively on their points are

(a) the designs $\mathscr{A}_n(2)$, and

(b) $\mathscr{L}^*(11)$ which, of course, is a 3-(12, 6, 2).

For every one of these designs the full automorphism group is even 3-transitive on the points. For the rest of this section we will concentrate on the special case

$\mathscr{L}^*(11)$ and establish some interesting combinatorial and group theoretic results.

First we note that the Hadamard 2-design $\mathscr{L}(11)$ is a symmetric 2-(11, 5, 2) or, in other words, a biplane with $k=5$. We have already discussed biplanes in Chapter 3 and from this discussion we can now establish some important properties of Aut $(\mathscr{L}(11))$.

Theorem 4.17

(a) Aut $(\mathscr{L}(11))$ has order $11 \cdot 10 \cdot 6$. It acts 2-transitively on the points of $\mathscr{L}(11)$ and, for any point X, $(\text{Aut}\,(\mathscr{L}(11)))_X \cong A_5$ (the alternating group of degree 5).

(b) Aut $(\mathscr{L}^*(11))$ acts 3-transitively on the points of $\mathscr{L}^*(11)$ and has order $12 \cdot 11 \cdot 10 \cdot 6$.

(c) $\mathscr{L}(11)$ admits a polarity.

Proof. (a) In Chapter 3 (see Example 3.7), we showed there was a unique biplane $\mathscr{B}$, with $k=5$, that Aut $\mathscr{B}$ was transitive on the points of $\mathscr{B}$ with $|\text{Aut}\,\mathscr{B}|=660$ and, for any point $X \in \mathscr{B}$, $(\text{Aut}\,\mathscr{B})_X \cong A_5$. Since $\mathscr{L}(11)$ is also a biplane with $k=5$, $\mathscr{L}(11)$ must be isomorphic to $\mathscr{B}$. Thus we need only establish the claim that Aut $(\mathscr{L}(11))$ acts 2-transitively on points. We will in fact show that Aut $(\mathscr{L}(11))$ acts 2-transitively on blocks and leave the reader to convince himself that this is sufficient. From Exercise 3.29 for any block x, $(\text{Aut}\,(\mathscr{L}(11)))_x$ is 2-transitive on the points of x. But, since there is a one-to-one correspondence between the pairs of points on x and the blocks of $\mathscr{L}(11)$ other than x, this means that $(\text{Aut}\,(\mathscr{L}(11)))_x$ is transitive on the remaining blocks. Thus Aut $(\mathscr{L}(11))$ is transitive on the blocks of $\mathscr{L}(11)$ and the stabiliser of any block is transitive on the remaining blocks; i.e. Aut $(\mathscr{L}(11))$ is 2-transitive on blocks.

(b) Since there is only one biplane with $k=5$, $(\mathscr{L}^*(11))_X \cong (\mathscr{L}^*(11))_Y$ for any pair of points X, Y of $\mathscr{L}^*(11)$. Thus by Corollary 4.7 Aut $(\mathscr{L}^*(11))$ is transitive on the points of $\mathscr{L}^*(11)$. Also Aut $(\mathscr{L}(11))$ is 2-transitive on the 11 points of $(\mathscr{L}^*(11))_X$, so Aut $(\mathscr{L}^*(11))$ is 3-transitive on the 12 points of $\mathscr{L}^*(11)$.

(c) By Exercise 2.9, $\mathscr{L}(11)$ has an abelian Singer group and hence, by Exercise 2.8, it has a polarity. □

We are now in a position to establish the distinguishing properties of the celebrated little Mathieu designs and groups. For $\lambda=2$ we will represent $\mathscr{K}$ by $\mathscr{M}_{11}$ and $\mathscr{K}^*$ by $\mathscr{M}_{12}$. $\mathscr{M}_{11}$ and $\mathscr{M}_{12}$ are known as the *little Mathieu designs*, while Aut $\mathscr{M}_{11}$ and Aut $\mathscr{M}_{12}$ are the *little Mathieu groups*.

Theorem 4.18. $\mathcal{M}_{11}$ is a 4-(11, 5, 1) design and $\mathcal{M}_{12}$ is a 5-(12, 6, 1). Aut $\mathcal{M}_{11}$ and Aut $\mathcal{M}_{12}$ are 4- and 5-transitive on the points of $\mathcal{M}_{11}$ and $\mathcal{M}_{12}$ respectively.

Proof. We shall need the following very elementary result from group theory:

(a) If G is a permutation group of a set $\mathcal{S}$ with $|\mathcal{S}|=m$ such that, for some set of s points, the subgroup of G fixing all s of the points has index $m(m-1)\cdots(m-s+1)$, then G is s-transitive on $\mathcal{S}$.

Now $\mathcal{M}_{11}$ is $\mathcal{K}$, and is isomorphic to $\mathcal{K}_0$. From Theorem 4.16, Aut $(\mathcal{L}^*(11))$ is an automorphism group of $\mathcal{K}_0 \cong \mathcal{M}_{11}$. We shall choose a set of four points of $\mathcal{K}_0$ and show that the subgroup H of Aut $(\mathcal{L}^*(11))$ fixing those four points is the identity.

Since, by Exercise 2.9, the additive cyclic group C_{11} is a Singer group of $\mathcal{L}(11)$ with difference set $D=\{1,3,4,5,9\}$, we can represent four points of $\mathcal{K}_0$ by the four parallel classes from $\mathcal{L}^*(11)$ given below (the two sets of six points on each line represent the two blocks of a parallel class):

$$\begin{array}{cccccc c cccccc}
\infty & 1 & 3 & 4 & 5 & 9 & \qquad & 0 & 2 & 6 & 7 & 8 & 10\\
\infty & 2 & 4 & 5 & 6 & 10 & & 1 & 3 & 7 & 8 & 9 & 0\\
\infty & 3 & 5 & 6 & 7 & 0 & & 2 & 4 & 8 & 9 & 10 & 1\\
\infty & 4 & 6 & 7 & 8 & 1 & & 3 & 5 & 9 & 10 & 0 & 2
\end{array}$$

We suppose $\gamma \in$ Aut $(\mathcal{L}^*(11))$ fixes each of the four parallel classes; then elementary calculations lead to $\gamma=1$. We only sketch this proof (see Exercise 4.15). If γ fixes ∞, then it fixes all eight blocks above, hence fixes their pairwise intersection. This quickly leads to the conclusion $\gamma=1$. If γ sends ∞ to 0, say, then γ interchanges the two blocks in the first, second and fourth classes, and fixes the two blocks in the third class. Then considering pairwise intersections, we are led to a contradiction. Alternatively, we might examine the point 4 and, by examining each of the four parallel classes in order, observe that its image under γ must be one of 2, 6, 7, 8, 10, one of 1, 3, 7, 8, 9, one of 1, 2, 8, 9, 10, and one of 2, 3, 5, 9, 10. But these four sets have empty intersection: a contradiction.

Proceeding in this fashion, we find $\gamma=1$. We now appeal to the group theoretic result (a). Then letting $G=$ Aut $(\mathcal{L}^*(11))$, G has degree 11 and the subgroup H fixing four points (pointwise) has order 1, which means it has index $12\cdot 11\cdot 10\cdot 6=11\cdot 10\cdot 9\cdot 8$ in G. Thus G is 4-transitive on the points of $\mathcal{M}_{11}$, and so Aut $\mathcal{M}_{11}$ is certainly 4-transitive as well.

Thus $\mathcal{M}_{11}$ is a 4-structure for $(11, 5, \mu)$. Since $\mathcal{M}_{11}$ has 66 blocks (note that it is a 2-structure for (11, 5, 12), from Lemma 4.14), we have

$$66 = \mu \frac{11 \cdot 10 \cdot 9 \cdot 8}{5 \cdot 4 \cdot 3 \cdot 2}$$

and hence $\mu = 1$. But then there are no repeated blocks (since there is only one block on four points), so $\mathscr{M}_{11}$ is a 4-(11, 5, 1) as claimed.

Aut $\mathscr{M}_{12}$ contains Aut ($\mathscr{L}^*(11)$) acting transitively (even 3-transitively) on its points (Theorem 4.17). But $\mathscr{M}_{12}$, defined from $\mathscr{M}_{11}$ by the complementation trick of Lemma 4.4, clearly has Aut $\mathscr{M}_{11}$ as its point-stabiliser. So Aut $\mathscr{M}_{12}$ is point-transitive with a 4-transitive stabiliser. Thus Aut $\mathscr{M}_{12}$ is 5-transitive on points, and, as above, we immediately see that $\mathscr{M}_{12}$ is a 5-(12, 6, 1). □

Corollary 4.19. The group Aut ($\mathscr{L}^*(11)$) occurs inside Aut $\mathscr{M}_{12}$ as:

(a) a subgroup 3-transitive on the 12 points of $\mathscr{M}_{12}$, and

(b) a subgroup 4-transitive on 11 points of $\mathscr{M}_{12}$, fixing the twelfth. □

Exercise 4.15.* Show that Aut $\mathscr{M}_{11} =$ Aut ($\mathscr{L}^*(11)$).

(Note the difficulties: an automorphism of $\mathscr{M}_{11}$ need not act on the points and blocks of $\mathscr{L}^*(11)$ (or, alternatively, the points and blocks of $\mathscr{L}(11)$); instead it acts on the objects of $\mathscr{K}_0$, or, alternatively, of $\mathscr{K}$.)

Aut $\mathscr{M}_{11}$ and Aut $\mathscr{M}_{12}$ are two of the famous Mathieu groups and are usually denoted by M_{11} and M_{12} respectively. We now do a little additional group theoretic analysis of them for which we need the result of Exercise 4.15.

Theorem 4.20. The groups Aut ($\mathscr{L}(11)$), M_{11} and M_{12} are all simple.

Proof. For this proof we need the following group theory result:

(b) Let G be a 2-transitive permutation group on a set $\mathscr{S}$ and let $N \neq 1$ be a normal subgroup of G. Then, for any $X \in \mathscr{S}$, N_X is a normal subgroup of G_X and, if $N_X = 1$, then $|N| = |\mathscr{S}|$ is a prime power and G_X induces an automorphism group of N which is transitive on the non-identity elements of N.

Let $G =$ Aut $\mathscr{L}(11)$. Then, by Theorem 4.17, G is 2-transitive on points and $G_X \cong A_5$ which, of course, is simple. Thus if $N \neq 1$ is a normal subgroup of G then, by (b), N_X is normal in $G_X \cong A_5$, which implies $N_X = 1$ or G_X. If $N_X = 1$ then $|N| = 11$ and A_5, of order 60, acts as an automorphism group of N. But this is impossible since the cyclic group of order 11 has only 10 automorphisms. Thus N_X must be G_X, which means that N must be G, i.e. that G is simple.

Now let $G=M_{11}=\text{Aut}\ \mathscr{L}^*(11)$. As above, since Aut $\mathscr{L}(11)$ is simple, if $N\neq 1$ is normal in G and $N\neq G$ then $N_X=1$. In this case $|N|=12$ which is a contradiction since 12 is not a prime power. Thus G is simple. (For the group theorist we note that here we are concentrating on the representation of M_{11} as a group of degree 12.)

Finally if we let $G=M_{12}$ then a repeat of the above argument shows M_{12} is also simple. □

We note, before proceeding, that the simple group Aut $\mathscr{L}(11)$ is the classical group $PSL(2, 11)$ acting on 11 letters. There are three other (larger) Mathieu groups which we will meet in Chapter 8.

Since $\mathscr{M}_{12}$ is a 5-(12, 6, 1) we know that, for any three points A, B, C, $(\mathscr{M}_{12})_{A,B,C}$ is a 2-(9, 3, 1), i.e. an affine plane. Thus we can construct it by starting with an affine plane of order 3, extending it to an inversive plane (with parameters 3-(10, 4, 1)), then extending to $\mathscr{M}_{11}$ and finally to $\mathscr{M}_{12}$. If we adopt this approach then it is natural to try to extend $\mathscr{M}_{12}$ to a 6-design. However, it is easy to prove that this is impossible.

Exercise 4.16. Show that $\mathscr{M}_{12}$ has no extension.

4.5 Some 4-designs and 5-designs

We begin this section by giving a theoretical construction for many families of t-designs. Although for most values of t it is impractical to attempt the construction, we do use the simplest case to get an infinite family of 2-designs. Furthermore, we give an indication of how Alltop constructed some new 4-designs, which are, in fact, always 5-designs.

Let K_n denote the complete graph on n vertices and let Γ be any subgraph of K_n. (In practice we shall want Γ to be very simple, as symmetric as possible, and to contain only a small number of vertices.) For any positive integer t, let $\Gamma_1, \Gamma_2, \ldots, \Gamma_s$ be the complete set of mutually non-isomorphic subgraphs of K_n such that each Γ_i is the union of exactly t isomorphic copies of Γ (i.e. each Γ_i is chosen from a different isomorphism class in K_n). Now let Δ be a subgraph of K_n which is the union of a number of isomorphic copies of each Γ_i and let n_i be the number of subgraphs of K_n which are isomorphic to Γ_i for $i=1, 2, \ldots, s$. Now Aut K_n is S_n, the symmetric group of degree n, and we can certainly find a subgroup G of S_n which is transitive on the subgraphs isomorphic to Γ and on the subgraphs isomorphic to Γ_i for all $i=1, 2, \ldots, s$.

Now let $\mathscr{D}$ be the structure whose points are all the subgraphs of K_n which are isomorphic to Γ and whose blocks are the subgraphs isomorphic to Δ, with inclusion as incidence. (Thus the points are the images of Γ under G and the blocks are the images of Δ under G.) The number of points of $\mathscr{D}$ is $|G: G_\Gamma|$ and the number of blocks of $\mathscr{D}$ is $|G: G_\Delta|$. If, for each i, we let α_i be the number of blocks on Γ_i then, by the transitivity of G, any subgraph isomorphic to Γ_i is also on α_i blocks. Similarly, if μ_i is the number of subgraphs of Δ isomorphic to Γ_i then each block of $\mathscr{D}$ contains μ_i isomorphic copies of Γ_i.

We now count the pairs (A, B), where $A \cong \Gamma_i$, $B \cong \Delta$ and $A \subseteq B$, in two different ways. Firstly there are n_i choices for A and, for each choice of A, there are α_i choices of B. Thus the number of such pairs is $n_i\alpha_i$. However, there are $x = |G: G_\Delta|$ choices for B and, for any choice of B, μ_i choices for A. Thus $n_i\alpha_i = x\mu_i$, so $\alpha_i = x\mu_i/n_i$. The structure $\mathscr{D}$ is a t-design if and only if each set of t points is on a constant number λ of blocks; that is, $\mathscr{D}$ is a t-design if and only if $\alpha_i = \lambda$ for all $i = 1, 2, \ldots, s$.

If we choose Γ carefully, then we can ensure that it is easy to determine each of the Γ_i and the n_i. Thus, in order to construct a t-design $\mathscr{D}$, it is only necessary to find a subgraph Δ such that the μ_i are in the correct ratios. The practical difficulty is that, for even the simplest choices of Γ and t, s tends to become very large. This forces the subgraph Δ to be very complicated and it becomes virtually impossible to make a suitable choice.

As an illustration we consider the case where Γ is the graph consisting of two vertices and the edge joining them, $t = 2$, and G is the full automorphism group S_n. In this case it is easy to see that $s = 2$ and that the possibilities for the Γ_i are: Γ_1 = two intersecting edges, and Γ_2 = two non-intersecting edges.

If $\Gamma_1 = \{(X, Y), (X, Z)\}$ then there are n choices for X. Given any choice of X there are $n-1$ choices for Y and then $n-2$ choices for Z. However, the order in which Y and Z are chosen is irrelevant and so $n_1 = (n(n-1)(n-2))/2$. Similarly if $\Gamma_2 = \{(X, Y), (Z, W)\}$ there are $(n(n-1))/2$ choices for (X, Y), and then $((n-2)(n-3))/2$ choices for (Z, W). Thus, since it does not matter in which order the edges are chosen, $n_2 = (n(n-1)(n-2)(n-3))/8$. The equation $n_1\mu_2 = \mu_1 n_2$ now becomes $n = (4\mu_2/\mu_1) + 3$ and so, if we can find a subgraph Δ with $n = (4\mu_2/\mu_1) + 3$, we will have constructed a 2-design with $v = (n(n-1))/2$ which admits S_n as an automorphism group. We also wish to choose Δ so that it is easy to compute μ_1 and μ_2. Perhaps the most natural choice to try for Δ might appear to be a complete subgraph of K_n. However, there is no proper complete subgraph which has μ_1 and μ_2 in the correct ratio.

Exercise 4.17. Show that if Δ is a complete subgraph with m vertices such that $n = (4\mu_2/\mu_1) + 3$ then $m = n$ so that $\mathscr{D}$ is trivial.

Suppose that we choose Δ to be the subgraph consisting of the set of edges joining consecutive points of the cycle $A_1, A_2, \ldots, A_m$, i.e. Δ is an m-sided polygon. Then $\mu_1 = m$ and $\mu_2 = (m(m-3))/2$ so that $(4\mu_2/\mu_1)+3 = 2m-3$. Thus if we choose m so that $n = 2m-3$ we have a 2-$((n(n-1))/2, (n+3)/2, \lambda)$ design. (Note that such a design must exist for all odd, n, $n \geqslant 3$.) Obviously we wish to compute λ.

Since $G = S_n$, the subgroup G_Δ restricted to Δ has order $2m$ (because it is the full automorphism group of Δ), and the subgroup fixing Δ pointwise has order $(n-m)!$ Thus $|G_\Delta| = (n-m)!(2m)$ so that $b = n!/(2m(n-m)!)$; then by Corollary 1.4 we can compute λ.

If we try the case $t=3$ and $G = S_n$, then $s=5$ and Δ is forced to be very complex.

Exercise 4.18. If $G = S_n$ and $t=3$ then show that $s=5$ and find the n_i for $1 \leqslant i \leqslant 5$; also determine the ratios of the μ_i necessary to give a 3-design.

In solving Exercise 4.18 the reader will see that in order to construct a 3-design n must be fairly large and that it appears extremely difficult to choose Δ.

Reverting back to the case $t=2$, another way to get a design is to keep the same choices for Γ_1, Γ_2 and Δ but to replace S_n by a smaller group G. All that is necessary is that G be transitive on the subgraphs isomorphic to Γ_1 and on those isomorphic to Γ_2. This will certainly be true if G is 4-transitive; hence the following exercise.

Exercise 4.19. Construct a 2-design on 55 points having M_{11} as an automorphism group transitive on points.

We now turn our attention to trying to construct t-designs with $t \geqslant 3$. If G is a t-homogeneous permutation group on a set $\mathscr{S}$ with v points, then G maps any set of k-subsets on t points onto the same number of k-subsets through any other set of t points. Thus, by choosing any given k-subset and all its images under G as blocks, we can use G to construct a t-design. However, since there are very few multiply transitive groups, this is not a very useful observation. We now look at some methods of constructing t-designs from permutation groups which are not t-homogeneous and use them to construct 4-designs and 5-designs from groups which are not 4-homogeneous.

Let G be a permutation group on a set $\mathscr{S}$. For any positive integer t let G split the t-subsets of $\mathscr{S}$ into m orbits $\mathbf{T}_1, \mathbf{T}_2, \ldots, \mathbf{T}_m$. Then we define a t-

proportionality vector of G on $\mathscr{S}$ to be the m-tuple $(|\mathbf{T}_1|, |\mathbf{T}_2|, \ldots, |\mathbf{T}_m|)$. (Note, for future reference, that

$$\sum_{i=1}^{m} |\mathbf{T}_i| = \binom{n}{t},$$

where $|\mathscr{S}|=n$.) If $\mathscr{U}$ is any subset of $\mathscr{S}$ then we call $(u_1, u_2, \ldots, u_m)$ a *t-proportionality vector of $\mathscr{U}$ with respect to G*, where u_i is the number of members of $\mathbf{T}_i$ contained in $\mathscr{U}$. If $\mathbf{U}$ denotes the orbit of $\mathscr{U}$ under G then, for any $i=1,\ldots, m$, each element of $\mathbf{T}_i$ is contained in the same number α_i of elements of $\mathbf{U}$.

If we define an incidence structure $\mathscr{D}$ whose points are the elements of $\mathscr{S}$ and whose blocks are the elements of $\mathbf{U}$ then $\mathscr{D}$ is a t-design if and only if $\alpha_1 = \alpha_2 = \cdots = \alpha_m$. But, as in the last section, we now count in two ways the pairs $(\mathscr{A}, \mathscr{B})$ with $\mathscr{A} \in \mathbf{T}_i$, and $\mathscr{B} \in \mathbf{U}$ and $\mathscr{A} \subseteq \mathscr{B}$ to get $\alpha_i|\mathbf{T}_i| = u_i|\mathbf{U}|$. Thus $\mathscr{D}$ is a t-design if and only if the t-proportionality vector of G on $\mathscr{S}$ and the t-proportionality vector of $\mathscr{U}$ with respect to G are *equivalent*, i.e. one is a multiple of the other.

Once we have chosen G, $\mathscr{S}$ and t, the t-proportionality vector of G on $\mathscr{S}$ is determined although, of course, it may be very difficult to compute it. If, however, we are able to determine the t-proportionality vector of G on $\mathscr{S}$ we then have the problem of finding a subset $\mathscr{U}$ with the correct t-proportionality vector. Naturally the smaller m is the easier it is to look for a suitable $\mathscr{U}$. If $m=1$ then G is t-homogeneous and this is the situation we are trying to avoid, thus we must look for a G, $\mathscr{S}$ and t such that $m=2$.

If $\mathscr{A}$ is the n-dimensional affine geometry over $GF(2)$ then, with the subspaces of any given dimension greater than one as blocks, $\mathscr{A}$ is a 3-design. Furthermore, with the planes as blocks, $\mathscr{A}$ is a 3-$(2^n, 4, 1)$ design (see Exercise 1.34). If G is the n-dimensional affine group acting on $\mathscr{A}$ then G acts as a 3-transitive permutation group on the points of $\mathscr{A}$. In particular, the stabiliser of any three points fixes the unique fourth point in the plane determined by the original three and is transitive on the remaining 2^n-4 points. Since any three points lie in a plane, a set of four points are either coplanar or consist of three coplanar points and one further point. Thus, by the preceding discussion, G splits the subsets of four points of $\mathscr{A}$ into exactly two orbits; i.e. if we take $\mathscr{S}$ to be the points of $\mathscr{A}$ and $t=4$ then we have $m=2$. Let $\mathbf{T}_1$ be the orbit of coplanar points then, clearly, $|\mathbf{T}_1|=$ the numbers of planes in $\mathscr{A}$. But any three points uniquely determine a plane and any given plane contains exactly four subsets of three points and so

$$|\mathbf{T}_1| = \binom{2^n}{3}\Big/4.$$

This gives

$$|\mathbf{T}_2| = \binom{2^n}{3}(2^{n-2}-1).$$

Now let $\mathscr{P}$ be any k-set of the points of $\mathscr{A}$ and let $\mathbf{B}$ be the orbit of $\mathscr{P}$ under Aut $\mathscr{A}$. If u_i, for $i=1,2$, is the number of members of $\mathbf{T}_i$ in $\mathscr{P}$ and if α_i is the number of members of $\mathscr{B}$ containing a fixed member of $\mathbf{T}_i$ then, from the above, $\alpha_i|\mathbf{T}_i| = u_i|\mathbf{B}|$. If $\mathscr{D}$ is the structure whose points are the points of $\mathscr{A}$ and whose blocks (as point sets) are the elements of $\mathbf{B}$, then $\mathscr{D}$ is a 4-design if and only if $\alpha_1 = \alpha_2$, i.e. if and only if $u_1/u_2 = |\mathbf{T}_1|/|\mathbf{T}_2| = 1/(2^n-4)$. Also, since any 4-set in $\mathscr{P}$ is in either $\mathbf{T}_1$ or in $\mathbf{T}_2$, we have

$$u_1 + u_2 = \binom{k}{4}.$$

Substituting, we find that $\mathscr{D}$ is a 4-design if and only if

$$u_2 = (2^n-4)\binom{k}{4}\Big/(2^n-3).$$

Thus a necessary condition for using this method to construct a 4-design $\mathscr{D}$ is that 2^n-3 should divide $\binom{k}{4}$.

Exercise 4.20.* Show that the smallest value of n for which we can choose k with $k \leqslant 2^{n-1}$ and $2^n-3 \mid \binom{k}{4}$ is $n=8$.

Whenever $2^n-3 \mid \binom{k}{4}$, this method will guarantee the existence of a 4-design with $v=2^n$ provided one can find a set of k points in $\mathscr{A}$ containing the appropriate number of planes. In [1] Alltop actually works through an example with $k=24$ and $n=8$. We will not give it here and the interested reader should consult the relevant paper.

It turns out that, whenever a 4-design is constructed in the above way, it is in fact even a 5-design. To see this, one merely has to perform similar calculations for the five subsets of $\mathscr{A}$ (i.e. determine the orbits under Aut $\mathscr{A}$, etc.), and see that the conditions for obtaining a 5-design are exactly the same as those for a 4-design. We leave it as a difficult exercise.

Exercise 4.21.* Show that any 4-design obtained in the above way is a 5-design.

As a result of Alltop's work, infinite families of 5-designs were discovered including, for example, designs with $v=2^n+2$ for every $n \geqslant 4$. At present the only 6-designs known have parameters 6-(20, 9, 112) and 6-(33, 8, 36).

4.6 Tight designs

In this section we prove an inequality for 4-designs analogous to Fisher's Inequality for 2-designs. This enables us to define a tight design, the analogue of a symmetric design, and we state, without proof, some results on tight designs. We begin by proving two simple lemmas about matrices.

Lemma 4.21. Let M be a square matrix with real entries satisfying:

(a) at least one column of M has column sum not zero;

(b) there are real constants a, b, c ($a \neq 0$) such that $M^2 = aI + bM + cJ$.

Then M is non-singular.

Proof. If M is singular, then there is a vector $\mathbf{u} \neq \mathbf{0}$ such that $\mathbf{u}M = \mathbf{0}$. If $\mathbf{u} \in \langle \mathbf{j} \rangle$ then $\mathbf{u}M = \mathbf{0}$ implies that the sum of the entries in every column of M is zero, violating (a). So $\mathbf{u} \notin \langle \mathbf{j} \rangle$ and, in particular, $\{\mathbf{u}, \mathbf{j}\}$ is a linearly independent set.

Now $\mathbf{u}M = \mathbf{0}$ implies (from (b)):

$$\begin{aligned} \mathbf{0} = \mathbf{u}M^2 &= a\mathbf{u}I + b\mathbf{u}M + c\mathbf{u}J \\ &= a\mathbf{u} + c\mathbf{u}J. \end{aligned}$$

If we write $|\mathbf{u}|$ for the sum of the coordinates of $\mathbf{u}$, then $\mathbf{u}J = |\mathbf{u}|\mathbf{j}$, and so

$$\mathbf{0} = a\mathbf{u} + c|\mathbf{u}|\mathbf{j}.$$

But this implies $a = c|\mathbf{u}| = 0$, since $\{\mathbf{u}, \mathbf{j}\}$ is linearly independent, violating (b). □

Now we recall the SR graph $T(n)$ of Example 3.12, and let D_n be its adjacency matrix. So the rows and columns of D_n are indexed by the 2-sets from the set $\{1, 2, \ldots, n\}$ with entry in row (i_1, j_1) and column (i_2, j_2) equal to 1 or 0 according as the set $\{i_1, j_1, i_2, j_2\}$ consists of exactly three elements or not. We have:

Lemma 4.22. $D_n^2 = 2(n-4)I + (n-6)D_n + 4J$.

Proof. See Example 3.12 and Corollary 3.43. □

We now prove a simple result about 4-designs which we will need to prove our theorem.

Lemma 4.23. Let $\mathscr{D}$ be a 4-design for (v, k, λ), with b blocks. Let C be the $\binom{v}{2} \times b$ matrix whose rows are indexed by the 2-sets of points of $\mathscr{D}$ and whose columns are indexed by the blocks of $\mathscr{D}$, with entry in positions $(\{P, Q\}, y)$ equal to 1 if P, Q are both in y, and 0 otherwise. Then

$$CC^{\mathrm{T}} = (\lambda_2 - \lambda)I + (\lambda_3 - \lambda)D_v + \lambda J.$$

Proof. A typical entry in CC^{T} is the inner product of row $\{P_1, Q_1\}$ with row $\{P_2, Q_2\}$, hence counts the number of blocks on the set of points $\{P_1, Q_1, P_2, Q_2\}$; note that this set can consist of two, three or four distinct points.

(a) If $\{P_1, Q_1\} = \{P_2, Q_2\}$, then the inner product of the two rows is λ_2, the numbers of blocks on two points. So the diagonal entries of CC^{T} are λ_2.

(b) If $|\{P_1, Q_1, P_2, Q_2\}| = 3$, then the inner product of the two rows is λ_3, the numbers of blocks on three points. So in those positions where D_v has 1, CC^{T} has λ_3.

(c) If $|\{P_1, Q_1, P_2, Q_2\}| = 4$, then the inner product of the two rows is λ. So in these positions, where D_v has 0, CC^{T} has λ.

Comparing this with the entries in $(\lambda_2 - \lambda)I + (\lambda_3 - \lambda)D_v + \lambda J$, the lemma is proved. □

These lemmas enable us to prove the following inequality.

Theorem 4.24. Let $\mathscr{D}$ be a 4-design for (v, k, λ), with $v \geqslant k + 2$ and with b blocks. Then $b \geqslant \binom{v}{2}$.

Proof. Let C be as in Lemma 4.23, and $M = CC^{\mathrm{T}}$. Write D for D_v. Then from Lemma 4.22, $D^2 = 2(v-4)I + (v-6)D + 4J$. Since $M = (\lambda_2 - \lambda)I + (\lambda_3 - \lambda)D + \lambda J$, we can compute M^2:

$$M^2 = (\lambda_2 - \lambda)^2 I + (\lambda_3 - \lambda)^2 D^2 + \lambda^2 J^2 + 2(\lambda_2 - \lambda)(\lambda_3 - \lambda)D$$
$$+ 2(\lambda_2 - \lambda)\lambda J + (\lambda_3 - \lambda)\lambda DJ + (\lambda_3 - \lambda)\lambda JD.$$

But it is easy to see that $DJ = JD = (v-2)J$ and, clearly, $J^2 = \binom{v}{2}J$. In addition,

$$D = \frac{1}{\lambda_3 - \lambda} M - \frac{\lambda_2 - \lambda}{\lambda_3 - \lambda} I - \frac{\lambda}{\lambda_3 - \lambda} J.$$

So $M^2 = xI + yM + zJ$ for real (even rational) x, y, z. Using the relations above,

$$x = 2(v-4)(\lambda_3 - \lambda)^2 - (v-6)(\lambda_2 - \lambda)(\lambda_3 - \lambda) - (\lambda_2 - \lambda)^2$$

and, by multiplying out, we see that:

$$x=[(\lambda_2-\lambda)+(v-4)(\lambda_3-\lambda)][2(\lambda_3-\lambda)-(\lambda_2-\lambda)].$$

So $x=0$ implies $\lambda_2-\lambda=2(\lambda_3-\lambda)$. But, using the expressions for λ_2 and λ_3 (see Theorem 1.2), this gives:

$$(v-2)(v-3)-2(v-3)(k-2)+(k-2)(k-3)=0$$

and, writing $w=v-2$, $l=k-2$, we have:

$$w^2-(2l+1)w+l^2+l=0$$

or

$$(w-l)(w-(l+1))=0.$$

Then $x=0$ implies $v=k$ or $v=k+1$, both impossible. So $x\neq 0$ and, since $\mathbf{j}M=\lambda_2\binom{k}{2}\mathbf{j}$, the column sums of M are $\lambda_2\binom{k}{2}\neq 0$. Then, by Lemma 4.21, M is non-singular. But M has $\binom{v}{2}$ rows (and columns), so the rank of C must be $\binom{v}{2}$, for $CC^{\mathrm{T}}=M$. Thus $b\geqslant\binom{v}{2}$, for C has b columns (compare the proof of Theorem 2.3). □

In fact, it is possible to prove the following:

Theorem 4.25. If $\mathscr{D}$ is a $2s$-design for (v, k, λ), with $v\geqslant k+s$, and with b blocks, then $b\geqslant\binom{v}{s}$.

(For the proof see [8].)

A $2s$-design for (v,k,λ) with b blocks and $b=\binom{v}{s}$ is called *tight*; thus a non-trivial 2-design is tight if and only if it is symmetric. Tight designs for $s>1$ are rare, and the following has been shown:

Theorem 4.26. If $\mathscr{D}$ is a tight 4-design with $v\geqslant k+2$, then $\mathscr{D}$ is a 4-(23, 7, 1) or the complement of a 4-(23, 7, 1).

(For the proof see [2, 5].)

We shall see in Chapter 8 that a 4-(23, 7, 1) not only exists, but is unique up to isomorphism; so there are exactly two non-trivial tight 4-designs. In fact, the situation for $s>2$ has been studied as well.

Theorem 4.27. If $\mathscr{D}$ is a 6-design for (v, k, λ) with $v\geqslant k+3$, then $\mathscr{D}$ is never tight.

(For the proof see [7].)

References

[1] is a reference for one of Alltop's papers on the construction of 5-designs and [6] contains the first known non-trivial 6-design. The original proof of Cameron's Theorem is in [3] while [2], [5], [6], [7] and [8] contain proofs of the quoted results on tight designs. Finally, more information on inversive planes can be found in [4].

[1] Alltop, W. O. '5-Designs in affine spaces'. *Pacific Journal of Maths.*, **39**, 3 (1971), 547–51.

[2] Bremner, A. 'The diophantine equation arising from tight 4-designs'. *Osaka J. Math.*, **16** (1979), 353–6.

[3] Cameron, P. J. 'Extending symmetric designs'. *J. Comb. Th.* (*A*), **14** (1973), 215–20.

[4] Dembowski, H. P. *Finite Geometries.* Berlin–Heidelberg–New York, Springer, 1968.

[5] Enomoto, S., Ito, W. & Noda, A. 'Tight 4-designs'. *Osaka J. Math.*, **16** (1979), 39–43.

[6] Nagliveras, S. S. & Leavitt, D. W. 'Simple 6-(33, 8, 36) designs from $P\Gamma L_2(32)$'. *Proceedings of the 1982 Durham Symposium on Computational Group Theory.*

[7] Petersen, C. 'On tight 6-designs'. *Osaka J. Math.*, **14** (1977), 417–55.

[8] Ray-Chaudhuri, D. K. & Wilson, R. M. 'On t-designs'. *Osaka J. Math.*, **12** (1975), 737–44.

5 Resolutions

5.1 Introduction

A resolution is a tactical decomposition with exactly one point class. Thus a parallelism is a particular, but very important, example of a resolution. In this chapter we look at resolutions in some detail.

After establishing some general properties of the parameters of resolvable designs, we look at strongly resolvable designs. These are a particularly important class of designs which include affine designs and their complements. We show that the only strongly resolvable 3-designs, and hence the only affine 3-designs, are the Hadamard 3-designs. We also establish the Orbit Theorem for strongly resolvable designs.

5.2 Resolutions

Resolutions were defined in Chapter 1. We recall that a resolution of a design $\mathscr{D}$ is a tactical decomposition in which there is only one point class. Thus a resolution is a partition of the blocks of $\mathscr{D}$ into classes such that the points of $\mathscr{D}$ and the blocks of any given class form a 1-design. Obviously if $\mathscr{D}$ is a 1-design then it admits a resolution with only one block class. However, this resolution is trivial and uninteresting and we shall always assume that a resolution has $c \geqslant 2$ block classes. If $\mathscr{B}_1, \mathscr{B}_2, \ldots, \mathscr{B}_c$ are the block classes of a resolution of a resolvable design $\mathscr{D}$ then we let σ_i and m_i $(i=1, 2, \ldots, c)$ be the parameters such that $\mathscr{B}_i$ is a 1-(v, k, σ_i) design with m_i blocks. There are some very simple restrictions on the possible values for the σ_i and m_i when $\mathscr{D}$ is a 1-design.

Exercise 5.1. Show that for any resolution of a t-(v, k, λ) design:

(a) $\sum_{i=1}^{c} m_i = b$;

(b) $\sum_{i=1}^{c} \sigma_i = r$;

(c) $v\sigma_i = m_i k$ for each i.

Exercise 5.2. Show that, for Example $\mathscr{K}$ of Chapter 1, $\mathscr{B}_1 = \{k_1, k_{13}\}$, $\mathscr{B}_2 = \{k_2, k_3, k_{11}, k_{14}\}$, $\mathscr{B}_3 = \{k_4, k_5, k_6, k_7, k_8, k_9, k_{10}, k_{12}\}$ gives a resolution, and determine $\sigma_1, \sigma_2, \sigma_3$.

Exercise 5.3. Find a resolution of Example $\mathscr{K}$ with $c = 7$ such that $\sigma_i = 1, m_i = 2$ for $i = 1, 2, \ldots, 7$.

Since being resolvable is not a very strong property we shall usually study resolutions in which the block classes have some extra property. A resolution is called a *σ-resolution* if each $\sigma_i = \sigma$ $(i = 1, \ldots, c)$. (Note from Exercise 5.1, that this automatically implies that the m_i are all equal. In this case we write $m_i = m$ $(i = 1 \ldots, c)$.) If a resolution has the property that any two blocks from the same class meet in ρ points, then we call it an *inner resolution* and call ρ the *inner constant*. Similarly if any two blocks from distinct classes meet in μ points then we say that we have an *outer resolution* and that μ is the *outer constant*. *Inner resolvable*, *outer resolvable* and *σ-resolvable* designs are defined in the obvious way as designs admitting the appropriate type of resolution. If a σ-resolution is both inner and outer then we call it a *strong resolution*.

Lemma 5.1. If $\mathscr{D}$ is a t-(v, k, λ) with a σ-resolution, then:

(a) there are $r\sigma^{-1}$ resolution classes;

(b) $b = rm\sigma^{-1}$.

Proof. If P is any point of $\mathscr{D}$ then P is on exactly σ blocks from each of the c resolution classes. Thus P is on $c\sigma$ blocks. Hence $r = \sigma c$ or $c = r\sigma^{-1}$. Each class contains exactly m blocks and each block is in exactly one class, so $b = mc = mr\sigma^{-1}$. □

Lemma 5.2. If $\mathscr{D}$ is a t-(v, k, λ) with an inner σ-resolution then $\rho = k(\sigma - 1)/(m - 1) = k^2(\sigma - 1)/(\sigma - k)$.

Proof. Let x be any block of $\mathscr{D}$. To prove the lemma we count, in two different ways, the flags (P, y) with P on x, and y in the same resolution class as x, $y \neq x$. There are k choices for P and then, for each choice of P, $\sigma-1$ choices for y. Alternatively there are $m-1$ choices for y and then, because the resolution is inner, ρ choices for P. Thus $k(\sigma-1)=\rho(m-1)$ or $\rho=k(\sigma-1)/(m-1)$. However, by Exercise 5.1, $v\sigma=mk$ which gives $\rho=k(\sigma-1)/(m-1)=k^2(\sigma-1)/(km-k)=k^2(\sigma-1)/(v-k)$. □

Exercise 5.4. If $\mathscr{D}$ is a t-(v, k, λ) with an outer σ-resolution show that $\mu=k(r-\sigma)/(b-m)=k^2/v$.

Historically, inner resolutions with $\rho=0$ have received more attention than general resolutions. As the next result shows, such inner resolutions are automatically 1-resolutions.

Theorem 5.3. An inner resolution of a t-(v, k, λ) $\mathscr{D}$ has $\rho=0$ if and only if it is a 1-resolution.

Proof. If $\rho=0$ and $\mathscr{B}_i$ is a resolution class of $\mathscr{D}$ then $\mathscr{B}_i$ is a 1-(v, k, σ_i) with the property that no two blocks of $\mathscr{B}_i$ intersect. But each point of $\mathscr{B}_i$ is on $\sigma_i \geqslant 1$ blocks of $\mathscr{B}_i$. Hence $\sigma_i=1$ and the resolution is a 1-resolution.

Conversely, if $\mathscr{B}_i$ is a 1-$(v, k, 1)$ then $\mathscr{B}_i$ cannot have two intersecting blocks, i.e. $\rho=0$. □

We remind the reader that we have already met 1-resolutions and that they are more commonly called *parallelisms*. As we saw in Exercise 5.3, Example $\mathscr{K}$ admits a parallelism. In fact, this example is merely one of a whole family which admit parallelisms.

Exercise 5.5. Show that any t-design in which the complement of a block is a block admits a parallelism with $m=2$.

We have also seen many other examples of designs with parallelisms.

Exercise 5.6. The blocks of $\mathscr{A}_{n,i}(q)$ are cosets $U+\mathbf{x}$ where U is an i-

dimensional subspace of $V_n(q)$ and $\mathbf{x}$ is a vector. Show that if $U_1+\mathbf{x} \sim U_2+\mathbf{y}$ if and only if $U_1=U_2$, then $\sim$ is a parallelism of $\mathscr{A}_{n,i}(q)$. Show further that this parallelism is an outer resolution if and only if $i=n-1$.

The parallelism of $\mathscr{A}_{n,i}(q)$ given in Exercise 5.6 is referred to as its *natural* one. We have already seen (in Exercises 5.2 and 5.3) that a design may admit different resolutions and we now set an exercise to show that $\mathscr{A}_{n,i}(q)$ can actually have more than on parallelism. Before posing the exercise, however, it is worth noting that taking any $i \neq n-1$ in Exercise 5.6 will give an inner resolution which is not outer.

Exercise 5.7. Let $\mathbf{H}$ be the set of all parallel classes of $\mathscr{A}_{3,1}(q)$ under the natural parallelism. If h_1, h_2 are any two intersecting lines with respective parallel classes $\mathscr{H}_1$ and $\mathscr{H}_2$ then, clearly, $\mathscr{H}_1 \neq \mathscr{H}_2$. If x is the plane defined by h_1 and h_2, denote by $\mathscr{H}'_1$ the set of lines of $\mathscr{H}_1$ not in x together with those of $\mathscr{H}_2$ in x. If $\mathscr{H}'_2$ is defined similarly, show that $\{\mathscr{H}'_1, \mathscr{H}'_2\} \cup \mathbf{H} \backslash \{\mathscr{H}_1, \mathscr{H}_2\}$ form the classes of a new parallelism of $\mathscr{A}_{3,1}(q)$.

Since we know that a t-(v, k, λ) design $\mathscr{D}$ with $t \geqslant 2$ has rank v (see Corollary 1.22), we know that the inequalities of Corollary 1.41 must hold for any tactical decomposition of $\mathscr{D}$. Thus we have the following stronger version of Fisher's Inequality for resolvable designs.

Theorem 5.4. If a 2-(v, k, λ) design $\mathscr{D}$ admits a resolution with c block classes then $b \geqslant v+c-1$.

Proof. Clearly, no proof is needed as this theorem is merely Corollary 1.41 with $v_1=1$ and $b_1=c$. However, we will give an alternative proof to illustrate a powerful 'trick' which is frequently useful for proving this type of result.

If $\mathscr{B}_1, \mathscr{B}_2, \ldots, \mathscr{B}_c$ are the classes of the resolution, let A be any incidence matrix of $\mathscr{D}$ with any labelling $P_1, \ldots, P_v$ of the points such that the first m_1 blocks form $\mathscr{B}_1$, the next m_2 form $\mathscr{B}_2$, and so on until the last m_c blocks form $\mathscr{B}_c$. We say that such an incidence matrix A is *partitioned* into the block classes. Now let B be the $v+c$ by $b+1$ matrix given by

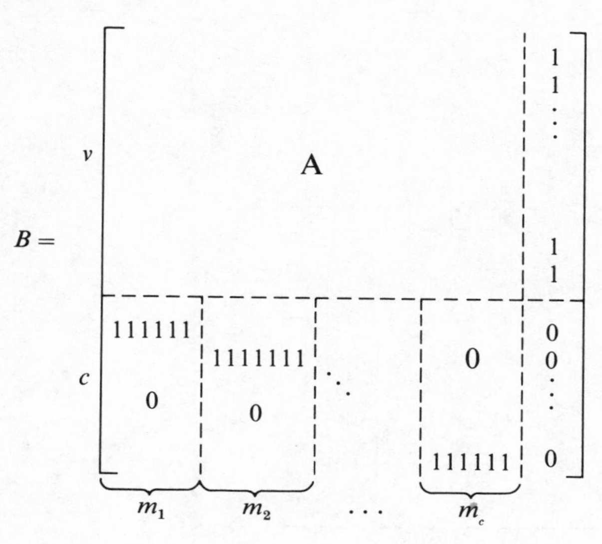

(Thus the $v+i$ row of B has 1s under the blocks of $\mathscr{B}_i$ and 0s everywhere else.)

Imitating the proof of Fisher's Inequality (Corollary 1.24) we compute BB^{T}. If

$$BB^{\mathrm{T}} = \begin{array}{c} \\ v \\ c \end{array}\!\!\begin{array}{c} \begin{array}{cc} v & c \end{array} \\ \left[\begin{array}{c|c} P & Q \\ \hline R & S \end{array}\right] \end{array}$$

then $P = AA^{\mathrm{T}} + J$ which, by Lemma 1.19, is equal to $nI_v + (\lambda+1)J_v$. The scalar product of the $(v+i)$th row of B with the jth row, $1 \leqslant j \leqslant v$, is equal to the number of columns which contain 1s in both rows. But this is precisely the number of blocks of class i which pass through the point P_j. Since this is σ_j, for all $j = 1, 2, \ldots, v$, it follows that

$$Q = \begin{bmatrix} \sigma_1 & \sigma_2 \ldots & \sigma_c \\ \sigma_1 & & \vdots \\ \vdots & & \\ \sigma_1 & \sigma_2 \ldots & \sigma_c \end{bmatrix}$$

Clearly, $R = Q^{\mathrm{T}}$. The (i, j)-entry of S is the scalar product of the $(v+i)$ and $(v+j)$ rows of B. If $i \neq j$ then this is 0, while if $i = j$ it is m_i. Thus

$$S = \begin{bmatrix} m_1 & & & 0 \\ & m_2 & & \\ & & \ddots & \\ 0 & & & m_c \end{bmatrix}$$

and

$$BB^{\mathrm{T}} = \left[\begin{array}{cccc|cccc} & & & & \sigma_1 & \sigma_2 & \dots & \sigma_c \\ & & & & \sigma_1 & & & \\ & nI_v + (\lambda+1)J_v & & & \vdots & & & \\ & & & & \sigma_1 & \sigma_2 & \dots & \sigma_c \\ \hline \sigma_1 & \sigma_1 & \dots & \sigma_1 & m_1 & & & \\ \sigma_2 & \sigma_2 & \dots & \sigma_2 & & m_2 & & 0 \\ \vdots & \vdots & & & & & \ddots & \\ \sigma_c & \sigma_c & \dots & \sigma_c & 0 & & & m_c \end{array}\right]$$

As in the proof of Fisher's Inequality, this is a non-singular square matrix. The proof of this is left as an exercise.

Exercise 5.8. Show that det $BB^{\mathrm{T}} = vn^{v-c} \prod_{i=1}^{c} m_i$ and deduce that BB^{T} has rank $v+c$.

Since BB^{T} has rank $v+c$, B itself must have rank at least $v+c$. Thus, as B has only $v+c$ rows, B has rank $v+c$ and, since it has $b+1$ columns, $v+c \leqslant b+1$, as required. □

In the next section we will consider the situation where $b+1=v+c$ and show that this occurs if and only if the resolution is strong. Meanwhile we state an obvious corollary.

Corollary 5.5. A symmetric design cannot admit a non-trivial resolution.

Proof. For a symmetric design $b=v$, so Theorem 5.4 becomes $v+1 \geqslant v+c$ or $c=1$. □

5.3 Strong resolutions

If a t-design $\mathscr{D}$ admits a strong resolution then its only intersection numbers are ρ and μ. If $t \geqslant 2$ then $\rho = \mu$ would imply that $\mathscr{D}$ was symmetric which, by Corollary 5.5, is impossible. Thus if $t \geqslant 2$ and $\mathscr{D}$ is strongly resolvable then $\mathscr{D}$ has exactly two distinct intersection numbers. This means that two blocks are in the same resolution class if and only if they intersect in ρ points, i.e. the strong resolution is unique. As we pointed out in Section 5.2, it is certainly possible for a design to admit two distinct resolutions. But the uniqueness is a consequence of the fact that the resolution is strong. (It is worth noting now that, as we shall see later, a design with two intersection numbers need not be strongly resolvable.) One immediate consequence of the uniqueness of a strong resolution is that it must be preserved by every automorphism of the design. This, in itself, makes strongly resolvability a remarkable property. We now discuss some of the other properties which make strong resolutions so interesting.

Theorem 5.6. Let $\mathscr{D}$ be a strongly resolvable 2-(v, k, λ). If A is an incidence matrix partitioned into the block classes then

$$A^{\mathrm{T}}A = \begin{bmatrix} \boxed{T} & & & & \\ & \boxed{T} & & & 0 \\ & & \boxed{T} & & \\ & & & \ddots & \\ & 0 & & & \boxed{T} \end{bmatrix} + \mu J_b$$

where $T = (k - \rho)I_m + (\rho - \mu)J_m$.

Proof. The (i, j)-entry of $A^{\mathrm{T}}A$ is the scalar product of the ith and jth rows of A^{T} which are, of course, the ith and jth columns of A. If $i = j$ then the inner product of the ith column of A with itself is the number of points on the ith block. Thus each diagonal entry of $A^{\mathrm{T}}A$ is k. If the ith and jth blocks are in the same class of the resolution then the inner product of the ith and jth columns is ρ, otherwise it is μ. The partitioning of A now establishes the theorem. $\square$

Strongly resolvable designs with $\rho = 0$ are called *affine* and have played a central role in the study of designs. One reason for their importance is, as we saw in Exercise 5.6, that the designs $\mathscr{A}_n(q)$ are affine designs. An affine plane is also an affine design.

Exercise 5.9. Show that a finite affine plane is an affine design.

If a design $\mathscr{D}$ admits a resolution then, clearly, $\mathscr{C}(\mathscr{D})$ must also admit a resolution (namely two blocks are in the same class of $\mathscr{C}(\mathscr{D})$ if and only if their complements are in the same class of the given resolution of $\mathscr{D}$). The resolutions of $\mathscr{D}$ and $\mathscr{C}(\mathscr{D})$ share certain properties.

Exercise 5.10. If a t-design $\mathscr{D}$ admits an inner (outer) resolution show that $\mathscr{C}(\mathscr{D})$ also admits an inner (outer) resolution.

As an immediate consequence of this exercise we see that the complement of a strongly resolvable design is also strongly resolvable. Thus, in particular, the complement of an affine design is strongly resolvable. However, it is not in general affine.

Exercise 5.11. Show that the complement of an affine design is affine if and only if $m=2$.

In order for a resolution to be strong it needs to be a σ-resolution with both inner and outer constants. We now prove a powerful theorem which shows that a resolution is strong if and only if $b=v+c-1$.

Theorem 5.7. Let $\mathscr{D}$ be a 2-(v, k, λ) which admits a resolution with c block classes. The resolution is strong if and only if $b+1=v+c$.

Proof. If $b+1=v+c$ then the matrix B in the proof of Theorem 5.4 is square and non-singular. Thus it has a two-sided inverse. This proof involves spotting the right inverse (by the use of results like Lemma 1.19) and then using the fact that it is a left inverse to get information about $A^{\mathrm{T}}A$ and hence about $\mathscr{D}$.

To simplify the writing of the proof we put

$$B = \begin{array}{c} \\ v \\ c \end{array}\!\!\begin{array}{c} \begin{array}{cc} b & 1 \end{array} \\ \left[\begin{array}{c|c} A & X \\ \hline Y & Z \end{array}\right] \end{array}$$

where X, Y, Z are as in Theorem 5.4.

We now define a new matrix

$$C = \begin{array}{c} \\ b \\ 1 \end{array}\!\!\begin{array}{c} \begin{array}{cc} v & c \end{array} \\ \left[\begin{array}{c|c} P & Q \\ \hline R & S \end{array}\right] \end{array},$$

where

$$P = \frac{1}{n} A^{\mathrm{T}} - \frac{k}{nv} J_{b,v},$$

$$R = \frac{1}{v} J_{1,v},$$

$$S = -\frac{k}{nv} J_{1,c},$$

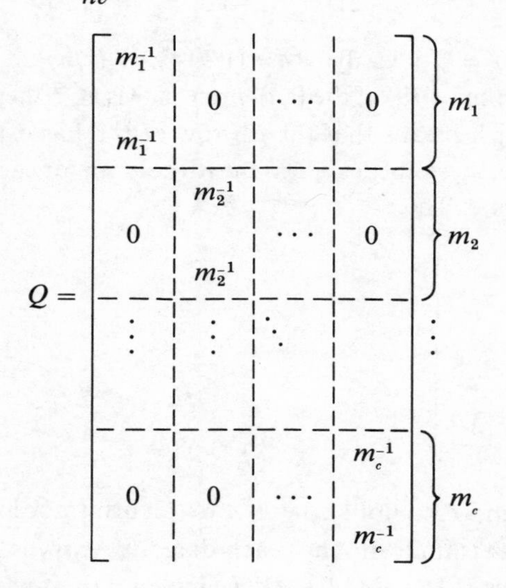

Then

$$BC = \left[\begin{array}{c|c} AP+XR & AQ+XS \\ \hline YP+ZR & YQ+ZS \end{array}\right]$$

However, we have enough information to simplify BC considerably. We begin with the top left-hand corner $AP + XR = (1/n)AA^{\mathrm{T}} - (k/nv)AJ_{b,v} + (1/v)J_v$. Hence, by Lemma 1.19 and Exercise 1.2, $AP + XR = I_v + (\lambda/n)J_v - (rk/nv)J_v + (1/v)J_v = I_v + ((\lambda v - rk + n)/nv)J_v$. But $n = r - \lambda$ and, by Corollary 1.4(c), $r(k-1) = \lambda(v-1)$. So $\lambda v - rk + n = 0$ and $AP + XR = I_v$.

Since $Z = O_{c,1}$ (where $O_{e,f}$ is the e by f matrix with each entry O), $ZR = O_{c,v}$. But $YP = Y(1/n)A^{\mathrm{T}} - (k/nv)YJ_{b,v}$. The (i, j)-entry of YA^{T} is equal to the number of blocks of class i which contain the jth point of $\mathscr{D}$. Thus the (i, j)-entry of YA^{T} is σ_i for each j. Similarly the (i,j)-entry of $YJ_{b,v}$ is m_i. Hence the (i, j)-entry of YP is $(\sigma_i/n) - (km_i/nv)$ which, by Exercise 5.1(c), is 0. Thus $YP + ZR = O_{c,v}$.

Similar computations, which the reader should check, give $BC = I_{v+c}$. Thus $CB = I_{v+c}$. But

$$CB = \begin{array}{c} \\ b \\ \\ 1 \end{array}\!\!\begin{array}{c} \begin{array}{cc} v & c \end{array} \\ \left[\begin{array}{c|c} P & Q \\ \hline R & S \end{array}\right] \end{array} \begin{array}{c} \begin{array}{cc} b & 1 \end{array} \\ \left[\begin{array}{c|c} A & X \\ \hline Y & Z \end{array}\right] \end{array}\!\!\begin{array}{c} \\ v \\ \\ c \end{array} = \begin{array}{c} \\ b \\ \\ 1 \end{array}\!\!\begin{array}{c} \begin{array}{cc} b & 1 \end{array} \\ \left[\begin{array}{c|c} PA+QY & PX+QZ \\ \hline RA+SY & RX+SZ \end{array}\right] \end{array}$$

Since $CB = I_{b+1}$, $PA + QY = I_b$. Clearly, $PA = (1/n)A^{\mathrm{T}}A - (k/nv)J_{b,v}A = (1/n)A^{\mathrm{T}}A - (k^2/nv)J_b$ (by Lemma 1.19). If the (i, j)-entry of Q is m_a^{-1} then the jth block of $\mathscr{D}$ is in class a which means that the ith row of QY has exactly m_a entries of m_a^{-1}. Furthermore, these entries are in the lth column for each l such that the lth block is in class a. Thus

$$QY = \begin{bmatrix} \boxed{X_1} & & & 0 \\ & \boxed{X_2} & & \\ & & \ddots & \\ 0 & & & \boxed{X_c} \end{bmatrix}$$

where $X_i = m_i^{-1}J_{m_i}$. Although we do not know $A^{\mathrm{T}}A$ for an arbitrary incidence matrix we do know, since $\mathscr{D}$ is uniform, that each diagonal entry is k. Thus, equating the diagonal entries of I_b and $PA + QY$, we have $1 = (k/n) - (k^2/nv) - m_i^{-1}$, for each i. But this means that the m_i are all equal. Putting $m_i = m$ for $i = 1, \ldots, c$ we have $I_b = (1/n)A^{\mathrm{T}}A - (b^2/nv)J_b + (1/m)X$, where X is the b by b matrix with blocks of J_m down the diagonal and zeros elsewhere. This rearranges to give $A^{\mathrm{T}}A = nI_b + (k^2/nv)J_b - (n/m)X$. However, the (i, j)-entry of $A^{\mathrm{T}}A$ gives the number of points common to blocks i and j. Hence two blocks in the same class meet $(k^2/v)-(n/m)$ times and two blocks from different classes meet k^2/v times. In other words, we have a strong resolution.

To prove the converse we use the fact that we know the form of $A^{\mathrm{T}}A$ to argue that $B^{\mathrm{T}}B$ is non-singular and deduce that B has rank $b+1$. Since we already know its rank is $v+c$ this completes the proof. We leave it as an exercise. □

As an alternative to the given matrix proof the following exercises give a counting proof of a similar result. Throughout the next five exercises it is to be assumed that $\mathscr{D}$ is a σ-resolvable 1-(v, k, r). For a fixed block d of $\mathscr{D}$ we label the other blocks d_i for $1 \leqslant i \leqslant b-1$ such that d is equivalent to d_i if and only if $1 \leqslant i \leqslant m-1$. Let ρ_i be the number of points common to d and d_i for $1 \leqslant i \leqslant m-1$ and μ_j be the number of points common to d and d_j for $m \leqslant j \leqslant b-1$.

Exercise 5.12. Prove $\sum_{i=1}^{m-1} \rho_i = k(\sigma-1)$.

Exercise 5.13. Prove $\sum_{j=m}^{b-1} \mu_j = k(r-\sigma)$.

We now let ρ and μ be the average values of the ρ_i and μ_j respectively.

Exercise 5.14. Show $\rho = k(\sigma-1)/(m-1)$ and $\mu = k(r-\sigma)/(b-m) = k^2/v$.

Exercise 5.15. Show $\sum_{i=1}^{m-1} \rho_i^2 + \sum_{j=m}^{b-1} \mu_j^2 = k(k-1)(\lambda-1) + k(r-1)$.

Exercise 5.16.* Show

$$\sum_{i=1}^{m-1} (\rho_i-\rho)^2 + \sum_{j=m}^{b-1} (\mu_j-\mu)^2 = \frac{(v-k)^2 kr[b-(v+c-1)]}{v(v-1)(b-c)}$$

and deduce that if a 1-(v, k, r) design $\mathscr{D}$ admits a σ-resolution with c block classes then the resolution is strong if and only if $b+1 = v+c$.

5.4 Strongly resolvable designs

As well as the standard parameters t, v, k, λ, etc., strongly resolvable designs have further ones: e.g. ρ, μ, σ and c. Not surprisingly, there are many relations between these parameters.

Theorem 5.8. Let $\mathscr{D}$ be a strongly resolvable 2-(v, k, λ); then $v = \mu m^2/\sigma^2$, $k = \mu m/\sigma$, $\lambda = (\mu m - \sigma)/(m-1)$, $b = m(\mu m^2 - \sigma^2)/\sigma^2(m-1)$, $r = (\mu m^2 - \sigma^2)/\sigma(m-1)$, $\rho = \mu m(\sigma - 1)/\sigma(m-1)$ and $c = (\mu m^2 - \sigma^2)/\sigma^2(m-1)$.

Proof. From Exercise 5.1 and Lemma 5.1, $v\sigma - mk$, $r = c\sigma$ and $b = mc$. Also, by Exercise 5.4, $k^2 = v\mu$. Thus $mk/\sigma = v = k^2/\mu$, so $k = m\mu/\sigma$ and $v = \mu m^2/\sigma$. From Theorem 5.11, $b = v + c - 1 = mc$, so $c = (v-1)/(m-1) = (\mu m^2 - \sigma^2)/\sigma^2(m-1)$. Hence $b = mc = m(\mu m^2 - \sigma^2)/\sigma^2(m-1)$, and $r = rc = (\mu m^2 - \sigma^2)/\sigma(m-1)$.

By Corollary 1.4, $\lambda = r(k-1)/(v-1) = (\mu m - \sigma)/(m-1)$ and, finally, from Lemma 5.2, $\rho = k(\sigma - 1)/(m-1) = \mu m(\sigma - 1)/\sigma(m-1)$. □

Corollary 5.9. If $\mathscr{D}$ is a strongly resolvable 2-(v, k, λ) then $r - \lambda = k - \rho$ and $\lambda(v-k) = (k-1)(k-\rho)$.

Proof. Obvious. □

Theorem 5.8 is very restrictive and shows that the number of strongly resolvable designs is limited. We now prove a theorem which characterises affine designs among the strongly resolvable designs.

Theorem 5.10. Let $\mathscr{D}$ be a strongly resolvable 2-(v, k, λ). Then

(a) $k \mid v$ if and only if $\mathscr{D}$ is affine or $k = v$;

(b) $\sigma \mid m$ if and only if $\mathscr{D}$ is affine or $k = v$.

Proof. If $\mathscr{D}$ is affine both cliams are immediate from Theorem 5.8 since $\sigma = 1$. If $d = (v, k)$ then $d \mid (v-k)$ and so, by Corollary 5.13, $d \mid (k-1)(k-\rho)$. However, since $d \mid k$, $(d, k-1) = 1$ so that $d \mid k - \rho$, i.e. $d \mid \rho$. If $d = k$ then $k \mid \rho$ and so, since $\rho \leqslant k$, either $\rho = 0$ or $\rho = k$. If $\rho = 0$, $\mathscr{D}$ is affine, while if $\rho = k$ there is only one block in each class which, since each point is on a block from each class implies there is only one block. If $\sigma \mid m$ then, since $v = \mu m^2/\sigma^2$ and $k = \mu m/\sigma$, $k \mid v$. □

Exercise 5.17. Show that, for a strongly resolvable design, $\rho < \mu$.

As we saw in Exercise 5.10, the complement of a strongly resolvable design is again strongly resolvable. We now look at the parameters of the complement.

Theorem 5.11. If $\mathscr{D}$ is a strongly resolvable 2-(v, k, λ) with $v \geqslant 2k$ and resolution constants σ, ρ, μ, m then $\mathscr{C}(\mathscr{D})$ is strongly resolvable with constants $m-\sigma$, $v-2k+\rho$, $v-2k+\mu$, m.

Proof. Clearly, there are the same number of blocks in a class of $\mathscr{C}(\mathscr{D})$ as in a class of $\mathscr{D}$. If P is any point of $\mathscr{D}$ and $\mathscr{B}_i$ is a resolution class, then P is on σ blocks of $\mathscr{B}_i$. Hence there are $m-\sigma$ blocks of $\mathscr{B}_i$ not containing P so that there are $m-\sigma$ blocks of $\mathscr{C}(\mathscr{B}_i)$ on P in $\mathscr{C}(\mathscr{D})$.

If x, y are two blocks in $\mathscr{B}_i$ then, since they intersect in ρ points, $|x \cup y| = 2k-\rho$. Then $|\mathscr{C}(x) \cap \mathscr{C}(y)| = v-2k+\rho$, i.e. two blocks in the same class of $\mathscr{C}(\mathscr{D})$ intersect in $v-2k+\rho$ points. The outer constant is evaluated in a similar way. □

Corollary 5.12. A strongly resolvable 2-design $\mathscr{D}$ with $m=2$ is affine.

Proof. Clearly, for either $\mathscr{D}$ or $\mathscr{C}(\mathscr{D})$, $v \geqslant 2k$. Let us assume that $\mathscr{D}$ has $v \geqslant 2k$. If the strong resolution of $\mathscr{D}$ is a σ-resolution then $\mathscr{C}(\mathscr{D})$ is $(2-\sigma)$-resolvable. Thus $\sigma \geqslant 1$ and $2-\sigma \geqslant 1$, i.e. $\sigma = 1$. Hence by Theorem 5.3, $\rho = 0$ and $\mathscr{D}$ is affine. But, since $m=2$, the complement of any block is a block so $\mathscr{C}(\mathscr{D}) = \mathscr{D}$ and the corollary is proved. □

As we have noted many times, a t-design is also an s-design for all $1 \leqslant s \leqslant t$. Obviously if $\mathscr{D}$ is a strongly resolvable t-design then, since the resolution merely depends on the intersection numbers, it is also a strongly resolvable s-design. However, we shall now show that there are no strongly resolvable t-designs with $t \geqslant 4$ and that the only strongly resolvable 3-designs have $m=2$. In Chapter 4 we saw that any Hadamard 2-$(4\lambda+3, 2\lambda+1, \lambda)$ had a unique extension to a Hadamard 3-$(4\lambda+4, 2\lambda+2, \lambda)$ in which the complement of a block is also a block. Furthermore, any block x of a Hadamard 3-$(4\lambda+4, 2\lambda+2, \lambda)$ intersects each block other than its complement x' in $\lambda+1$ points. It is a simple exercise to show that, taking each block and its complement as a complete resolution class, a Hadamard 3-design is affine with $m=2$.

Exercise 5.18. Show that any Hadamard 3-design is affine and find its resolution constants.

In fact, Hadamard 3-designs are the only strongly resolvable designs with $m=2$ and are the only affine 3-designs.

Theorem 5.13. If $\mathscr{D}$ is a non-trivial strongly resolvable 2-(v, k, λ) with $m=2$ then $\mathscr{D}$ is a Hadamard 3-design.

Proof. By Corollary 5.12, $\mathscr{D}$ is affine. Thus $\sigma=1$ and by Theorem 5.8 $\mathscr{D}$ is a 2-$(4\mu, 2\mu, 2\mu-1)$ with $b=2(4\mu-1)$, $r=4\mu-1=c$.

For any point X of $\mathscr{D}$ the contraction of $\mathscr{D}$ at X, $\mathscr{D}_X$, is a 1-$(4\mu-1, 2\mu-1, 2\mu-1)$ with $4\mu-1$ blocks (see Theorem 1.14(c)). Thus $\mathscr{D}_X$ is a square 1-design. However, since $\mathscr{D}$ is affine, any two blocks of $\mathscr{D}_X$ intersect in $\mu-1$ points. This means that the dual of $\mathscr{D}_X$ is a symmetric 2-design. But, by Theorem 1.26, a 2-design is symmetric if and only if its dual is also a 2-design. Thus $\mathscr{D}_X$ is a 2-$(4\mu-1, 2\mu-1, \mu-1)$ or, putting $\lambda=\mu-1$, a 2-$(4\lambda+3, 2\lambda+1, \lambda)$. In other words, $\mathscr{D}_X$ is a Hadamard 2-design and, since any Hadamard 2-design is uniquely extendable, $\mathscr{D}$ is a Hadamard 3-design. □

Theorem 5.14. If $\mathscr{D}$ is an affine 3-design then $\mathscr{D}$ is a Hadamard 3-design.

Proof. Since $\mathscr{D}$ is affine $\sigma=1$ and so, from Theorem 5.8, $\mathscr{D}$ is a 3-$(\mu m^2, \mu m, \lambda)$ with $b=m(\mu m^2-1)/(m-1)$, $r=(\mu m^2-1)/(m-1)$ and $\lambda_2=(\mu m-1)/(m-1)$.

Let P be a point of $\mathscr{D}$. Then $\mathscr{D}_P$ is a 2-$(\mu m^2-1, \mu m-1, \lambda)$ with $(\mu m^2-1)/(m-1)$ blocks. However, since $\mathscr{D}$ is affine, any two blocks of $\mathscr{D}_P$ intersect in $\mu-1$ points. Thus $\mathscr{D}_P$ is a 2-design whose dual is also a 2-design which, by Theorem 1.14, means that $\mathscr{D}_P$ is symmetric. Hence $\mu m^2-1=(\mu m^2-1)/(m-1)$ or $m=2$ and, by Theorem 5.13, this implies that $\mathscr{D}$ is Hadamard. □

We now show that Hadamard 3-designs are the only strongly resolvable 3-designs.

Theorem 5.15. The following statements are equivalent for a non-trivial 2-design $\mathscr{D}$:

(a) $\mathscr{D}$ is a strongly resolvable 3-design;
(b) $\mathscr{D}$ is a Hadamard 3-design;
(c) $\mathscr{D}$ is strongly resolvable with $m=2$.

Proof. We have already shown (in Theorem 5.13 and Exercise 5.16) that (c) implies (b) and (b) implies both (a) and (c). Thus we have only to show that (a) implies either (b) or (c).

If $\mathscr{D}$ is a strongly resolvable 2-$(\mu m^2/\sigma^2, \mu m/\sigma, \mu m\sigma/(m-1))$, which is also a 3-design, then, for any point A of $\mathscr{D}$, $\mathscr{D}_A$ is a 2-$((\mu m^2/\sigma^2)-1, (\mu m/\sigma)-1, \lambda)$ for some λ, with $(\mu m^2-\sigma^2)/\sigma(m-1)$ blocks. (Note that, for $\mathscr{D}$, $r=(\mu m^2-\sigma^2)/\sigma(m-1)$; see Theorem 5.8.) So, by Fisher's Inequality, $(\mu m^2-\sigma^2)/\sigma(m-1) \geqslant (\mu m^2-\sigma^2)/\sigma^2$ or $\sigma \geqslant m-1$. However, since $\mu < v = \mu m^2/\sigma^2$, $\sigma < m$. Thus $\sigma = m-1$ and, by Theorem 5.11, $\mathscr{C}(\mathscr{D})$ is strongly 1-resolvable, i.e. $\mathscr{C}(\mathscr{D})$ is affine. But, by Exercise 1.39, $\mathscr{C}(\mathscr{D})$ is a 3-design. Theorem 5.14 now completes the proof. □

Having shown that all affine 3-designs have $m=2$ it is straightforward to prove that there are no strongly resolvable t-designs with $t \geqslant 4$.

Theorem 5.16. There are no strongly resolvable t-designs with $t \geqslant 4$.

Proof. Obviously it is sufficient to prove that there are no strongly resolvable 4-designs. If $\mathscr{D}$ is a strongly resolvable 4-design then, since it is also a 3-design, Theorem 5.15 implies it is a Hadamard 3-design. Thus it is a 4-$(4x+4, 2x+2, \lambda)$ for some λ, with $\lambda_3 = x$ and $8x+6$ blocks. From Theorem 1.2

$$\lambda = \frac{(8x+6)(2x+2)(2x+1)2x(2x-1)}{(4x+4)(4x+3)(4x+2)(4x+1)} = \frac{x(2x-1)}{4x+1}.$$

Since $x \geqslant 1$, $x(2x-1)/(4x+1)$ is never an integer and the theorem is proved. □

We have seen that a strongly resolvable 2-design is a 3-design if and only if $m=2$, and that there are no strongly resolvable 4-designs. It is now natural to ask when a strongly resolvable 1-design is a 2-design. To answer this question we will use a counting argument similar to the dual of the one outlined in Exercises 5.12–5.16. The reader should also try to find a matrix proof.

Theorem 5.17. Let $\mathscr{D}$ be a strongly resolvable 1-(v, k, r) with b blocks and c resolution classes. Then $\mathscr{D}$ is a 2-design if and only if $b = v + c - 1$.

Proof. For a fixed point P of $\mathscr{D}$, label the other points $P_1, \ldots, P_{v-1}$ and let w_i be the number of blocks common to P and P_i $(i=1, \ldots, v-1)$. Counting flags (P_i, d), where P is on d, in two ways gives

$$\sum_{i=1}^{v-1} w_i = r(k-1). \tag{1}$$

If λ is the average value of w_i then

$$\lambda = \frac{r(k-1)}{v-1}. \tag{2}$$

We note that $\mathscr{D}$ is a 2-design if and only if for every P, $\lambda = w_i$ for each i and the value of λ must also be independent of the choice of P.

Next we count triples (P_i, d_1, d_2), where P and P_i are in $d_1 \cap d_2$. If $d_1 = d_2$ then there are $r(k-1)$ such triples. If d_1 and d_2 are in the same resolution class but $d_1 \neq d_2$ then, for each of the r choices of d_1 there are $\sigma - 1$ choices of d_2 and then $\rho - 1$ choices for P_i. Hence there are $r(\sigma-1)(\rho-1)$ such triples with equivalent, but distinct, blocks. Similarly there are $r(r-\sigma)(\mu-1)$ triples with blocks from different classes. Alternatively for any given P_i there are w_i choices for d_1 and w_i choices for d_2. Thus there are $\sum_{i=1}^{v-1} w_i^2$ such triples and we have

$$\sum_{i=1}^{v-1} w_i^2 = r[(k-1)+(\sigma-1)(\rho-1)+(r-\sigma)(\mu-1)]. \tag{3}$$

Next, as usual, we calculate $\sum_{i=1}^{v-1} (\lambda - w_i)^2$. Clearly, $\sum_{i=1}^{v-1} (\lambda - w_i)^2 = \sum_{i=1}^{v-1} w_i^2 - 2\lambda \sum_{i=1}^{v-1} w_i + (v-1)\lambda^2$ and thus from (1), (2) and (3) we have

$$\sum_{i=1}^{v-1} (\lambda - w_i)^2 = r[(k-1)+(\sigma-1)(\rho-1)+(r-\sigma)(\mu-1)] - \frac{r^2(k-1)^2}{v-1}. \tag{4}$$

From Exercises 5.1 and 5.4, $\sigma = mk/v$ and $\mu = k^2/v$. Substituting these values for σ and μ in (4) and simplifying gives

$$\sum_{i=1}^{v-1} (\lambda - w_i)^2 = \frac{r}{kv(v-1)}[kr(k^2v - k^3 - v^2 + v - vk^2 + 2vk - v) + (m-1)\rho(v-1)(v\rho - k^2) + k^2(v-1)(v-k)]. \tag{5}$$

But, from Lemma 5.2, $\rho(m-1) = k(\sigma-1) = k(mk-v)/v$, so that $v\rho(m-1) = k(mk-v)$ or $(v\rho - k^2)(m-1) = k(k-v)$. Making this substitution in (5) and rearranging gives

$$\sum_{i=1}^{v-1} (\lambda - w_i)^2 = \frac{r(v-k)}{v(v-1)}[k(v-1) - r(v-k) - \rho(v-1)]. \tag{6}$$

By Corollary 1.4(b), $bk = vr$ so that $k(v-1) - r(v-k) - \rho(v-1) = k(v-1) + k\sigma - kb + k(r-c) - \rho(v-1) = k(v+c-b-1) - \rho(v-1) + k(r-c)$. But, from Exercise 5.1, $r = \sigma c$ so $k(r-c) - kc(\sigma-1) = c(m-1)$ (by Lemma 5.2). Thus

$$\sum_{i=1}^{v-1} (\lambda - w_i)^2 = \frac{r(v-k)(k-\rho)(v+c-b-1)}{v(v-1)}.$$

Clearly, $\sum_{i=1}^{v-1} (\lambda - w_i)^2 \geqslant 0$ and thus, since $r(v-k)(b-\rho)/(v(v-1)) \geqslant 0$, $v+c \geqslant b+1$ with equality if and only if $\lambda = w_i$ for each i. Since the value of $\sum_{i=1}^{v-1} (\lambda - w_i)^2$ is independent of the choice of P the theorem is proved. □

Every automorphism of a strongly resolvable design preserves the strong resolution, and we have an orbit theorem for strongly resolvable designs:

Theorem 5.18. Let G be an automorphism group of a strongly resolvable 2-$(\mu m^2/\sigma^2, \mu m/\sigma, (\mu m - \sigma)/(m-1))$. If G has v_1 point orbits, b_1 block orbits and c_1 orbits of block classes then $b_1 + 1 = v_1 + c_1$.

Proof. Let $\mathscr{P}_i$ $(1 \leqslant i \leqslant v_1)$, x_j $(1 \leqslant j \leqslant b_1)$ and $\mathscr{C}_l$ $(1 \leqslant l \leqslant c_1)$ denote the point, block and resolution class orbits of G respectively. Let $(\mathscr{P}_i, x_j)$ and $(x_j, \mathscr{P}_i)$ be as defined in Chapter 1 and let $(x_j, \mathscr{C}_l)$ denote the number of blocks of x_j which are in any given resolution class of $\mathscr{C}_l$.

For any given i, $\sum_{j=1}^{b_1} (\mathscr{P}_i, x_j)(x_j, \mathscr{P}_i) = r + \lambda(|\mathscr{P}_i| - 1) = r - \lambda + \lambda|\mathscr{P}_i|$. (To see this, note that, for any fixed P in $\mathscr{P}_i$ and for any fixed j, $(\mathscr{P}_i, x_j)(x_j, \mathscr{P}_i)$ is the number of flags (Q, x) with Q in $\mathscr{P}_i$, x in x_j and P on x. So to evaluate $\sum_{j=1}^{b_1} (\mathscr{P}_i, x_j)(x_j, \mathscr{P}_i)$ we merely have to count configurations (P, Q, x) with P, Q in $\mathscr{P}_i$ and P, Q on x. If $P = Q$ there are r such flags, while if $P \neq Q$ there are $|\mathscr{P}_i| - 1$ choices for Q and each choice gives λ choices for x.) Thus

$$\sum_{i=1}^{v_1} \sum_{j=1}^{b_1} (\mathscr{P}_i, x_j)(x_j, \mathscr{P}_i) = (r - \lambda)v_1 + \lambda v. \tag{1}$$

A similar counting argument to that use for $\sum_{j=1}^{b_1} (\mathscr{P}_i, x_j)(x_j, \mathscr{P}_i)$ gives

$$\sum_{i=1}^{v_1} (\mathscr{P}_i, x_j)(x_j, \mathscr{P}_i) = k + \mu\left(|x_j| - \sum_{l=1}^{c_1} (x_j, \mathscr{C}_l)\right) + \rho\left(\sum_{l=1}^{c_1} (x_j, \mathscr{C}_l) - 1\right).$$

(Here we note that the number of blocks in x_j which are in the same orbit as a given block x in x_j is $\sum_l (x_j, \mathscr{C}_l)$. Thus for any given x there are $\sum_l (x_j, \mathscr{C}_l) - 1$ blocks in x_j intersecting it in ρ points and $|x_j| - \sum_l (x_j, \mathscr{C}_l)$ which intersect it in μ points.) But, since $\sum_j (x_j, \mathscr{C}_l) = m$ for any l, $\sum_{j,l} (x_j, \mathscr{C}_l) = mc_1$ and so

$$\sum_{i,j} (\mathscr{P}_i, x_j)(x_j, \mathscr{P}_i) = \mu b + (\rho - \mu)mc_1 + (k - \rho)b_1. \tag{2}$$

Equating (1) and (2) we have

$$(r - \lambda)v_1 + \lambda v = \mu b + (\rho - \mu)mc_1 + (k - \rho)b_1.$$

But we can use Theorem 5.8 to get

$$\left(\frac{\mu m^2 - \sigma^2}{\sigma(m-1)} - \frac{(\mu m - \sigma)\sigma}{(m-1)\sigma}\right)v_1 + mc_1\left(\frac{\mu\sigma(m-1)}{\sigma(m-1)} - \frac{\mu m(\sigma - 1)}{\sigma(m-1)}\right)$$
$$= b_1\left(\frac{\mu m(m-1)}{\sigma(m-1)} - \frac{\mu m(\sigma-1)}{\sigma(m-1)}\right) + \left(\frac{\mu m(\mu m^2 - \sigma^2)}{\sigma^2(m-1)} - \frac{(\mu m - \sigma)\mu m^2}{(m-1)\sigma^2}\right),$$

which, on cancelling $\mu m(m-\sigma)/(\sigma(m-1))$, gives the required result. □

Exercise 5.19.* Use Block's Lemma (Theorem 1.40) to give an alternative proof for Theorem 5.18.

5.5 Examples of strongly resolvable designs

We have already seen that Hadamard 3-designs, the designs $\mathscr{A}_n(q)$ and affine planes are all examples of affine designs. Thus, of course, both they and their complements are strongly resolvable. Each of these three families are discussed in considerable detail in other chapters. However, in this section we give a construction for an infinite number of strongly resolvable designs which are neither affine nor the complements of affine designs. At one time it was conjectured that a 2-design was strongly resolvable if and only if it (or its complement) was affine. The construction given here gives counter examples to this conjecture.

Exercise 5.20. If $\mathscr{D}$ is a strongly resolvable 2-$(\mu m^2/\sigma^2, \mu m/\sigma, (\mu m - \sigma)/(m-1))$ with $\sigma \neq 1$ (i.e. not affine), show that $m \geqslant 7$ and $\sigma \geqslant 3$. Show, further, $m = 7$ if and only if $\sigma = 3$.

We shall construct a strongly resolvable 2-design with m blocks in a class for any integer m such that

(a) there exists an affine design with m blocks in a parallel class, and
(b) there exists a symmetric 2-design with m points.

Although this is a severe restriction on m there are still many possibilities. If m

is any prime power then there is certainly an affine design, namely an affine geometry, with m blocks in a parallel class. For certain of these values of m the Payley designs are symmetric designs with m points. There are also many examples of projective planes with a prime-power number of points, for prime powers that are not primes. But there are no known affine designs where m is not a prime power, and no strongly resolvable designs either.

Let $\mathscr{A}$ be any affine 2-$(wm^2, wm, (wm-1)/(m-1))$ with $r=c=(wm^2-1)/(m-1)$ parallel classes. In addition let $\mathscr{D}_1, \mathscr{D}_2, \ldots, \mathscr{D}_c$ be c, not necessarily isomorphic, symmetric 2-(m, σ, α) designs with the same parameters and each having m points. For each distinct parallel class $\mathscr{C}_i$ of $\mathscr{A}$ $(1 \leqslant i \leqslant c)$ let π_i be any bijection from the blocks of the design $\mathscr{D}_i$ onto the blocks of the parallel class $\mathscr{C}_i$.

We define a structure $\mathscr{D}$ as follows. The points of $\mathscr{D}$ are the points of $\mathscr{A}$. For any point X in $\mathscr{D}_i$ and any of the bijections π_i we define a block $d(X, \pi_i)$ of $\mathscr{D}$ by $d(X, \pi_i) = \bigcup_{X \in x} x^{\pi_i}$ (where, of course, we are identifying blocks with the set of points incident with them).

Theorem 5.19. $\mathscr{D}$ is a 2-$(wm^2, wm\sigma, \lambda)$, where $\lambda = (wm\sigma - 1)\sigma/(m-1)$.

Proof. Since $\mathscr{D}$ has the same point set as $\mathscr{A}$, $\mathscr{D}$ certainly has wm^2 points. For any point X of $\mathscr{D}$, there are σ blocks of $\mathscr{D}_i$ which contain it. Thus $d(X, \pi_i)$ is the union of σ parallel blocks of $\mathscr{A}$. Since these blocks are parallel they have no common points, so each block of $\mathscr{D}$ contains wm points. (Note that, since each π_i is a bijection and no two points in any $\mathscr{D}_i$ are incident with the same block set, $\mathscr{D}$ cannot have any repeated blocks.)

Let P, Q be any two points of $\mathscr{D}$ (and hence also of $\mathscr{A}$). They are on exactly $(wm-1)/(m-1)$ blocks of $\mathscr{A}$. But, since there are σ points incident with each block of $\mathscr{D}_i$, each block of $\mathscr{A}$ occurs σ times in a block of $\mathscr{D}$. Thus these $(wm-1)/(m-1)$ blocks of $\mathscr{A}$ give rise to $\sigma(wm-1)/(m-1)$ blocks of $\mathscr{D}$ which are common to P and Q. However, there are also $(wm^2-1)/(m-1)-(wm-1)/(m-1) = wm$ blocks of $\mathscr{A}$ which are incident with P but not Q. For each of these wm blocks of $\mathscr{A}$ there is precisely one parallel block through Q, and this pair of blocks is in exactly α blocks of $\mathscr{D}$. (Note that since each $\mathscr{D}_i$ is symmetric any two blocks of $\mathscr{D}_i$ intersect in α points.) Hence any two points of $\mathscr{D}$ are on $\lambda = \sigma(wm-1)/(m-1) + wm\alpha$ blocks of $\mathscr{D}$. But, since each $\mathscr{D}_i$ is symmetric, by Corollary 1.4(a) we have $\alpha(m-1) = \sigma(\sigma-1)$. Substituting this in the expression for λ gives $\lambda = (wm\sigma - 1)\sigma/(m-1)$. □

Theorem 5.20. $\mathscr{D}$ is strongly resolvable.

Proof. We define an equivalence relation $\sim$ on the blocks of $\mathscr{D}$ by $d \sim g$ if and only if there exist points X and Y in the same $\mathscr{D}_i$ such that $d = \bigcup_{X \in x} x^{\pi_i}$ and $g = \bigcup_{Y \in x} x^{\pi_i}$; i.e. they are parallel if and only if they 'come from' the same parallel class of $\mathscr{A}$. It is easy to see that $\sim$ is an equivalence relation, that there are c equivalence classes and that, since each point lies on one block from each parallel class of $\mathscr{A}$ and each block of $\mathscr{A}$ occurs in σ blocks of $\mathscr{D}$, each point of $\mathscr{D}$ is on σ blocks from each class. Thus $\sim$ is a σ-resolution. But $\mathscr{D}$ and $\mathscr{A}$ have the same number of blocks and points. Thus, for $\sim$ on $\mathscr{D}$, $b = v + c - 1$ so that $\sim$ is a strong resolution. □

Clearly, there is considerable freedom of choice in this construction so there are many examples of non-isomorphic strongly resolvable 2-designs with the same parameters.

References

[1] is a survey article on tactical decompositions and provides suitable further reading for anyone interested in the material of this chapter.

[1] Beker, H. J., Mitchell, C. J. & Piper, F. C. 'Tactical decompositions of designs'. *Aquationes Mathematicae*, **23** (1982), 123–52.

6 Other 2-designs

6.1 Introduction

In the earlier chapters we have developed quite a bit of general theory about symmetric 2-designs and constructed several interesting classes. There are, of course, many other 2-designs which are not symmetric but there is much less general theory about them. So far we have met a few of these designs (e.g. affine geometries), and in Section 4.5 gave Alltop's method of constructing more. In this chapter we give further examples of non-symmetric 2-designs. We begin by choosing two classes which have been widely studied for their intrinsic interest, and finish with a large class that have interesting connections with group and field theory (and which generalise both the first class of this chapter as well as the Hadamard 2-designs $L(q)$ of Chapter 3).

Section 6.2 deals with Steiner Triple Systems, which are merely 2-designs with $k=3$ and $\lambda=1$. Certain projective and affine geometries give us easy examples (and in the more detailed theory not included here, these play the role of 'classical' Steiner Triple Systems), but we give others, as well as showing that two Steiner Triple Systems can be 'pasted' together to build a new design of the family. In Section 6.3 we define unitals and construct their classical examples, from unitary polarities of projective planes $\mathscr{P}_2(q^2)$. Other examples are known as well (but never with parameters different from the classical unitals), and there are many unsolved problems about the existence and the classification of unitals. Finally in Section 6.4 we give a more general construction of 2-designs, which is interesting (among other reasons) because of the particular algebraic flavour of the techniques.

6.2 Steiner triple systems

A Steiner Triple System (abbreviated 'STS') is merely a 2-design for $(v, 3, 1)$. We shall include the trivial 2-(3, 3, 1) as an STS. We begin our discussion by establishing a simple restriction on the possible values of v.

Lemma 6.1. If $\mathscr{S}$ is an STS with v points, then $v \equiv 1$ or 3 (mod 6).

Proof. Let $\mathscr{S}$ be an STS, i.e. a 2-$(v, 3, 1)$. Then b and r are given by

$$r = (v-1)/2,$$
$$b = v(v-1)/6.$$

Hence $v \equiv 1$ (mod 2) and $v(v-1) \equiv 0$ (mod 3), which together imply that $v \equiv 1$ or 3 (mod 6). □

In fact, at least one STS exists for every value of $v > 1$ permitted by Lemma 6.1. We shall not attempt to prove this (but see, for example, [4]); however, we notice:

Theorem 6.2

(a) The design $\mathscr{P}_{n,1}(2)$ of points and lines in a projective geometry over $GF(2)$ is an STS.

(b) The design $\mathscr{A}_{n,1}(3)$ of points and lines in an affine geometry over $GF(3)$ is an STS.

Proof. In either case two points are on exactly one line and the line size is 3. (Alternatively see Exercises 1.31 and 1.33.) □

But these are not the only STSs, as our next theorem shows.

Theorem 6.3. If q is a prime power, $q \equiv 1$ (mod 6), then there is an STS with q points.

Proof. Let K be the field $GF(q)$, and K^* its multiplicative group. Since K^* is cyclic of order $q-1 \equiv 0$ (mod 3), K^* contains exactly two elements of order 3; let these be a, a^2, and let $A = \langle a \rangle$ be the subgroup of order 3 in K^*. Furthermore, let B be the set of all elements $a^i - a^j$, $a^i \neq a^j$, so

$$B = \{1-a, 1-a^2, a-1, a^2-1, a-a^2, a^2-a\}.$$

Let $A' = \langle -1, a \rangle = \langle -a \rangle = \{\pm 1, \pm a, \pm a^2\}$ be the subgroup of order 6 in K^*. We make a number of simple observations:

(a) If $b_1, b_2 \in B$, then there is a unique $c \in A'$ such that $cb_1 = b_2$ (and hence c is even unique in K).

This is easily seen; e.g. multiply $1-a$ by the six elements of A'.

(b) The six elements of B are distinct.

To see this, note that the six (distinct) elements of A', multiplied by, say, $1-a$, give six distinct products, so B has six distinct elements.

(c) If $x, y \in K^*$, then $xB \cap yB \neq \phi$ if and only if $xy^{-1} \in A'$ (and thus if and only if $xB = yB$).

For if $xb_1 = yb_2$, for $b_1, b_2 \in B$, then $xy^{-1}b_1 = b_2$. Since there is $c \in A'$ such that $cb_1 = b_2$, it follows that $xy^{-1} = c$.

(d) The distinct sets xB, $x \in K^*$, partition K^*, and there are $(q-1)/6$ such sets.

This is an immediate consequence of (c), and of the fact that, for $k \in K^*$, k must lie in the set $k(1-a)^{-1}B$.

Now choose elements $s_1, s_2, \ldots, s_t$, where $t = (q-1)/6$, such that $s_1B, s_2B, \ldots, s_tB$ partition K^*. We will construct $\mathscr{S}$ as follows:

the points of $\mathscr{S}$ are symbols (x), for all $x \in K$;
the blocks of $\mathscr{S}$ are symbols $[s_iA + y]$ for all $i = 1, 2, \ldots, t$, and all $y \in K$;
incidence in $\mathscr{S}$ is the 'natural' rule: (x) is on $[s_iA + y]$ if $x \in s_iA + y$.

Certainly $\mathscr{S}$ has q points, and every block has three points. For any $d \in K$, the mapping

$$\alpha_d : (x) \to (x+d)$$
$$[s_iA + y] \to [s_iA + y + d]$$

is clearly an automorphism of $\mathscr{S}$, and the set of all α_d, which is a group isomorphic to $(K, +)$, is transitive on the points of $\mathscr{S}$. So to prove that $\mathscr{S}$ is a 2-structure it suffices to show that the number of blocks on (0) and (x) $(x \neq 0)$ is a constant independent of x. The block $[s_iA + y]$ contains (0) and (x) if and only if 0 and $x \in s_iA + y$, which is equivalent to:

$$-y \in s_iA \quad \text{and} \quad x - y \in s_iA. \tag{1}$$

But $x \in s_iB$ for a unique choice of i, so $x = s_ia^j - s_ia^m$; hence with $y = -s_ia^m$, equation (1) is satisfied. Conversely, if $y = -s_ja^u$ is a solution of (1), then $x \in s_jA - s_ja^u$ implies $x \in s_jB$, so $j = i$ and $a^u = a^m$.

Thus (0) and (x) are on exactly one common block and so, since it has no repeated blocks, $\mathscr{S}$ is a 2-$(q, 3, 1)$. □

Exercise 6.1

(a) Construct an STS with seven points as in Theorem 6.3 and show that it is isomorphic to $\mathscr{P}_{2,1}(2)$.

(b) Construct an STS with 19 points as in Theorem 6.3.

Exercise 6.2. Using Theorem 6.3, construct an STS with 13 points and find its full automorphism group.

We now give a purely combinatorial trick that enables us to 'paste' two STSs together.

Theorem 6.4. Let $\mathscr{S}_1$ and $\mathscr{S}_2$ be STSs with v_1 and v_2 points respectively. Then $\mathscr{S}$, defined below, is an STS with $v_1 v_2$ points.

The points of $\mathscr{S}$ are the symbols (X_1, X_2), where X_i is any point of $\mathscr{S}_i$ for $i=1, 2$, and the blocks of $\mathscr{S}$ are the following point sets:

(a) if X is a point of $\mathscr{S}_1$ and X_2, Y_2, Z_2 are three points of a block in $\mathscr{S}_2$, then (X, X_2), (X, Y_2), (X, Z_2) are the three points of a block in $\mathscr{S}$;

(b) the same as in (a), with the roles of $\mathscr{S}_1$ and $\mathscr{S}_2$ interchanged;

(c) if X_i, Y_i, Z_i are the three points of a block in $\mathscr{S}_i$ for $i=1, 2$, then the six 3-sets (X_1, A), (Y_1, B), (Z_1, C), where A, B, C is any permutation of X_2, Y_2, Z_2, are blocks of $\mathscr{S}$.

We leave the proof as a straightforward but interesting exercise.

Exercise 6.3. Prove Theorem 6.4. □

Remembering that there is an STS with three points, the results of this section enable us to construct STSs for many small values of v:

Exercise 6.4. Show that there is an STS with every permissible value of $v < 150$, with at most 10 exceptions.

Notice that from parametric considerations alone, any STS is a candidate to have an extension to a 3-design for $(v+1, 4, 1)$. They do not, in fact, all have extensions, but it is true that for every (permissible) value of v there is at least one STS which does extend to a 3-design. The interested reader will find more on this (and on STSs) in [2, 3].

6.3 Unitals

A *unital* is defined to be a 2-design for $(n^3+1, n+1, 1)$, with $n>1$. The name is derived from the fact that the first (and perhaps most 'classical') examples were constructed from the set of absolute points and non-absolute lines of a unitary polarity of a projective plane. We shall construct these unitals (for the planes $\mathscr{P}_2(q^2)$), but first we need some results from the theory of finite fields.

Theorem 6.5. Let K be the field $GF(q^2)$ and K_0 its subfield $GF(q)$. Then:

(a) $x \to x^q$ is an automorphism of K, $x^q = x$ if and only if $x \in K_0$, and $x^{q^2} = x$ for all $x \in K$.

(b) there is an element $t \in K$, $t \neq 0$, such that $t^q = -t$;

(c) if $t^q = -t$, then the set of elements $y \in K$ such that $y^q = -y$ is exactly the set tK_0;

(d) if $y \in K_0$ $(y \neq 0)$ then $x^{q+1} = y$ has exactly $q+1$ solutions $x \in K$.

Proof. (a) is well known and we omit a proof. To prove (b), first suppose q is even; then $-1 = +1$, so $t = 1$ satisfies the condition. If q is odd, then, since 2 divides $q+1$, $2(q-1)$ must divide $(q+1)(q-1) = q^2 - 1$. So the cyclic group K^* has an element t of order $2(q-1)$, and t^{q-1} has order 2; the only element of order 2 in K is -1, so $t^{q-1} = -1$, and thus $t^q = -t$.

For (c) it is certainly true that an element tk_0, for $k_0 \in K_0$, satisfies $(tk_0)^q = t^q k_0^q = -tk_0$. Conversely, if $x^q = -x$, then $(xt^{-1})^q = x^q t^{-q} = (-x)(-t^{-1}) = xt^{-1}$, so $xt^{-1} \in K_0$, by (a).

Finally, the cyclic subgroup K_0^* of the cyclic group K^* has order $q-1$ and index $q+1$ in K^*, so (d) is an easy consequence of the most elementary group theory. □

Now let $\mathscr{P} = \mathscr{P}_2(q^2)$ be the projective plane of order q^2 over the field K. As in Chapter 3, we represent the points of $\mathscr{P}$ by $\langle \mathbf{v} \rangle$ and the lines of $\mathscr{P}$ by $\langle \mathbf{w}^{\mathrm{T}} \rangle$, for non-zero vectors $\mathbf{v}$, $\mathbf{w}$ in $V_3(K)$ (having fixed a basis for $V_3(K)$). So $\langle \mathbf{v} \rangle$ is on $\langle \mathbf{w}^{\mathrm{T}} \rangle$ if and only if $\mathbf{v} \cdot \mathbf{w}^{\mathrm{T}} = 0$. If $\mathbf{v} = (x, y, z)$, we write $\mathbf{v}^q = (x^q, y^q, z^q)$.

Theorem 6.6. The mapping σ given by

$$\sigma : \langle \mathbf{v} \rangle \to \langle (\mathbf{v}^q)^{\mathrm{T}} \rangle$$
$$\langle \mathbf{w}^{\mathrm{T}} \rangle \to \langle \mathbf{w}^q \rangle$$

is a polarity of $\mathscr{P}$.

Proof. $\langle \mathbf{v} \rangle$ is on $\langle \mathbf{w}^{\mathrm{T}} \rangle$ if and only if $\mathbf{v} \cdot \mathbf{w}^{\mathrm{T}} = 0$, which is equivalent to $\mathbf{w} \cdot \mathbf{v}^{\mathrm{T}} = 0$, and this last is equivalent to $\mathbf{w}^q \cdot (\mathbf{v}^q)^{\mathrm{T}} = 0$. Thus, since σ is obviously a bijection of order 2, it is a polarity. □

We recall that a point P (line y) of $\mathscr{P}$ is absolute if it is on P^σ (y^σ) and is non-absolute otherwise, and establish an important property of the absolute points of σ.

Lemma 6.7. The line joining two distinct absolute points is non-absolute, and contains $q+1$ absolute points.

Proof. Let $P_1=\langle \mathbf{v}_1\rangle$ and $P_2=\langle \mathbf{v}_2\rangle$ be distinct absolute points. The points on the line joining these two points are exactly the points $\langle x\mathbf{v}_1+y\mathbf{v}_2\rangle$, for $x, y\in K$, not both x, y zero, and such a point is absolute if and only if

$$(x\mathbf{v}_1+y\mathbf{v}_2)\cdot(x^q(\mathbf{v}_1^q)^{\mathrm{T}}+y^q(\mathbf{v}_2^q)^{\mathrm{T}})=0.$$

Writing $a=\mathbf{v}_1\cdot(\mathbf{v}_2^q)^{\mathrm{T}}$, and remembering that P_1 and P_2 are absolute, this is equivalent to:

$$axy^q+a^qx^qy=0. \tag{1}$$

If $a=0$, then every point on the line through P_1 and P_2 is absolute. But $a=0$ implies $P_1=\langle \mathbf{v}_1\rangle$ is on $\langle(\mathbf{v}_2^q)^{\mathrm{T}}\rangle$ and $P_2=\langle \mathbf{v}_2\rangle$ is on $\langle(\mathbf{v}_1^q)^{\mathrm{T}}\rangle$. Together with P_1 on $\langle(\mathbf{v}_1^q)^{\mathrm{T}}\rangle$ and P_2 on $\langle(\mathbf{v}_2^q)^{\mathrm{T}}\rangle$, this means that

$$\langle(\mathbf{v}_1^q)^{\mathrm{T}}\rangle=\langle(\mathbf{v}_2^q)^{\mathrm{T}}\rangle,\quad \text{so } \langle \mathbf{v}_1\rangle=\langle \mathbf{v}_2\rangle.$$

This is a contradiction, so $a\neq 0$.

If the line on P_1 and P_2 is absolute, then its image under σ is an absolute point on the same line, so we can, without loss of generality, assume that it is P_1 itself. But then $P_2=\langle \mathbf{v}_2\rangle$ is on $P_1^\sigma=\langle(\mathbf{v}_1^q)^{\mathrm{T}}\rangle$ which implies that $a=0$. So the line is not absolute.

Equation (1) can be written:

$$az+a^qz^q=0, \tag{2}$$

where $z=x/y$, as long as $y\neq 0$. But $y=0$ corresponds to the point P_1, so any other absolute points on the line through P_1 and P_2 correspond to solutions of (2). Since (2) is merely an equation $t^q=-t$, with $t=az$, there are q solutions $t\in K$. So our line has $q+1$ absolute points on it, and is non-absolute, as claimed. □

Now we define $\mathscr{U}=\mathscr{U}(q)$ to be the structure whose points are the absolute points in $\mathscr{P}$ and whose blocks are the lines of $\mathscr{P}$ which contain at least two absolute points. We have already proved:

Lemma 6.8. $\mathscr{U}$ is a 2-design for $(v, q+1, 1)$, where v is the number of absolute points in $\mathscr{P}$. □

Our only remaining problem is to find v.

Exercise 6.5. Show that the number of points $\langle(x, y, z)\rangle$ in $\mathscr{P}$ whose coordinates satisfy

$$x^{1+q}+y^{1+q}+z^{1+q}=0$$

is q^3+1. Thus show that $\mathscr{U}$ is a 2-$(q^3+1, q+1, 1)$, and that every non-absolute line in $\mathscr{P}$ is a block of $\mathscr{U}$.

The solution of Exercise 6.5 is elementary but fussy. Using the same sort of reasoning as in the proof of Lemma 6.7, we give an alternate proof that every non-absolute line is a block of $\mathscr{U}$.

Lemma 6.9. Every non-absolute line of $\mathscr{P}$ contains $q+1$ absolute points.

Proof. Let L be a non-absolute line and $P_1=\langle \mathbf{v}_1\rangle$ a non-absolute point on L. The line P_1^σ meets L in a point $P_2\neq P_1$; if P_2 were absolute, then, since P_2 is on P_1^σ, P_2^σ would be on P_1 and certainly on P_2, so P_2^σ would be L, contradicting the fact that L is non-absolute. Let $P_2=\langle \mathbf{v}_2\rangle$. We have $\mathbf{v}_1\cdot(\mathbf{v}_2^q)^{\mathrm{T}}=0$, so also $\mathbf{v}_2\cdot(\mathbf{v}_1^q)^{\mathrm{T}}=0$.

An arbitrary point Q on L has the form $Q=\langle x\mathbf{v}_1+y\mathbf{v}_2\rangle$, and if $Q\neq P_1$, then $y\neq 0$. Q is absolute if and only if

$$(x\mathbf{v}_1+y\mathbf{v}_2)\cdot(x^q(\mathbf{v}_1^q)^{\mathrm{T}}+y^q(\mathbf{v}_2^q)^{\mathrm{T}})=0.$$

Writing $a_i=\mathbf{v}_i\cdot(\mathbf{v}_i^q)^{\mathrm{T}}$ for $i=1, 2$, we see that neither a_i is zero, since neither P_i is absolute. Then the condition that Q be absolute is $a_1x^{1+q}+a_2y^{1+q}=0$. If $\mathbf{v}_1=(u, w, z)$, then $a_1=u^{1+q}+w^{1+q}+z^{1+q}$, and hence $a_1^q=a_1$ (from Theorem 6.5(a)); so both a_i are in K_0. Thus the condition that Q be absolute is equivalent to

$$(x/y)^{1+q}=-a_2/a_1,$$

where $-a_2/a_1\in K_0$. Thus, by Theorem 6.5(d), there are $q+1$ solutions x/y and, consequently, $q+1$ absolute points on L. □

This gives us an alternative way of showing that $\mathscr{U}$ has the parameters stated in Exercise 6.5.

Theorem 6.10. $\mathscr{U}$ is a 2-$(q^3+1, q+1, 1)$.

Proof. The number of points in $\mathscr{U}$ is the number of absolute points, v, say. So there are v absolute lines, hence q^4+q^2+1-v non-absolute lines, all of which are blocks of $\mathcal{U}$, by Lemma 6.9. This gives us:

$$b=q^4+q^2+1-v=\frac{v(v-1)}{(q+1)q},$$

which simplifies to:

$$(v-(q^3+1))(v-q(q^2+q+1))=0.$$

Thus $v=q^3+1$ or $v=q(q^2+q+1)$. But $v=q(q^2+q+1)$ implies that in $\mathcal{U}$, $r=(v-1)/q=q^2+q+1-1/q$, which is not an integer. Thus $v=q^3+1$. □

Exercise 6.6. Show that there is only one unital, up to isomorphism, with $n=2$, and that it is isomorphic to the affine plane of order 3. Hence conclude that the projective plane of order 4 contains a subdesign isomorphic to the affine plane of order 3.

The unitals $\mathcal{U}(q)$ that we have constructed are not the only ones known (but no 2-$(n^3+1, n+1, 1)$ has ever been found in which n is not a prime power). The following exercise shows how to construct some more.

Exercise 6.7.* Show that if $\mathscr{P}$ is any projective plane of order n^2 and if σ is a polarity with n^3+1 absolute points then the absolute points and non-absolute lines form a unital with parameters 2-$(n^3+1, n+1, 1)$.

There are many examples of planes admitting such polarities and the interested reader should consult [1].

6.4 Groups and more 2-designs

In this section we generalise the construction of STSs given in Theorem 6.3 to a wider class of 2-designs. This class will also include the Paley–Hadamard 2-designs $\mathscr{L}(q)$ of Chapter 3. The method used involves more group theory than in the construction used in Theorem 6.3, but we will give elementary proofs wherever possible. First we need a technical result:

Lemma 6.11. Let K be the finite field $GF(q)$, A a subgroup of the multiplicative group K^*, $A \neq 1$. Then if $s \in K^*$ and $y \in K$ satisfy $sA + y = A$, it follows that $y = 0$ and $s \in A$.

Proof. We write $\mu_{b,c}$, for $b \in K^*$ and $c \in K$, for the permutation $x\mu_{b,c} = bx + c$ of the elements of K. The group G of all such permutations is 2-transitive on K and only the identity fixes two symbols of K. The element $\mu_{s,y}$ has the property $A\mu_{s,y} = A$ and, since certainly $A\mu_{b,0} = A$, for all $b \in A$, the group F generated by all $\mu_{b,0}$, for $b \in A$ and by $\mu_{s,y}$, fixes A and is transitive on A. If we write A_0 for the subgroup of F consisting of all elements $\mu_{b,0}$ with $b \in A$, then $A_0 \cong A$ and A_0 is transitive on A.

Clearly, $F = A_0 F_1$, where F_1 is the subgroup of F fixing the symbol $1 \in A$. Only the identity in F can fix as many as two symbols in A (since the whole group G has that property acting on all of K), so F is a Frobenius group and A_0 is its 'Frobenius kernel', i.e. A_0 is normal in F.

If the conclusions of the lemma do not hold, then $F \neq A_0$; i.e. there is an element $\mu_{t,u}$ in F_1 ($\mu_{t,u} \neq$ identity). So

$$1\mu_{t,u} = t + u = 1,$$

hence $u = -t + 1$. It is straightforward to see that $\mu_{t,u}^{-1} = \mu_{t^{-1}, -ut^{-1}}$ and, since $\mu_{t,u}$ must normalise A_0, if we choose $a \in A$ $(a \neq 1)$, then

$$\mu_{t,u}^{-1}\mu_{a,0}\mu_{t,u} = \mu_{b,0}$$

for some $b \in A$. Using $\mu_{x,y}\mu_{v,w} = \mu_{xv, yv+w}$, this yields

$$\mu_{a,u-au} = \mu_{b,0},$$

hence $a = b$, $u - au = 0$. So $u = au$, which, since $a \neq 1$, implies $u = 0$ and then $t = 1$. Thus $F = A_0$ and so $s \in A$, $y = 0$. □

(The reader with some group theoretic background might like to find more elegant proofs of this lemma.) Next we need a combinatorial result, which has considerable interest beyond our use of it here.

Theorem 6.12. Suppose T is a group and there are subsets $A_1, A_2, \ldots, A_n$ of T such that for any $t \in T$ $(t \neq 1)$, there are exactly μ pairs a_{ij}, a_{ik}, where a_{ij}, a_{ik} are in the same set A_i, such that $t = a_{ij}a_{ik}^{-1}$ (here i is not a constant, but takes on different values for a fixed t). Then the structure $\mathscr{S}$, whose points are the symbols (t), for all $t \in T$, and whose blocks are the symbols $[A_i s]$, for $i = 1, 2, \ldots, n$ and $s \in T$, with incidence given by: (t) is in $[A_i s]$ if $t \in A_i s$, is a 2-structure with μ blocks on two points. $\mathscr{S}$ is uniform if and only if all the A_i have the same number of elements.

Proof. First we note that T induces (faithfully) an automorphism group of $\mathscr{S}$, since the mapping $(t) \to (tb)$, $[A_i s] \to [A_i sb]$, for any $b \in T$, is an automorphism of $\mathscr{S}$. Thus to check that $\mathscr{S}$ is a 2-structure, it suffices to check that (1) and (t), for $t \neq 1$, are on μ blocks. It is easily seen that the μ pairs a_{ij}, a_{ik} such that $t = a_{ij} a_{ik}^{-1}$ give us the μ blocks, where $s = a_{ik}^{-1}$. □

Now we proceed to our construction. Let K be the field $GF(q)$, and let A be a subgroup of K^*, of order m, where $2 < m < q-1$. The group $(K, +)$ will play the role of T in Theorem 6.12 and we shall choose elements $s_1, s_2, \ldots, s_n$ in K^* such that $As_1, As_2, \ldots, As_n$ play the role of the A_i in Theorem 6.12. Since T is written additively, the condition of Theorem 6.12 is that for any $t \in K$ there should be μ choices of $a_i, a_k \in A$ and of s_j such that $t = a_i s_j - a_k s_j$.

Let B be the set of all elements $a_i - a_j$, with $a_i, a_j \in A$ $(a_i \neq a_j)$. (We want to consider B as a sort of 'multiset': if there are repeated elements of B, we include them as often as they occur, so B consists of $m(m-1)$ elements.) Let A' be the subgroup of K^* that results from adjoining -1 to A; so

(a) if q is even, $A' = A$;
(b) if q is odd and m is even, $A' = A$;
(c) if both q and m are odd, then $A' = \langle -1, A \rangle$ has order $2m$.

Now let $A's_1, A's_2, \ldots, A's_n$ be the distinct cosets of A' in K^*, so that $n = (q-1)/m$ if q is even or m is even, and $n = (q-1)/(2m)$ if both q and m are odd.

Lemma 6.13. If $h \in A'$, then $Bh = B$.

The proof is trivial, and we omit it. □

Lemma 6.14. If $x \in K^*$ and $A's_i x = A's_j$, then $Bs_i x = Bs_j$.

Proof. $A's_i x = A's_j$ if and only if $s_i x = hs_j$ for some $h \in A'$. So $Bs_i x = Bhs_j = Bs_j$. □

Lemma 6.15. If $x \in K^*$ and x occurs μ times in the union of the sets Bs_i, then every element of K^* occurs μ times in the union of the sets Bs_i.

Proof. If $y \in K^*$, then $y = xt$ for some $t \in K^*$. Since x occurs μ times in the union of the Bs_i, $xt=y$ occurs μ times in the union of the Bs_it. But from Lemma 6.14, the sets Bs_it, as i varies, are the sets Bs_j, permuted in some order. □

Lemma 6.16. $As_i+y=As_j+z$ if and only if $s_i=s_j$ and $y=z$.

Proof. From $As_i+y=As_j+z$ we deduce $As_is_j^{-1}+(y-z)s_j^{-1}=A$, so, from Lemma 6.11, we have $y=z$ and $s_is_j^{-1}\in A$. But $s_is_j^{-1}\in A$ implies $s_is_j^{-1}\in A'$, so $A's_i=A's_j$, and thus $s_i=s_j$. □

Theorem 6.17. Let $\mathscr{S}$ be a structure whose points are (x), for all $x\in K$, whose blocks are $[As_i+y]$ for all $i=1, 2, \ldots, n$, and for all $y\in K$, with the natural incidence rule: (x) is on $[As_i+y]$ if $x\in As_i+y$. Then $\mathscr{S}$ is a 2-design for $(q, m, m-1)$ if q or m is even, and is a 2-design for $(q, m, (m-1)/2)$ if both q and m are odd.

Proof. The conditions of Theorem 6.12 are satisfied, with $T=(K, +)$ and $A_i=As_i$, by Lemma 6.15. Clearly, $\mathscr{S}$ is uniform with block size m. Lemma 6.16 implies that $\mathscr{S}$ has no repeated blocks. Obviously the number of blocks in $\mathscr{S}$ is nq and the value of λ can be computed from this. □

Corollary 6.18. If $q\equiv 3 \pmod 4$ and $m=(q-1)/2$, then $\mathscr{S}$ is a Hadamard 2-design. □

Exercise 6.8. Show that the Hadamard 2-design of Corollary 6.18 is isomorphic to $\mathscr{L}(q)$ (see Chapter 3).

Corollary 6.19. If $q\equiv 1 \pmod 6$ and $m=3$, then $\mathscr{S}$ is an STS. □

Exercise 6.9. Show that the STS of Corollary 6.19 is isomorphic to the STS constructed in Theorem 6.3.

Exercise 6.10. Construct 2-designs with parameters (9, 4, 3), (13, 6, 5), (16, 3, 2) and (29, 7, 3).

Exercise 6.11. Show that the design $\mathscr{S}$ of Theorem 6.17 has an automorphism group of order mq.

Exercise 6.12. Let $\mathscr{S}$ be the design of Theorem 6.17, with $q \equiv 1 \pmod 4$ and $m = (q-1)/2$.

(a) Show that Aut ($\mathscr{S}$) is 2-transitive on the points of $\mathscr{S}$ and transitive on its blocks, but never 2-transitive on the blocks.

(b) Show that $\mathscr{S}$ always has an extension to a 3-design $\mathscr{S}^*$, but that $\mathscr{S}^*$ never has an extension to a 4-design unless $q = 9$.

Exercise 6.13.* As in Exercise 6.12, let $q = 9$, $m = 4$.

(a) Find the full group Aut ($\mathscr{S}$).

(b) Show that there is an extension $\mathscr{S}^*$ of $\mathscr{S}$ such that Aut ($\mathscr{S}^*$) is 3-transitive on points, and such that $\mathscr{S}^*$ has an extension to a 4-design.

(Compare with Exercise 4.7; the design $\mathscr{S}$ of Exercise 6.13 is, in fact, isomorphic to the design $\mathscr{S}_{X,Y}$ of Exercise 4.7.)

References

Although it does not contain the original proof [4] is a suitable reference for a proof of the fact that there exists an STS for every admissible value of v. [2] and [3] contain discussion of STSs and the problem of extending them to Steiner quadruple systems. Finally [1] provides an excellent survey of unitals in projective planes.

[1] Ganley, M. J. 'On correlations and collineations of projective planes'. Ph.D. Thesis, University of London, 1971.

[2] Hanani, H. 'On quadruple systems'. *Canad. J. Math.*, **12** (1961), 145–57.

[3] Hanani, H. 'The existence and construction of balanced incomplete block designs'. *Ann. Math. Statist.*, **32** (1961), 361–8.

[4] Moore, E. H. 'Concerning triple systems'. *Math. Ann.*, **43** (1961), 271–85.

7 Some 1-designs

7.1 Introduction

In general, 1-designs are less interesting than t-designs with $t > 1$; it is possible to construct them easily and there do not seem to be many deep theorems about them. But with certain extra properties imposed upon them, 1-designs can become complicated and important objects, in particular with crucial connections to group theory and geometry.

Among the most important 1-designs are *generalised quadrangles*, which are studied in Section 7.3 as special members of a class of 1-designs called *Γ_α-geometries*, introduced in Section 7.2. Using SR graphs we prove some elementary results about Γ_α-geometries, and we give some infinite families of examples. In particular, we develop two infinite families of generalised quadrangles, one classical (in the sense that it comes from a polarity of a projective geometry), the other not. The other classical generalised quadrangles involve deeper projective geometry and algebra, and we do not include them. (The surprisingly rich and complex theory of generalised quadrangles, and their connections to group theory as well as geometry, is beyond the scope of this book; some of the flavour of the subject is all that we can impart.) There are many unsolved problems (about existence, non-existence, and structure) in the area of Γ_α-geometries.

In Section 7.4 we study *semisymmetric designs*, which are 1-designs that generalise (and include) symmetric designs. Besides a number of examples, the section contains results about upper and lower bounds on the number of points in a semisymmetric design, and touches upon the many open questions in this area. In Section 7.5 we consider *divisible* semisymmetric designs (historically often called *group divisible*, a term we avoid: besides being unnecessary, it is certainly misleading, since it has nothing to do with groups). Here there are much sharper bounds on the number of points, and also a Bruck–Ryser–Chowla-type non-existence theorem. But again, there are many unsolved and interesting problems about divisible semisymmetric designs.

7.2 Semilinear spaces and Γ_α-geometries

A connected structure $\mathcal{S}$ which satisfies:

(SL1) $\mathcal{S}$ contains at least two points and at least two blocks;
(SL2) two distinct points of $\mathcal{S}$ are on at most one common block;
(SL3) every block contains at least two points and every point is on at least two blocks;

is called a *semilinear space.* We shall call the blocks of a semilinear space *lines*, and say that two points on a common line are *collinear.*

Lemma 7.1. If $\mathcal{S}$ is a semilinear space, then $\mathcal{S}$ has no repeated lines and $\mathcal{S}^{\mathrm{T}}$ is also a semilinear space.

Proof. If y and z are lines of $\mathcal{S}$, containing the distinct points P and Q, then from (SL2) above we have $y=z$. This implies that $\mathcal{S}$ has no repeated lines, and that $\mathcal{S}^{\mathrm{T}}$ is a semilinear space as well. □

A semilinear space is called *linear* if any two distinct points are on exactly one common line.

A semilinear space $\mathcal{S}$ which is both regular and uniform is a *semilinear design.* (This is a slight abuse of our terminology, since $\mathcal{S}$ is more than a design, it is a 1-design as well.) If $\mathcal{S}$ is a semilinear design with $s+1$ points on every line and $t+1$ lines on every point, then we say that $\mathcal{S}$ has *order* (s, t). Among the semilinear designs, an interesting and important subclass are the Γ_α-*geometries*: a semilinear design $\mathcal{S}$ is a Γ_α-geometry if for any point P and any line y not on P, there are exactly α points on y which are collinear with P. Obviously in a Γ_α-geometry, we always have $\alpha>0$, since otherwise $\mathcal{S}$ would only have one line.

Exercise 7.1

(a) Let $\mathcal{S}$ be the structure of points and lines in a projective geometry $\mathcal{P}(n, q)$. Observe that $\mathcal{S}$ is a linear space and show that $\mathcal{S}^{\mathrm{T}}$ (which is always semilinear) is a linear space if and only if $n=2$.

(b) Let $\mathcal{S}$ be any set of points and lines in $\mathcal{P}(n, q)$. What conditions are necessary for $\mathcal{S}$ to be a semilinear space?

(c) Let $\mathcal{P}=\mathcal{P}(n, q)$ and let H, P be, respectively, a fixed hyperplane and a fixed point in $\mathcal{P}$. Let $\mathcal{S}$ be the structure whose points are the points of

$\mathscr{P}$ not in H and not equal to P and whose lines are the lines of $\mathscr{P}$ not in H and not containing P. Show that $\mathscr{S}$ is a semilinear design and find its order.

(d) Let $\mathscr{S}$ be the structure of (c) above. Show that $\mathscr{S}$ is a Γ_α-geometry if and only if $n=2$ and P is on H, and in this case find α. (Note that then $\mathscr{S}$ is merely an affine plane with one parallel class removed.)

(e) Let $\mathscr{N}$ be a net of order n with m parallel classes, $m>1$. Show that $\mathscr{N}$ is a Γ_{m-1}-geometry of order $(n-1, m-1)$.

As we shall see, nets are the only Γ_t-geometries of order (s, t), and this, of course, will characterise the Γ_s-geometries of order (s, t) as well.

Theorem 7.2. If $\mathscr{S}$ is a Γ_α-geometry of order (s, t), then $\mathscr{S}$ has v points and b lines, where

$$v=(s+1)(st+\alpha)/\alpha,$$

$$b=(t+1)(st+\alpha)/\alpha.$$

Proof. Since $\mathscr{S}^{\mathrm{T}}$ is a Γ_α-geometry of order (t, s), it suffices to prove that v has the indicated form. Now let Γ be the point-adjacency graph of $\mathscr{S}$ (see Sections 1.7 and 3.7). We shall prove that Γ is strongly regular.

If P is a point of $\mathscr{S}$, then each of the $t+1$ lines through P carries s points other than P, so P is adjacent to $s(t+1)$ vertices in Γ, hence Γ is regular of degree $s(t+1)$. Suppose P and Q are distinct collinear points of $\mathscr{S}$ (which is equivalent to saying that P and Q are adjacent in Γ). Then there are $s-1$ points X on the line which joins P and Q. For any line y on P which does not contain Q, there are α points on y joined to Q, one of which is P; so for each of the t lines y (on P but not containing Q) there are $\alpha-1$ points $X\neq P$ on y, but collinear with Q. Thus P and Q are joined to exactly $s-1+t(\alpha-1)$ points X.

If P and Q are points of $\mathscr{S}$ which do not lie on a common line, then for each of the $t+1$ lines y on P there are α points X on y which are collinear with Q, so P and Q are joined to $\alpha(t+1)$ common points. Now Γ is SR with parameters $(v, s(t+1), s-1+t(\alpha-1), \alpha(t+1))$. From Theorem 3.39 we have:

$$s(t+1)[s(t+1)-s+1-t(\alpha-1)-1]=\alpha(t+1)[v-s(t+1)-1].$$

Solving this for v, we find $v=(s+1)(st+\alpha)/\alpha$. □

Now if $\alpha=s+1$ then, clearly, $\mathscr{S}$ is merely a 2-design for $(v, s+1, 1)$, and conversely. Furthermore, the graph Γ is trivial in that case. In fact:

Lemma 7.3. The graph Γ of Theorem 7.2 is a trivial SR graph if and only if $\alpha = s$ or $s+1$. If $\mathscr{S}$ is a Γ_s-geometry of order (s, t) then the relationship $\sim$ given by: '$P \sim Q$ if and only if $P = Q$ or P and Q are not collinear', is an equivalence relation on the points, and each equivalence class consists of $t+1$ points.

Proof. Clearly Γ is always connected, so Γ is a trivial SR graph if and only if its complement Γ^c is not connected. Now the complement of an SR graph with parameters (v, k, λ, μ) has parameters $(v, v-k-1, \lambda', \mu')$, where $\mu' = v - 2k + \lambda$. Then it is immediate that Γ^c is not connected if and only if $\mu' = 0$, which is equivalent to

$$\frac{(s+1)(st+\alpha)}{\alpha} - 2s(t+1) + s - 1 + t(\alpha - 1) = 0.$$

Simplifying, this last equation becomes:

$$t\alpha^2 - t(2s+1)\alpha + t(s^2+s) = 0$$

and thus:

$$\alpha^2 - (2s+1)\alpha + s(s+1) = 0.$$

The zeros of this equation are $\alpha = s$ and $s+1$.

Now suppose $\mathscr{S}$ is a Γ_s-geometry of order (s, t). The fact that the graph Γ^c is not connected means that if a point X is collinear with P, then X is collinear with every point Q which is *not* collinear with P (check this!). Now suppose that $P \sim Q$, $P \sim R$ $(Q \neq R)$; choose a pair of lines y, z such that y is on Q and z is on R. Then Q is the unique point on y not collinear with P, and R is the unique point on z not collinear with P. Hence any point $X \neq Q$ on y must be collinear with R, so Q is the unique point on y not collinear with R. Thus $Q \sim R$. The rest of the proof that $\sim$ is an equivalence relation is trivial.

Let $\mathscr{C}$ be an equivalence class under $\sim$. Then if y is any line of $\mathscr{S}$, y must meet $\mathscr{C}$ exactly once: for if $P \in \mathscr{C}$, then either y is on P, or there is a unique point Q on y such that $Q \sim P$, so $Q \in \mathscr{C}$. Then we count pairs (P, y), where $P \in \mathscr{C}$ and y is on P. If m is the number of points in $\mathscr{C}$, then there are m choices of P and $t+1$ choices of y on P; on the other hand, there are b choices of y, and one point P for each y. Substituting the value of b from Theorem 7.2, we have $m = t+1$. □

The reader might notice that an alternate proof of the second part of Lemma 7.3 could be given without using the graph Γ, perhaps most conveniently by dualising and noting that then if P is a point and y a line, P not on y, there must exist exactly one line on P that does not meet y.

Theorem 7.4. If $\mathscr{S}$ is a Γ_α-geometry of order (s, t), then $s+t-\alpha+1$ divides $st(s+1)(t+1)/\alpha$.

Proof. If the graph Γ is trivial, then α is s or $s+1$, and it is easy to see that the theorem is (more or less trivially) true. If Γ is not trivial we can use Theorem 3.44 and, in particular, the polynomial $f(x)$ which gives us the eigenvalues.

Using the parameters of Γ from the proof of Theorem 7.2, we compute $f(x)$:

$$f(x)=x^2-(s-1+t(\alpha-1)-\alpha(t+1))x-(s(t+1)-\alpha(t+1)),$$

which simplifies to $f(x)=(x-(s-\alpha))(x+(t+1))$, and so the eigenvalues of Γ are $s-\alpha$, $-(t+1)$. The multiplicities m_1, m_2 must satisfy

$$m_1+m_2=(s+1)(st+\alpha)/\alpha-1, \quad (s-\alpha)m_1-(t+1)m_2=-s(t+1).$$

Multiply the first of these equations by $t+1$ and add to the second; this gives

$$(s+t-\alpha+1)m_1=st(s+1)(t+1)/\alpha$$

and, since m_1 is an integer, this proves the theorem. □

Now we can use Lemma 7.3 to characterise Γ_α-geometries of order (s, t) if $\alpha=s$ or t.

Theorem 7.5. If $\mathscr{S}$ is a Γ_t-geometry of order (s, t), then $\mathscr{S}$ is a net; so if $\mathscr{S}$ is a Γ_s-geometry of order (s, t), then $\mathscr{S}$ is the dual of a net.

Proof. Suppose $\mathscr{S}$ is a Γ_t-geometry of order (s, t). Then, dualising the result of Lemma 7.3, the lines of $\mathscr{S}$ can be partitioned into equivalence classes of $s+1$ lines each such that two lines in the same class do not meet and two lines from different classes meet in one point. From Theorem 7.2, $\mathscr{S}$ has $(s+1)^2$ points and $(s+1)(t+1)$ lines, so the number of equivalence classes of lines is $t+1$. Now (see Section 3.4) it follows immediately that $\mathscr{S}$ is a net of order $s+1$ with $t+1$ parallel classes. The other half of the theorem is the dual. □

Suppose $\mathscr{S}$ is a non-empty subdesign of the points and lines in a projective geometry $\mathscr{P}$, with the property that any line of $\mathscr{P}$ which contains one point of $\mathscr{S}$ must contain exactly k points of $\mathscr{S}$, where $k>1$. Then $\mathscr{S}$ is called a *saturated subdesign* of $\mathscr{P}$. (If $n=2$, then these are called *maximal* (v, k)*-arcs* in the literature, where v is the number of points of $\mathscr{S}$.)

Exercise 7.2. Show that the set $\mathscr{S}$ of points and lines of an affine geometry $\mathscr{A}(n, q)$ forms a saturated subdesign in $\mathscr{P}(n, q)$.

Exercise 7.3. Let $\mathscr{S}$ be a subset of the points and lines in $\mathscr{P}(n, q)$, and let r be the number of lines of $\mathscr{P}(n, q)$ on a point of $\mathscr{P}(n, q)$ (so $r=(q^n-1)/(q-1)$). Show that the following are equivalent:

(a) $\mathscr{S}$ is a saturated subdesign of $\mathscr{P}(n, q)$;
(b) $\mathscr{S}$ is a 2-design for $((k-1)r+1, k, 1)$, with $k>1$;
(c) $\mathscr{S}$ is a 1-design for (v, k, r), for some v, with $k>1$.

Theorem 7.6. Suppose $\mathscr{P}=\mathscr{P}(n, q)$ and that H is a fixed hyperplane in $\mathscr{P}$. Let $\mathscr{S}$ be a saturated subdesign in $H \cong \mathscr{P}(n-1, q)$. Define $\mathscr{D}=\mathscr{D}(\mathscr{S})$ to be the structure of all points of $\mathscr{P}$ not on H and of all lines of $\mathscr{P}$ not in H but containing a point of $\mathscr{S}$. If $\mathscr{S}$ has parameters 2-$(v, k, 1)$ then $\mathscr{D}$ is a Γ_{k-1}-geometry of order $(q-1, v-1)$.

Proof. Certainly the lines of $\mathscr{D}$ contain q points, since each has 'lost' the point where it met the hyperplane H. A point of $\mathscr{D}$ is joined to the v points of $\mathscr{S}$ by v distinct lines, since each line not in H meets H only once. So the order of $\mathscr{D}$ is $(q-1, v-1)$.

If y is a line of $\mathscr{D}$ and P is a point of $\mathscr{D}$ not on y, then the plane U generated by P and y meets H in a line z and, since y must contain a point X of $\mathscr{S}$, it follows that z contains X as well. So z is a line of $\mathscr{S}$ and contains $k-1$ points $X_1, X_2, \ldots, X_{k-1}$ of $\mathscr{S}$ other than X. All the X_i lie in U, so the lines PX_i are all in U and each meets y (for PX_i and y are two lines in the projective plane U). Since any line on P that meets y must lie in U and must meet $\mathscr{S}$, it follows that there are exactly these $k-1$ lines on P which meet y. □

In fact, as the next theorem shows, the most interesting saturated subdesigns in projective geometries occur only in projective planes, and are thus maximal (v, k)-arcs.

Theorem 7.7. Let $\mathscr{S}$ be a saturated subdesign in $\mathscr{P}(n, q)$, with $n>2$. Then $\mathscr{S}$ is the set of points and lines of an affine geometry $\mathscr{A}(n, q)$ contained in $\mathscr{P}(n, q)$, or $\mathscr{S}$ contains all the points and lines of $\mathscr{P}(n, q)$.

Proof. Let $\mathscr{P}=\mathscr{P}(n, q)$, with $n>2$, and let $\mathscr{S}$ be a saturated subdesign of $\mathscr{P}$, with k points on a line. It is easy to see that if $k=q+1$, then $\mathscr{S}$ contains all points and lines of $\mathscr{P}$, so let us suppose $k \leqslant q$. If H is a hyperplane of $\mathscr{P}$ and H contains a point X of $\mathscr{S}$, then, since every line of H which lies on X must be in $\mathscr{S}$, it follows immediately that the points and lines of $H \cap \mathscr{S}$ form a saturated subdesign of $H \cong \mathscr{P}(n-1, q)$, which also has k points on a line. To finish the proof we use induction and show (a) if $k=q$ then $\mathscr{S}$ is the set of points and lines of $\mathscr{P}$ lying in some affine geometry $\mathscr{A}(n, q)$ in $\mathscr{P}$, and (b) if $n=3$, then $k=q$.

(a) Suppose $k=q$. Then it follows from Exercise 7.3 that $\mathscr{S}$ has $v=(k-1)r+1$ points, i.e. that $v=(q-1)(q^n-1)/(q-1)+1=q^n$. Any hyperplane H which meets $\mathscr{S}$ must meet it in a saturated subdesign with q points on a line. Let us count flags (X, H), where X is in $\mathscr{S}$ and H is a hyperplane on X. On the one hand, the number of such flags is $q^n(q^n-1)/(q-1)$, since there are $(q^n-1)/(q-1)$ hyperplanes on a point. On the other hand, if N is the number of hyperplanes that meet $\mathscr{S}$, then the number of such flags is Nq^{n-1}. Solving for N, we have $N=q^n+q^{n-1}+\cdots+q=(q^{n+1}-1)/(q-1)-1$. Since $\mathscr{P}$ contains $(q^{n+1}-1)/(q-1)$ hyperplanes, there is a unique hyperplane not meeting $\mathscr{S}$ and, clearly, $\mathscr{S}$ is the set of points and lines not in this hyperplane; i.e. $\mathscr{S}$ is the set of points and lines in an $\mathscr{A}(n, q)$ in $\mathscr{P}$.

(b) Now let $n=3$ and suppose $\mathscr{S}$ is a saturated subdesign in $\mathscr{P}=\mathscr{P}(3, q)$, with k points on a line. Each point of $\mathscr{P}$ is on q^2+q+1 lines and on q^2+q+1 planes of $\mathscr{P}$. So $\mathscr{S}$ has $v=(q^2+q+1)(k-1)+1$ points. If H is a plane of $\mathscr{P}$ which meets $\mathscr{S}$, then H contains $v^*=(q+1)(k-1)+1$ points of $\mathscr{S}$, since $H \cap \mathscr{S}$ gives a saturated subdesign in H, and each point of H is on $q+1$ lines of H.

Now we count flags (X, H), where X is in $\mathscr{S}$ and H is a plane on X. On the one hand, the number of flags is $v(q^2+q+1)$, and, on the other hand, it is $M_2[(q+1)(k-1)+1]$, where M_2 is the number of planes of $\mathscr{P}$ that meet $\mathscr{S}$. So:

$$(q+1)(k-1)+1 \text{ divides } (q^2+q+1)[(q^2+q+1)(k-1)+1]. \tag{1}$$

Now choose a point Q not in $\mathscr{S}$ and count flags (X, H), where X is in $\mathscr{S}$ and H is a plane on Q and X. A plane on Q and X is on the line QX, and there are $q+1$ planes on a line. Hence if M_1 is the number of planes on Q which meet $\mathscr{S}$, we have $v(q+1)=M_1[(q+1)(k-1)+1]$, and thus:

$$(q+1)(k-1)+1 \text{ divides } (q+1)[(q^2+q+1)(k-1)+1]. \tag{2}$$

But $q+1$ and q^2+q+1 are relatively prime, so (1) and (2) together imply

$$(q+1)(k-1)+1 \text{ divides } (q^2+q+1)(k-1)+1. \tag{3}$$

Since $(q^2+q+1)(k-1)+1=q^2(k-1)+(q+1)(k-1)+1$, (3) implies

$$(q+1)(k-1)+1 \text{ divides } (k-1)q^2. \tag{4}$$

Now taking congruences modulo $(q+1)(k-1)+1$, we have $(k-1)q \equiv -(k-1)-1$, so $(k-1)q^2 \equiv -(k-1)q-q \equiv -(q-k)$. Thus:

$$(q+1)(k-1)+1 \text{ divides } q-k. \tag{5}$$

Now if $q=k$, then, from (a), $\mathscr{S}$ is the set of points and lines of an affine geometry $\mathscr{A}(3, q)$. If $q>k$, then (5) implies that $(q+1)(k-1)+1 \leqslant q-k$, or $(k-2)q \leqslant -2k<0$, which is impossible, since $k \geqslant 2$, so $0 \leqslant (k-2)q$.

Hence, by induction on n, the theorem is proved. □

Exercise 7.4

(a) Let $\mathscr{S}$ be the saturated subdesign of points and lines of an $\mathscr{A}(n-1, q)$ in $\mathscr{P}(n-1, q)$, and $\mathscr{D}=\mathscr{D}(\mathscr{S})$, as in Theorem 7.6. Find the parameters of $\mathscr{D}$.

(b) Let $\mathscr{S}$ be a saturated subdesign with k points on a line, in the projective plane $\mathscr{P}$ of order n. Show that k divides n.

Exercise 7.5.* In the following exercise, let $q=2^n$, and $\mathscr{P}=\mathscr{P}_2(q)$.

(a) Let $\mathscr{O}'$, be the set of points $\langle (x, y, z) \rangle$ in $\mathscr{P}$ which satisfy the equation $xy+yz+zx=0$. Show that $\mathscr{O}'$ consists of exactly $q+1$ points.

(b) Show that a line of $\mathscr{P}$ meets $\mathscr{O}'$ in at most two points.

(c) If $\langle (a, b, c) \rangle$ is a point of $\mathscr{O}'$, then show that every line, but exactly one, on $\langle (a, b, c) \rangle$ meets $\mathscr{O}'$ again; the unique line not meeting $\mathscr{O}'$ again is called the *tangent line to* $\mathscr{O}'$ at $\langle (a, b, c) \rangle$. (Hint: show that the tangent line has coordinates $\langle (b+c,\ a+c,\ a+b)^{\mathrm{T}} \rangle$, and that if a point $\langle (x, y, z) \rangle$ is on both $\mathscr{O}'$ and this last line, then $x\sqrt{(b+c)}= y\sqrt{(a+c)}=z\sqrt{(a+b)}$, and so $\langle (x, y, z) \rangle$ is uniquely determined and must be $\langle (a, b, c) \rangle$.)

(d) Show that $\langle (1, 1, 1) \rangle$ is on every tangent line to $\mathscr{O}'$, and hence $\mathscr{O}=\mathscr{O}' \cup \langle (1, 1, 1) \rangle$ is a set of $q+2$ points in $\mathscr{P}$, no three collinear.

(e) Show that $\mathscr{O}$, together with the lines on any two of its points, is a saturated subdesign of $\mathscr{P}$ with $k=2$. (In a projective plane of even order, the point set of a saturated subdesign with $k=2$ is a *hyperoval*.)

The point of the lengthy exercise above is to show that hyperovals exist in $\mathscr{P}_2(q)$, if q is even (from Exercise 7.4(b), they do not exist if q is odd). For from hyperovals we can construct Γ_α-geometries in more than one way, and including a very interesting class of Γ_1-geometries.

Theorem 7.8

(a) Let H be a fixed plane in $\mathscr{P}(3, q)$, where q is even, let $\mathscr{O}$ be a hyperoval in $H \cong \mathscr{P}_2(q)$ and $\mathscr{S}$ the saturated subdesign in H consisting of the points of $\mathscr{O}$ and the lines on two points of $\mathscr{O}$. Then $\mathscr{D}(\mathscr{S})$, as in Theorem 7.6, is a Γ_1-geometry of order $(q-1, q+1)$.

(b) Let $\mathcal{O}$ be a hyperoval in a projective plane $\mathcal{P}$ of even order q, and $\mathcal{E}$ the structure of points of $\mathcal{P}$ not on $\mathcal{O}$ and lines of $\mathcal{P}$ which meet $\mathcal{O}$. Then $\mathcal{E}$ is a $\Gamma_{(q-2)/2}$-geometry of order $(q-2, q/2)$.

Proof. (a) is an immediate corollary of Theorem 7.6. For (b), note that each line of $\mathcal{E}$ contains $q+1-2=q-1$ points, while any point of $\mathcal{E}$ is joined to the $q+2$ points of $\mathcal{O}$ by lines which meet $\mathcal{O}$ in two points (since every line through a point of $\mathcal{O}$ contains one more point of $\mathcal{O}$), so a point of $\mathcal{E}$ is on $(q+2)/2=q/2+1$ lines of $\mathcal{E}$. Hence the order of $\mathcal{E}$ is $(q-2, q/2)$.

Now if y is a line of $\mathcal{E}$ and X is a point of $\mathcal{E}$, X not on y, then y meets $\mathcal{O}$ in two points A and B. The point X is joined to A and B by lines which meet $\mathcal{O}$, i.e. lines of $\mathcal{E}$ (but these lines do not meet y in $\mathcal{E}$, since A and B are not in $\mathcal{E}$). The other $q/2-1$ lines of $\mathcal{E}$ on X meet y in points not in $\mathcal{O}$, i.e. points in $\mathcal{E}$, so X is joined to y by $(q-2)/2$ lines. □

Theorem 7.8 shows that Γ_α-geometries of order (s, t) exist not only for $\alpha=1$, s, $s+1$, t and $t+1$, but also for other values. There exist other saturated subdesigns in projective planes, which give yet more values of α, as the next exercise illustrates.

Exercise 7.6. Let $\mathcal{O}$ be a hyperoval in a projective plane $\mathcal{P}$ of even order n, and let $\mathcal{S}$ be the structure of all points of $\mathcal{P}$ not in $\mathcal{O}$ and of all lines of $\mathcal{P}$ which do not meet $\mathcal{O}$.

(a) Show that $\mathcal{S}^{\mathrm{T}}$ is a saturated subdesign in the dual projective plane $\mathcal{P}^{\mathrm{T}}$ with $k=n/2$ points on a line.

(b) Suppose n is a prime power and that $\mathcal{P} \cong \mathcal{P}^{\mathrm{T}} \cong \mathcal{P}_2(n)$. Find the parameters of the Γ_α-geometry $\mathcal{D}=\mathcal{D}(\mathcal{S}^{\mathrm{T}})$ constructed in $\mathcal{P}(3, n)$ as in Theorem 7.6.

(c) Show that the construction of Theorem 7.8(b) can be generalised to construct a Γ_α-geometry $\mathcal{E}$ and find the parameters of $\mathcal{E}$.

Exercise 7.7. Let $\mathcal{Y}$ be the set $\{1, 2, 3, 4, 5, 6\}$. Let $\mathcal{S}$ be the structure whose points are the 15 2-sets from $\mathcal{Y}$ and whose blocks are given by arbitrary partitions of $\mathcal{Y}$ into three disjoint 2-sets. For example, $(1, 2)$, $(3, 4)$, $(5, 6)$ are three points on a block, and so are $(1, 2)$, $(3, 5)$, $(4, 6)$. Show that $\mathcal{S}$ is a Γ_1-geometry of order $(2, 2)$.

7.3 Generalised quadrangles

A *grid* is a Γ_1-geometry of order $(s, 1)$; $\mathcal{N}$ is a grid in Exercise 7.1(e) if $m=2$

(and, clearly, all grids are of this form). The Γ_1-geometries have a special name: they are called *generalised quadrangles* (or *generalised 4-gons*), which we shall abbreviate to 'GQ'. (A grid is sometimes called a trivial GQ.) Note that a GQ is merely a semilinear design in which any point not on a line is joined to exactly one point on that line. Hence a GQ can contain no 'triangles', but it does contain 'quadrangles': if P and Q are non-collinear points, then, for any choice of a pair of lines y, z on P, there is a unique pair y', z' of lines on Q such that y' meets y and z' meets z.

Corollary 7.9. A GQ of order (s, t) has $(s+1)(st+1)$ points and $(t+1)(st+1)$ lines; furthermore $s+t$ divides $st(s+1)(t+1)$.

Proof. This is merely Theorems 7.2 and 7.8 with $\alpha=1$. □

We have already constructed an infinite family of GQs in Theorem 7.8(a), but the 'classical' GQs are of a different form, constructed out of certain polarities of projective geometries. We can construct one family of classical GQs fairly easily, using only a little projective geometry (and some matrix theory). First some geometry.

Lemma 7.10. The projective geometry $\mathscr{P}=\mathscr{P}(3, K)$ (where K is any field) has a polarity σ with the property that all points in $\mathscr{P}$ are absolute. (Such a σ is called a *null-polarity*.)

Proof. Suppose first that the field K does not have characteristic 2, and let A be any 4-by-4 non-singular skew-symmetric matrix over K. Let $V=V_4(K)$ be the 4-dimensional vector space over K consisting of all 4-tuples from K, so that the points of $\mathscr{P}$ are the 1-dimensional subspaces $\langle \mathbf{v} \rangle$ in V, and the planes can be represented by 1-dimensional subspaces $\langle \mathbf{w}^{\mathrm{T}} \rangle$ of the dual space V^{T}. Then $\langle \mathbf{v} \rangle$ is on $\langle \mathbf{w}^{\mathrm{T}} \rangle$ if and only if $\mathbf{v}\cdot\mathbf{w}^{\mathrm{T}}=0$. (The lines of $\mathscr{P}$, which are the 2-dimensional subspaces of V, do not have such simple representations.) Define a mapping σ on the points and planes of $\mathscr{P}$ by

$$\sigma : \langle \mathbf{v} \rangle \to \langle A\mathbf{v}^{\mathrm{T}} \rangle$$
$$\langle \mathbf{w}^{\mathrm{T}} \rangle \to \langle \mathbf{w}A^{-1} \rangle.$$

It is easy to see that σ is a polarity, inducing a mapping from the lines to the lines.

If $P=\langle \mathbf{v} \rangle$ is a point, then P is on P^{σ} if and only if $\mathbf{v}A\mathbf{v}^{\mathrm{T}}=0$. But the 'one-by-one matrix' $\mathbf{v}A\mathbf{v}^{\mathrm{T}}$ equals its own transpose, so

$$\mathbf{v}A\mathbf{v}^{\mathrm{T}}=(\mathbf{v}A\mathbf{v}^{\mathrm{T}})^{\mathrm{T}}=\mathbf{v}A^{\mathrm{T}}\mathbf{v}^{\mathrm{T}}=-\mathbf{v}A\mathbf{v}^{\mathrm{T}},$$

since $A^{\mathrm{T}}=-A$. Hence $2(\mathbf{v}A\mathbf{v}^{\mathrm{T}})=0$ and thus $\mathbf{v}A\mathbf{v}^{\mathrm{T}}=0$. So P is absolute.

If the characteristic of K is 2, let A be the matrix

$$\begin{bmatrix} 0 & 1 & 0 & 0 \\ 1 & 0 & 0 & 0 \\ 0 & 0 & 0 & 1 \\ 0 & 0 & 1 & 0 \end{bmatrix}$$

(noting of course that $-1=1$, so skew-symmetry is equivalent to symmetry). Then the proof can be followed through, but in more detail, to show that σ, defined as above, has the property that all points are absolute. (See Exercise 7.8.)

Exercise 7.8. Finish the proof of Lemma 7.10 for the case of characteristic 2. □

The polarity σ of Lemma 7.10 is called *symplectic*. If a line y of $\mathscr{P}$ has the property that $y=y^{\sigma}$, then y will be called *absolute* (in the fullness of the theory, such a line would be called *totally isotropic*).

Theorem 7.11. The structure $\mathscr{W}(q)$ of all the points of $\mathscr{P}=\mathscr{P}(3,q)$ and of all the absolute lines of a symplectic polarity σ is a GQ of order (q,q).

Proof. We prove this in steps.

(a) If y is a line on the point P, and y is in P^{σ}, then y is absolute.

Choose a plane $h\neq P^{\sigma}$ such that h contains y (i.e. $y=h\cap P^{\sigma}$). There is a point H such that $H^{\sigma}=h$, and H is on h (because σ is symplectic). But h is on P, so $H=h^{\sigma}$ is on P^{σ}. Hence H lies in $P^{\sigma}\cap h=y$. Since y is on H and P, y^{σ} is in $H^{\sigma}=h$ and in P^{σ}. Thus $y^{\sigma}=h\cap P^{\sigma}=y$. This proves (a).

(b) If y is a line on the point P, and y is not in P^{σ}, then y is not absolute.

Suppose, to the contrary, that $y^{\sigma}=y$. Then for all points X on y, X^{σ} contains $y=y^{\sigma}$, so X^{σ} contains P. But then X is in P^{σ}, so y is in P^{σ}. This proves (b).

(c) Each point of $\mathscr{W}(q)$ is on $q+1$ absolute lines (and of course each absolute line contains $q+1$ points of $\mathscr{W}(q)$).

The second statement is trivial. The only absolute lines on a point P, from (a) and (b), are the lines of the plane P^σ which contain P, and, since P^σ is a projective plane of order q, there are $q+1$ such lines. This proves (c).

Now we finish the proof. Certainly $\mathscr{W}(q)$ is a semilinear design of order (q, q), from (c). Let y be an absolute line and P a point not on y. Then P^σ does not contain $y^\sigma = y$, so the plane P^σ meets the line y in one point X. The line z on P and X is absolute, by (a), and z meets y. No other absolute line on P can meet y, since z is the only line on P in the plane P^σ which meets y. □

Exercise 7.9. Show that $\mathscr{W}(2)$ is isomorphic to the GQ $\mathscr{S}$ of Exercise 7.7.

Exercise 7.10. Let $\mathscr{S}$ be a GQ of order (s, t). For each point X in $\mathscr{S}$, define $X^\perp$ to be the union of X and all points collinear with X. If X and Y are distinct points, define $(X, Y)^\perp = X^\perp \cap Y^\perp$. (The symbol '$\perp$' is read as 'perp'.)

(a) If X and Y are collinear, then show that $(X, Y)^\perp$ consists of all the points on the line through X and Y.

(b) If X and Y are not collinear, then show that $(X, Y)^\perp$ is a set of $t+1$ points no two of which are collinear.

Exercise 7.11.* If X and Y are distinct non-collinear points of $\mathscr{W}(q)$, show that $(X, Y)^\perp$ consists of all the points on some non-absolute line of $\mathscr{P}$, and that every non-absolute line of $\mathscr{P}$ equals $(X, Y)^\perp$ for some pair of non-collinear points X and Y in $\mathscr{W}(q)$. (Hence, in fact, $\mathscr{P}$ can be reconstructed from $\mathscr{W}(q)$: this makes an interesting exercise as well.)

7.4 Semisymmetric designs

A class of 1-designs of considerable interest are the *semisymmetric designs* (abbreviated 'SSD'). These are defined as follows: let λ be a positive integer and $\mathscr{S}$ a structure with more than λ points, satisfying

(SS1) two distinct points of $\mathscr{S}$ are on 0 or λ common blocks;

(SS2) two distinct blocks of $\mathscr{S}$ are on 0 or λ common points;

(SS3) $\mathscr{S}$ is connected;

(SS4) if $\lambda = 1$, then there is a constant k such that every block of $\mathscr{S}$ contains k points and every point of $\mathscr{S}$ is on k blocks.

Then we say that $\mathscr{S}$ is an SSD for λ, or merely an SSD. Note that a symmetric design with parameters (v, k, λ) is an SSD for λ. The explanation for the exceptional axiom (SS4) is that any semilinear space (e.g. an arbitrary set of points in a projective geometry, together with the lines that contain at least two of them) satisfies (SS1), (SS2), (SS3), with $\lambda=1$, and need not be uniform or regular. But the other SSDs satisfy the following lemma, so we impose it as an axiom if $\lambda=1$.

Lemma 7.12. If $\mathscr{S}$ is an SSD for λ, $\lambda>1$, then there is an integer k such that every block of $\mathscr{S}$ contains k points and every point of $\mathscr{S}$ is on k blocks; so $\mathscr{S}$ is regular and uniform.

Proof. Let r_X be the number of blocks on the point X of $\mathscr{S}$ and k_w the number of points on the block w of $\mathscr{S}$. Now suppose P is a point on the block y. We shall count flags (X, z), where X is a point on y, but $X\neq P$, and z is a block on P, but $z\neq y$. There are k_y-1 choices of X, and since each is joined to P by y, it is joined to P by $\lambda-1$ blocks $z\neq y$. So the number of such flags is $(\lambda-1)(k_y-1)$. On the other hand, there are r_P-1 blocks $z\neq y$ on P, and, since each meets y once in P, it meets y again in $\lambda-1$ points $X\neq P$; hence the number of such flags is $(\lambda-1)(r_P-1)$. Thus, since $\lambda>1$, $k_y=r_P$ for any point P on y. Since $\mathscr{S}$ is connected, this proves the lemma (e.g. $k_z=r_P$ for all blocks z on a point P on y, then $r_Q=k_z$ for all points Q on such blocks z, etc.). □

Lemma 7.13. An SSD is square and is a design.

Proof. If v and b are the numbers of points and blocks, respectively, then counting flags we have $vk=bk$, so $b=v$. If an SSD for λ, with block size k, has repeated blocks y and z, say, then y and z meet in k points, so $k=\lambda$; hence every block meeting y must meet it in $\lambda=k$ points, so from connectivity it follows that all blocks of $\mathscr{S}$ contain the k points of y, and no other points. So $\mathscr{S}$ has $k=\lambda$ points, contradicting the definition. □

If $\mathscr{S}$ is an SSD for λ, with v points and k points on a block, then we say that $\mathscr{S}$ has parameters $(v, k, [\lambda])$; if $\mathscr{S}$ is not a symmetric design, then we say that it is a *proper* SSD. We now give some examples of proper SSDs.

Let $\mathscr{P}$ be a finite projective plane of order n.

Example 7.1. Let A be an elation group of $\mathscr{P}$ with axis z, centre Z; i.e. Z is on z and every element of A fixes all the points on z and all the lines through Z. Let $|A|=m$. Then m divides n, and the n^2 points of $\mathscr{P}$ not on z are decomposed into A-orbits of size m, with n/m such orbits on each line y on Z, if $y \neq z$; similarly the n^2 lines of $\mathscr{P}$ not on Z are decomposed into A-orbits of size m, with n/m such orbits on each point Y on z, if $Y \neq Z$. Let $\mathscr{S}=\mathscr{S}(\mathscr{P}, A)$ be the structure whose points are the n^2/m A-orbits of points not on Z, whose blocks are the n^2/m A-orbits of lines not on Z, and whose incidence is induced from $\mathscr{P}$ as follows: if $\mathscr{B}$ is an A-orbit of points and $\mathscr{C}$ is an A-orbit of lines, then $\mathscr{B}$ is on $\mathscr{C}$ in $\mathscr{S}$ if B is on c for some $B \in \mathscr{B}$ and some $c \in \mathscr{C}$. Then $\mathscr{S}$ is an SSD with parameters $(n^2/m, n, [m])$.

Example 7.2. Let A be a homology group of $\mathscr{P}$ with axis z, centre Z; i.e. Z is not on z and every element of A fixes all the points on the line z and all the lines through the point Z. Let $|A|=m$. Now the n^2-1 points of $\mathscr{P}$ not equal to Z and not on z are broken into $(n^2-1)/m$ A-orbits, $(n-1)/m$ A-orbits lying on any line y through Z; the lines are similarly broken into $(n^2-1)/m$ A-orbits, $(n-1)/m$ of which lie on a point Y on z. Proceeding as in Example 7.1, we construct $\mathscr{S}=\mathscr{S}(\mathscr{P}, A)$ which is an SSD with parameters $((n^2-1)/m, n, [m])$.

Example 7.3. Let A be a group of Baer automorphisms with Baer subplane $\mathscr{P}_0$; i.e. $\mathscr{P}_0$ is a subplane of order $\sqrt{n}$, and every element of A fixes all the points and lines of $\mathscr{P}_0$. We can imitate Example 7.1 and use the $n^2-\sqrt{n}$ points (and $n^2-\sqrt{n}$ lines) of $\mathscr{P}$ not in $\mathscr{P}_0$ to construct $\mathscr{S}=\mathscr{S}(\mathscr{P}, A)$, which is an SSD for $((n^2-\sqrt{n})/m, n, [m])$, where $m=|A|$.

These examples can certainly be applied to the projective planes $\mathscr{P}=\mathscr{P}_2(q)$, since elation groups exist in $\mathscr{P}$ for any prime power m dividing q, homology groups of order m exist for any m dividing $q-1$ and, if q is a square, Baer subgroups exist of order $m=2$. But notice that $m=1$ is also permitted in all three cases.

Exercise 7.12.* Prove that the three structures defined in Examples 7.1, 7.2 and 7.3 are indeed SSDs with the given parameters.

Exercise 7.13. Show that for any fixed $\lambda>0$, there exist infinitely many proper SSDs for λ. (Hint: for $\lambda=1$, all three examples give such designs; for $\lambda>1$, we can use Example 7.2 and the fact that for any positive integer m there are infinitely many primes p such that $p \equiv 1 \pmod{m}$.)

Theorem 7.14. Let $\mathscr{S}$ be an SSD for $(v, k, [\lambda])$. Then λ divides $k(k-1)$, and

$$\frac{k(k-1)}{\lambda} + 1 \leqslant v,$$

with equality if and only if S is a symmetric design. If $\lambda > 1$, then

$$v \leqslant 2^{k-1}(k-\lambda) \Big/ \binom{k-2}{\lambda-1},$$

and so there are only a finite number of SSDs with fixed $\lambda > 1$, and with fixed block size k.

Proof. Let $\mathscr{S}$ be an SSD for $(v, k, [\lambda])$, and P a fixed point in $\mathscr{S}$. We count flags (X, y), where $X \neq P$, and y is on P. If we let m be the number of points of $\mathscr{S}$ which are joined to P then each of the m choices of X gives rise to λ choices of Y. Thus the number of flags is $m\lambda$. On the other hand, there are k blocks y on P, each containing $k-1$ choices of X, so the number of flags is $k(k-1)$. Hence $m = k(k-1)/\lambda$, and λ divides $k(k-1)$; since P is not yet counted, we have $v \geqslant k(k-1)/\lambda + 1$. Clearly, equality implies that P is joined to all points of $\mathscr{S}$, and then the same count applied at any point Q also implies that Q is joined to all the points, so $\mathscr{S}$ is symmetric. If $\mathscr{S}$ is symmetric then, trivially, equality holds.

To prove the second inequality, we suppose $\lambda > 1$ and consider the incidence graph Γ of $\mathscr{S}$. We choose a point P and let $\mathscr{M}_1$ be the set of blocks on P, $\mathscr{M}_2$ the set of points $\neq P$ which are on blocks of $\mathscr{M}_1$, $\mathscr{M}_3$ the set of blocks not in $\mathscr{M}_1$ which are on points of $\mathscr{M}_2$, etc. That is, $\mathscr{M}_0 = \{P\}$ and $\mathscr{M}_{i+1}$ is the set of elements of $\mathscr{S}$ which are not in $\mathscr{M}_{i-1}$ but are incident with at least one element of $\mathscr{M}_i$. All the elements of a fixed $\mathscr{M}_i$ are either points or blocks, and in Γ all edges from $\mathscr{M}_i$ go either to $\mathscr{M}_{i-1}$ or to $\mathscr{M}_{i+1}$. Let $M_i = |\mathscr{M}_i|$. We shall prove:

(a) $M_0 = 1$, $M_1 = k$, $M_2 = k(k-1)/\lambda$;

(b) for $i > 2$,

$$M_i \leqslant \frac{k(k-1)(k-\lambda)(k-\lambda-1)\cdots(k-\lambda-i+3)}{\lambda(\lambda+1)\cdots(\lambda+i-2)};$$

(c) an element in $\mathscr{M}_i$ is joined to at least $c_i = \lambda + i - 2$ elements in $\mathscr{M}_{i-1}$ and to at most $k - c_i = k - \lambda - i + 2$ elements in $\mathscr{M}_{i+1}$, if $i > 1$;

(d) $\mathscr{M}_i$ is empty if $i > k - \lambda + 2$.

(See Figure 7.1.)

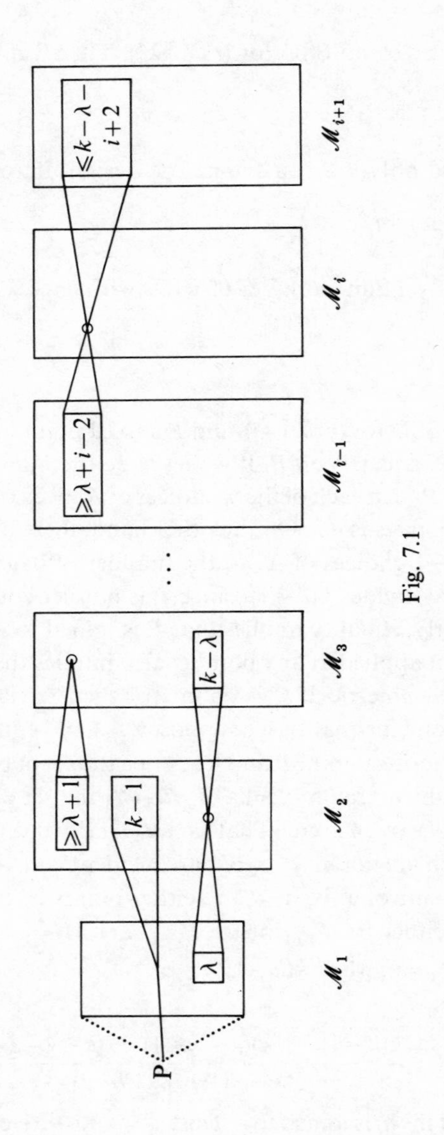
P
λ
k−1
≥λ+1
k−λ
≥λ+i−2
≤k−λ−i+2
$\mathscr{M}_1$
$\mathscr{M}_2$
$\mathscr{M}_3$
. . .
$\mathscr{M}_{i-1}$
$\mathscr{M}_i$
$\mathscr{M}_{i+1}$

Fig. 7.1

Now $M_0=1$ and $M_1=k$ are obvious. M_2 is just the integer m of the first part of the proof of the theorem, so $M_2=k(k-1)/\lambda$. Thus we have proved (a).

We prove (c) next. Clearly, $c_2=\lambda$ (see Figure 7.1). Suppose (c) is true for i; we demonstrate it for $i+1$. We can assume that $\mathcal{M}_{i+1}$ consists of points, and let $X\in\mathcal{M}_{i+1}$. Then X is on at least one block y of $\mathcal{M}_i$ and y contains at least c_i points of $\mathcal{M}_{i-1}$, so X is joined to each of these points in $\mathcal{M}_{i-1}$ $\lambda-1$ more times. Count flags (Y,z), where Y is on y, $Y\in\mathcal{M}_{i-1}$, and where z is on X (see Figure 7.2.). The number of such flags is at least $c_i(\lambda-1)$, and is also equal to $\lambda-1$ times the number of choices of z. So there are at least c_i choices of z. But, counting y as well, we see that X is on at least c_i+1 blocks of $\mathcal{M}_i$. This proves (c).

Now (d) follows from (c), since certainly $\lambda+i-2\leqslant k$.

Finally we show (b). First, for $i=3$ we count flags (Z, y), where $Z\in\mathcal{M}_2$ and $y\in\mathcal{M}_3$. There are $M_2=k(k-1)/\lambda$ choices of Z in $\mathcal{M}_2$ and $k-\lambda$ choices of y in $\mathcal{M}_3$ for each Z, so $k(k-1)(k-\lambda)/\lambda$ such flags. For each of the M_3 choices of y in $\mathcal{M}_3$ there are at least $c_3=\lambda+1$ choices of Z in $\mathcal{M}_2$, hence

$$\frac{k(k-1)(k-\lambda)}{\lambda}\geqslant M_3(\lambda+1),$$

and thus,

$$M_3\leqslant\frac{k(k-1)(k-\lambda)}{\lambda(\lambda+1)}.$$

Inductively, suppose M_{i-1} satisfies (b), and consider flags (X, Y), where $X\in\mathcal{M}_{i-1}$, $Y\in\mathcal{M}_i$. The number of X is M_{i-1} and the number of Y in $\mathcal{M}_i$ joined to X is at most $k-\lambda-i+3$. So the number of such flags is at most $M_{i-1}(k-\lambda-i+3)$. On the other hand, for each of the M_i choices of Y, there are at least $c_i=\lambda+i-2$ choices of $X\in\mathcal{M}_{i-1}$, so the number of such flags is at least $M_i(\lambda+i-2)$. Hence $M_i(\lambda+i-2)\leqslant M_{i-1}(k-\lambda-i+3)$ and, substituting the known upper bound for M_{i-1}, we find the desired inequality for M_i.

Now $2v=\sum M_i$. For $i\geqslant 3$, let us write $\bar{M}_i$ for the right-hand side of (b):

$$\bar{M}_i=\frac{k(k-1)(k-\lambda)(k-\lambda-1)\cdots(k-\lambda-i+3)}{\lambda(\lambda+1)\cdots(\lambda+i-2)}.$$

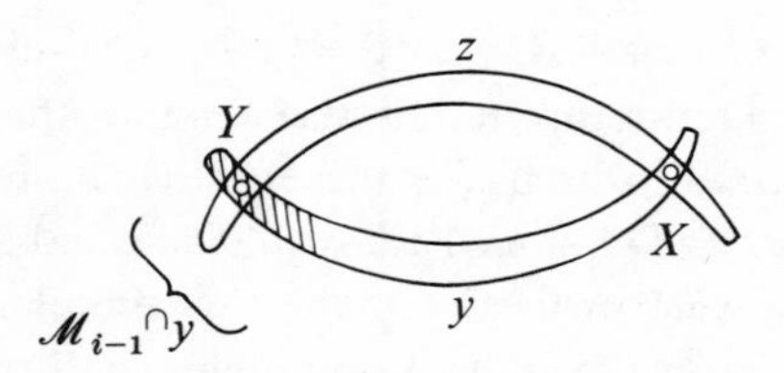

Fig. 7.2

Then it is straightforward that:

$$\bar{M}_i = \frac{k-\lambda}{\binom{k-2}{\lambda-1}} \binom{k}{\lambda+i-2}.$$

From this it follows that:

$$2v \leqslant 1 + k + \frac{k(k-1)}{\lambda} + \frac{k-\lambda}{\binom{k-2}{\lambda-1}} \left[\binom{k}{\lambda+1} + \binom{k}{\lambda+2} + \cdots + \binom{k}{k-\lambda+2} \right].$$

It is now an elementary (but fussy and tedious) job to see that the right-hand side of this last inequality is bounded by

$$2^k(k-\lambda) \Big/ \binom{k-2}{\lambda-1},$$

from which the inequality for v follows. □

Exercise 7.14. Show that the right-hand side of the last inequality in the proof of Theorem 7.14 is indeed bounded by

$$2^k(k-\lambda) \Big/ \binom{k-2}{\lambda-1}.$$

In fact, as the next exercise illustrates, this bound for v above is very rough in general.

Exercise 7.15. In the proof of Theorem 7.14 we could have stated

$$2v \leqslant 1 + k + \frac{k(k-1)}{\lambda} + [\bar{M}_3] + [\bar{M}_4] + \cdots + [\bar{M}_{k-\lambda-2}],$$

where $[x]$ is the greatest integer in x. Compute the upper bound given this way for v when $k=15$ and $\lambda=7$, and compare it with the bound on v given by Theorem 7.14. (The two answers should be, respectively, 60 and 76.)

Note that the proof of Theorem 7.14 breaks down if $\lambda=1$ (where?) and, as we shall see, there is no upper bound in terms of k in that case. On the other hand, when $\lambda=2$, Theorem 7.14 is sharp. Before discussing these cases in more detail, we introduce some terminology: an SSD for $\lambda=1$ is a *semiplane*, and an SSD for $\lambda=2$ is a *semibiplane*. Clearly, every projective plane is a semiplane, and every biplane is a semibiplane. First we examine semibiplanes a little more deeply.

Corollary 7.15. If $\mathscr{B}$ is a semibiplane with parameters $(v, k, [2])$, then

$$\binom{k}{2}+1 \leqslant v \leqslant 2^{k-1}. \quad \square$$

Proof. This is merely Theorem 7.14 with $\lambda=2$. $\square$

The existence of any single semibiplane enables us to construct an infinite family:

Theorem 7.16. If $\mathscr{B}$ is a semibiplane with parameters $(v, k, [2])$, then there is a semibiplane $\mathscr{B}^2$, called the *double* of $\mathscr{B}$, with parameters $(2v, k+1, [2])$.

Proof. Let A be an incidence matrix for $\mathscr{B}$. Then both AA^{T} and $A^{\mathrm{T}}A$ have k everywhere on their main diagonal, and all other entries are 0 or 2. Let

$$A_2=\begin{pmatrix} A & I \\ I & A^{\mathrm{T}} \end{pmatrix},$$

so A_2 is a $2v$ by $2v$ matrix, all entries 0 or 1, with $k+1$ 1s in every row and every column. We have

$$A_2A_2^{\mathrm{T}}=\begin{pmatrix} A & I \\ I & A^{\mathrm{T}} \end{pmatrix}\begin{pmatrix} A^{\mathrm{T}} & I \\ I & A \end{pmatrix}=\begin{pmatrix} AA^{\mathrm{T}}+I & 2A \\ 2A^{\mathrm{T}} & A^{\mathrm{T}}A+I \end{pmatrix}.$$

But $AA^{\mathrm{T}}+1$ and $A^{\mathrm{T}}A+I$ have $k+1$ on the main diagonal and 0 or 2 elsewhere, while $2A$ and $2A^{\mathrm{T}}$ have 0 or 2 in all positions. So if $\mathscr{B}^2$ is the structure with incidence matrix A_2, then $\mathscr{B}^2$ is a regular and uniform structure with the property that any two points meet in 0 or two blocks; clearly, the same argument applied to $A_2^{\mathrm{T}}A_2$ proves that any two blocks of $\mathscr{B}^2$ meet in 0 or two points. From the form of A_2 it is easy to see that $\mathscr{B}^2$ is connected, so $\mathscr{B}^2$ is a semibiplane with the required parameters. $\square$

Corollary 7.17. If there exists a semibiplane with parameters $(v, k, [2])$, then there exists a semibiplane with parameters $(2^s v, k+s, [2])$ for any positive integer s. In particular, there is a semibiplane with parameters $(2^{k-1}, k, [2])$ for all $k>2$.

Proof. The first sentence is immediate from Theorem 7.16: double the initial semibiplane s times. For the second sentence, note that there is a semibiplane with parameters (4, 3, [2]) (in fact, a semibiplane for (4, 3, [2]) is a symmetric design, and is trivial). □

Thus the upper bound on v given by Theorem 7.14 is best possible when $\lambda=2$. (The reader might like to check that this upper bound does *not* depend on the computations in Exercise 7.14 when $\lambda=2$.) These 'maximal' semibiplanes are even unique (up to isomorphism), although this is not easy to show (see [3]).

Exercise 7.16. Let A be an incidence matrix for a semibiplane $\mathscr{B}$ with parameters $(v, k, [2])$, $k>4$. Suppose A satisfies: $A=A^{\mathrm{T}}$, and all entries on the main diagonal of A are 1 (which is equivalent to $\mathscr{B}$ possessing a polarity with all points absolute). Then show that $A-I$ is the incidence matrix for a semibiplane with parameters $(v, k-1, [2])$. Show, furthermore, that the biplane $\mathscr{B}_2$ of Example 3.3, with parameters (16, 6, 2), has such an incidence matrix A, so there exists a (proper) semibiplane with parameters (16, 5, [2]).

Exercise 7.17. Show that there exist semibiplanes with the following values of v and k:

k	v
3	4, and no other value of v
4	7, 8, and no other values of v
5	11, 12, 14, 16
6	16, 22, 24, 28, 32
7	24, 32
8	32
9	37, 39, 40

Exercise 7.18. Show that the semibiplanes with the following values of v and k are unique up to isomorphism: $(v, k)=(4, 3), (7, 4), (8, 4), (11, 5), (12, 5)$. (Some of these uniquenesses have been established already!)

Now we give an exercise which establishes that for semiplanes the upper bound of Theorem 7.14 does not hold, and indeed no upper bound exists at all.

Exercise 7.19. Let Z_v be the additive group of integers modulo v, and suppose D is a subset of Z_v, $|D|>2$, satisfying:

(a) the set of differences d_i-d_j, for $d_i, d_j \in D$, $d_i \neq d_j$, are all distinct;

(b) $1=d_1-d_2$ for some pair $d_1, d_2 \in D$.

Let $\mathscr{S}$ be the structure whose points are x for all $x \in Z_v$, and whose blocks are $D+y$ for all $y \in Z_v$, with 'natural' incidences: x is on $D+y$ if $x \in D+y$.

(i) Show that $\mathscr{S}$ is a semiplane with v points and $k=|D|$ points on every block. (Remember that $\mathscr{S}$ must be shown to be connected.)

(ii) Show that if $D=\{2^i \mid 0 \leqslant i < k\}$ and if $v \geqslant 2^k-1$, then D satisfies (a) and (b) above. Conclude that for a fixed value of k there is a semiplane with block size k for all $v \geqslant 2^k-1$.

We conclude this section with a long exercise which uses semibiplanes to construct a 3-design for (22, 6, 1). This 3-design will arise again in Chapter 8, where it is constructed in a completely different way, but is shown to be unique.

Exercise 7.20.* Let $\mathscr{L}$ be the biplane $\mathscr{L}(11)$ of Chapter 3, with parameters (11, 5, 2). We recall that $\mathscr{L}$ is unique (up to isomorphism) with those parameters (see Example 3.2). Let $\mathscr{B}$ be the double of $\mathscr{L}$, so $\mathscr{B}$ has parameters (22, 6, [2]).

(a) Show that $\mathscr{B}$ has a partition into two subbiplanes, each with parameters (11, 5, 2), the one naturally isomorphic to $\mathscr{L}$, the other to $\mathscr{L}^{\mathrm{T}}$. (Of course, $\mathscr{L} \cong \mathscr{L}^{\mathrm{T}}$, but we do not need that here.) We identify these two subbiplanes as $\mathscr{L}$ and $\mathscr{L}^{\mathrm{T}}$. Show that a block of $\mathscr{B}$ contains five points of $\mathscr{L}$ (or of $\mathscr{L}^{\mathrm{T}}$) and one point of $\mathscr{L}^{\mathrm{T}}$ (or of $\mathscr{L}$).

(b) Define an *oval* $\mathscr{O}$ in $\mathscr{L}$ to be a maximal set of points of $\mathscr{L}$ such that no three points of $\mathscr{O}$ are on a block. Show that an oval contains three points, that $\mathscr{L}$ contains 55 ovals, and that any pair of distinct points of $\mathscr{L}$ are in three ovals.

(c) If $\mathscr{O}$ is an oval in $\mathscr{L}$, show that on each point of $\mathscr{O}$ there is a unique block which contains no other points of $\mathscr{O}$; these three blocks are called the *tangent* blocks to $\mathscr{O}$.

(d) An *oval* in $\mathscr{B}$ is defined analogously (see (b)). Show that an oval in $\mathscr{B}$ has six points, three in each of $\mathscr{L}$ and $\mathscr{L}^{\mathrm{T}}$; each of these 3-sets being an oval in $\mathscr{L}$ or $\mathscr{L}^{\mathrm{T}}$, respectively, and the three tangent blocks to this oval in $\mathscr{L}(\mathscr{L}^{\mathrm{T}})$ contain, besides their five points in $\mathscr{L}(\mathscr{L}^{\mathrm{T}})$, one point each of the oval in $\mathscr{L}^{\mathrm{T}}(\mathscr{L})$.

(e) Show that $\mathscr{B}$ has 55 ovals.

(f) Show that three points of $\mathscr{L}(\mathscr{L}^{\mathrm{T}})$ are in a unique block of $\mathscr{B}$ or in a unique oval of $\mathscr{B}$.

(g) If X, Y are points of $\mathscr{L}$, y a block of $\mathscr{L}$, then show that:

 (i) if X is on y, Y not on y, then there is a unique oval $\mathscr{O}$ in $\mathscr{L}$ such that X, Y are in $\mathscr{O}$ and y is a tangent block to $\mathscr{O}$;

 (ii) if neither X nor Y is on y then there is a unique oval $\mathscr{O}$ in $\mathscr{L}$ such that X, Y are in $\mathscr{O}$ and y is a tangent block to $\mathscr{O}$.

(h) From (f) and (g) show that three points of $\mathscr{B}$ are either on a unique block of $\mathscr{B}$ or in a unique oval of $\mathscr{B}$.

(j) If $\mathscr{M}$ is the structure whose points are the points of $\mathscr{B}$ and whose blocks are (i) the blocks of $\mathscr{B}$, and (ii) the ovals of $\mathscr{B}$, then show that $\mathscr{M}$ is a 3-design for (22, 6, 1).

NB. In Chapter 8 we not only demonstrate that a 3-(22, 6, 1) is unique up to isomorphism, but that its automorphism group is 3-transitive on points and is non-soluble. Our construction above, while elementary and interesting, does not permit an easy proof of these additional facts. But it reveals what the construction in Chapter 8 does not: that a 3-(22, 6, 1) contains a sub–semi-biplane with parameters $(22, 6, [2])$ and a subbiplane with parameters $(11, 5, 2)$.

Exercise 7.21. Show that there exists an SR graph with parameters $(77, 16, 0, 4)$. (Hint: see Theorem 3.46, and Exercise 7.20 above.)

7.5 Divisible semisymmetric designs

Let $\mathscr{S}$ be a proper SSD for $(v, k, [\lambda])$. Then a natural generalisation of the concept of a resolution (see Chapter 5) for $\mathscr{S}$ is that the blocks of $\mathscr{S}$ be partitioned into classes $\mathscr{B}_1, \mathscr{B}_2, \ldots, \mathscr{B}_m$ such that two blocks of $\mathscr{S}$ are on λ_1 or λ_2 common points, where $\lambda_1 \neq \lambda_2$, according as the blocks are in, or are not in, the same class $\mathscr{B}_i$. Then certainly λ_1, λ_2 must be 0, λ in some order. We say that $\mathscr{S}$ is *block-divisible* if such a decomposition of its blocks can be made, and we call the $\mathscr{B}_i$ the (block-) *divisibility classes.* Clearly a dual definition can be made for the points of $\mathscr{S}$, in which case we say that $\mathscr{S}$ is *point-divisible*, and refer to the (point-) *divisibility classes.*

Lemma 7.18. If $\mathscr{S}$ is a point-divisible SSD for λ, then the divisibility class of the point X consists of X and all points not joined to X. Thus $\mathscr{S}$ is point-divisible if and only if the relationship '$P \sim Q$ if and only if $P = Q$ or P and Q are not joined' is an equivalence relation.

Proof. If X and Y are in the same class ($X \neq Y$) and are joined by λ blocks, then neither X nor Y can be joined to any point in any other class. Since $\mathscr{S}$ is connected, this is impossible. □

We can include symmetric designs in our definition of point or block-divisibility by defining each class to consist of one element only; then a symmetric design is both point- and block-divisible, and Lemma 7.18 is true for symmetric designs as well.

Lemma 7.19. Suppose $\mathscr{S}$ is a point-divisible SSD for $(v, k, [\lambda])$, and y, z are a pair of blocks that do not meet. Then, for each point P on y there is a unique point Q on z such that $P \sim Q$ and every block on P, excepting y, meets z.

Proof. Let c be the number of blocks on P which meet z, and let d be the number of points on z which are joined to P. Let us count flags (X, w), where X is on z and w is on P; for each of c choices of w there are λ points X, and for each of the d choices of X there are λ blocks w. So $c\lambda = d\lambda$, hence $c = d$. Now $c < k$ since y does not meet z. But if $d < k-1$, then there are at least two points X, Y on z not joined to P. Then $P \sim X$, and $P \sim Y$, so $X \sim Y$; this is impossible since X and Y are joined by the block z. Thus $c = d = k-1$, which proves the lemma. □

Theorem 7.20. Suppose $\mathscr{S}$ is an SSD. Then $\mathscr{S}$ is point-divisible if and only if it is block-divisible.

Proof. Let $\mathscr{S}$ be point-divisible. We must show that the relationship on blocks given by '$y \sim z$ if and only if $y = z$ or y and z do not meet' is an equivalence relation. Clearly, what is required is: if y and w do not meet z then $y = w$ or y and w do not meet. Suppose that there is a point P on y and w; then from Lemma 7.19, every block on P, excepting y, meets z and so, since w does not meet z, we must have $w = y$. Hence $\mathscr{S}$ is block-divisible, and thus the converse holds as well. □

In view of Theorem 7.20, we refer to a point- or block-divisible SSD as a *divisible* SSD. Before proceeding, we mention one aspect of divisibility that is not like resolvability: the blocks of a class need not cover the points of the SSD, so we do not necessarily have a resolution. As an illustration we set the following exercise.

Exercise 7.22. Consider the set $D = \{0, 1, 2, 5, 10\}$ in the additive group Z_{12} of integers modulo 12.

(a) Show that every $x \in Z_{12}$, $x \neq 0, 6$ has exactly two representations in the form $x = d_1 - d_2$, for $d_1, d_2 \in D$.

(b) If $\mathscr{S}$ is the structure whose points are the elements of Z_{12} and whose blocks are the subsets $D + y$, for all y in Z_{12}, show that S is an SSD for $(12, 5, [2])$.

(c) Show that $\mathscr{S}$ is divisible, and find all its point and block classes; if $\mathscr{B}_i$ is a block class, show that there are exactly two points of $\mathscr{S}$ which lie in no block of $\mathscr{B}_i$.

(NB. This SSD has been met before: see Example 7.2 and Exercises 7.17 and 7.18.)
We now establish some properties of the parameters of a divisible SSD.

Theorem 7.21. Suppose $\mathscr{S}$ is a divisible SSD for $(v, k, [\lambda])$. Then the size of every point (or block) divisibility class is $c = v - k(k-1)/\lambda$. In addition, c divides v and $v \leqslant k^2/\lambda$.

Proof. If P is a point of $\mathscr{S}$ then P is joined to $k(k-1)/\lambda$ points of $\mathscr{S}$. Thus the set consisting of P and all points not joined to P has $c = v - k(k-1)/\lambda$ points, which is then the (common) class size. Clearly, c divides v since the classes partition the points.

Now let $\mathscr{C}$ be a point class, and X a point not in $\mathscr{C}$. Then X is joined to every point in $\mathscr{C}$ (otherwise it would be in $\mathscr{C}$), and no block on X contains more than one point in $\mathscr{C}$ (for two points of $\mathscr{C}$ cannot lie on a common block). We count flags (Y, y), where Y is in $\mathscr{C}$ and y is on X. There are c choices of Y and λ choices of y for each Y, hence the number of flags is $c\lambda$. But there are at most k choices of y, and one choice of Y for each permitted y, so the number of flags is at most k. Thus $c\lambda \leqslant k$, so $c \leqslant k/\lambda$. Then:

$$v = c + k(k-1)/\lambda \leqslant k/\lambda + k(k-1)/\lambda = k^2/\lambda. \quad \square$$

Exercise 7.23. Show that all the SSDs of Examples 7.1, 7.2 and 7.3 are divisible, and find the class size c for each.

Exercise 7.24. Let $\mathscr{S}$ be an SSD for $(v, k, [\lambda])$, with $v = k(k-1)/\lambda + 2$. Show that $\mathscr{S}$ is divisible with class size 2, and that v is even.

Exercise 7.25. Let $\mathscr{D} = \mathscr{D}(H)$ be the generalised Hadamard design of Exercise 3.25, where H is a Hadamard matrix of order $n > 4$. Then show that $\mathscr{D}$ is an SSD for $(2n, n, [n/2])$, and that $\mathscr{D}$ is divisible.

The conditions of Exercise 7.24 are quite special and imply that if $v - k(k-1)/\lambda = 1$ or 2, then $\mathscr{S}$ is divisible. If $v - k(k-1)/\lambda = 3$ this is not necessarily so: there are three semibiplanes for $(18, 6, [2])$, and one is indeed divisible with class size 3, while the other two are not divisible.

Theorem 7.22. Let $\mathscr{S}$ be a divisible SSD for $(v, k, [\lambda])$, with class size c. Then $\bar{\mathscr{S}}$, defined below, is a symmetric design for $(v/c, k, \lambda c)$, unless $k=\lambda c$, in which case it is a trivial structure for (k, k, k) with k (repeated) blocks.

The points of $\bar{\mathscr{S}}$ are the point-divisibility classes, and the blocks of $\bar{\mathscr{S}}$ are the block-divisibility classes; the point class $\mathscr{C}$ is on the block class $\mathscr{B}$ if P is on y for some $P \in \mathscr{C}$ and some $y \in \mathscr{B}$.

Proof. If P is on y in $\mathscr{S}$, then the c points in the class of P are on c different blocks of the class of y; i.e. one each on each block in the class of y. So incidence is well defined. Let us write $\langle X \rangle$ (or $\langle y \rangle$) for the class of X (or y). If $\langle X \rangle$ and $\langle Y \rangle$ are distinct point classes, then X is joined to each of the c points in $\langle Y \rangle$ by λ blocks, and hence to all the points in $\langle Y \rangle$ by λc blocks: none of these can coincide, since no block can contain more than one point in $\langle X \rangle$ or more than one point in $\langle Y \rangle$. These λc blocks are in λc different block classes, since any two of them meet in X. If $X_1 \in \langle X \rangle$ then, be Lemma 7.19, λc blocks on X_1 which meet $\langle Y \rangle$ give exactly the same λc block classes. Hence $\bar{\mathscr{S}}$ is a square 2-structure with v/c points, and k points on a block (for the k points on a block of $\mathscr{S}$ are in k different point classes); so $\bar{\mathscr{S}}$ has parameters $(v/c, k, \lambda c)$, and is a symmetric design unless it has repeated blocks. Suppose $\langle y \rangle \neq \langle z \rangle$, but that $\langle y \rangle$ and $\langle z \rangle$ are incident with the same point classes. Then z contains k points, all of them on some block of $\langle y \rangle$; since z is not in $\langle y \rangle$, z meets every block of y in λ points. So z contains λc points, thus $k=\lambda c$.

Conversely, if $k=\lambda c$, then $v/c=k(k-1)/\lambda c+1=\lambda c$, so $\bar{\mathscr{S}}$ is a 2-structure for (k, k, k) with k blocks. □

Exercise 7.26. Consider the three divisible SSDs of Examples 7.1, 7.2 and 7.3. Show:

(a) in 7.1, $\bar{\mathscr{S}}$ is a 2-structure for (n, n, n);
(b) in 7.2, $\bar{\mathscr{S}}$ is a (trivial) symmetric design for $(n+1, n, n-1)$;
(c)* in 7.3, $\bar{\mathscr{S}}$ is the complement of a projective plane of order $\sqrt{n}$.

We can derive a non-existence theorem for divisible SSDs from Theorem 7.22, using the Bruck–Ryser–Chowla Theorem (Theorem 2.3) on $\bar{\mathscr{S}}$. Note that if $k=\lambda c$, when $\bar{\mathscr{S}}$ is not a symmetric design, the conclusion of the following theorem is trivially satisfied.

Theorem 7.23. Let $\mathscr{S}$ be a divisible SSD for $(v, k, [\lambda])$, with class size c.

(a) If v/c is even then $k-\lambda c$ is a square;

(b) if v/c is odd, then there are integers x, y, z, not all zero, such that

$$x^2=(k-\lambda c)y^2+(-1)^{(m-1)/2}\lambda c z^2,$$

where $m=v/c$ (the number of point classes).

Proof. Apply Theorem 2.3. □

Exercise 7.27. Show that there is no SSD with parameters: $(44, 7, [1])$, $(58, 8, [1])$, $(68, 12, [2])$, $(44, 15, [5])$.

Exercise 7.28. Let $\mathscr{S}$ be a divisible SSD for $(v, k, [\lambda])$, with class size c and $m=v/c$ point (or block) classes.

(a) Show that there is an incidence matrix A for $\mathscr{S}$ such that

$$AA^{\mathrm{T}}=\begin{bmatrix} kI_c & & & \\ & kI_c & & \lambda \\ & & \ddots & \\ & \lambda & & kI_c \end{bmatrix}$$

that is, with m matrices kI_c down the main diagonal, and λ elsewhere.

(b) Show that $\det(AA^{\mathrm{T}})=\det D$, where

$$D=\begin{bmatrix} kI_c+\lambda(m-1)J_c & & & * \\ & kI_c-\lambda J_c & & \\ & & kI_c-\lambda J_c & \\ 0 & & \ddots & \\ & & & kI_c-\lambda J_c \end{bmatrix}$$

(with $m-1$ blocks $kI_c-\lambda J_c$ indicated by a brace)

that is, with the indicated matrices down the main diagonal, and 0 elsewhere below them.

(c) Show that

$$\det D=k^{m(c-1)}(k-\lambda c)^{m-1}(k+\lambda c(m-1)),$$

and that $k+\lambda c(m-1)=k^2$, so

$$\det D=k^{m(c-1)+2}(k-\lambda c)^{m-1}.$$

(d) Conclude that:

(i) if $k > \lambda c$, then $\mathscr{S}$ has rank v;
(ii) $(k - \lambda c)^{m-1} k^{v-m}$ is a square;
(iii) if v is even and k is odd, then k is a square.

(Hint: For parts (b) and (c), see the trick used in the proof of Lemma 1.21. As well, use Theorem 7.23.)

Exercise 7.29. Show that there is no SSD with parameters (30, 8, [2]).

These non-existence theorems (Theorem 7.23 and Exercise 7.28) are special cases of a more general non-existence theorem for divisible SSDs, called the Bose–Connor Theorem (see [1] for more). In fact, the Bose–Connor Theorem applies to more general structures, *partially symmetric designs*, where the expression '0 or λ' in axioms (SS1) and (SS2) at the beginning of Section 7.3 is replaced by 'λ_1 or λ_2'.

References

Many articles have been written on Generalised Quadrangles and [2] is a recent survey article about them. [3] is the standard reference for semibiplanes and [1] contains the Bose–Connor Theorem.

[1] Bose, R. C. & Connor, W. S. 'Combinatorial properties of group divisible incomplete block designs'. *Ann. Math. Statist.*, **23** (1952), 367–83.

[2] Thas, J. A. 'Combinatorics of finite generalised quadrangles: A survey'. *Annals of Discrete Math.*, **14** (1982), 57–76.

[3] Wild, P. R. 'On semibiplanes'. Ph.D. Thesis, University of London, 1980.

8 The large Mathieu designs

8.1 Introduction

In this chapter we discuss the large Mathieu designs and their automorphism groups. There are three large Mathieu designs, denoted by $\mathcal{M}_{22}$, $\mathcal{M}_{23}$ and $\mathcal{M}_{24}$, and these are the unique (as we shall see) designs with parameters 3-(22, 6, 1), 4-(23, 7, 1) and 5-(24, 8, 1) respectively. Their automorphism groups contain the celebrated Mathieu groups M_{22}, M_{23} and M_{24}. Notice that we have already constructed $\mathcal{M}_{22}$ in Exercise 7.20, although its uniqueness is yet to be established.

In Section 8.2 we assume that a 5-(24, 8, 1) exists and show that it must be constructed from $\mathscr{P}_2(4)$, the unique projective plane of order 4, in a very special way. Then in Section 8.3 we show that the construction 'works', i.e. that $\mathcal{M}_{24}$ exists: furthermore, we establish its uniqueness. The existence of $\mathcal{M}_{22}$ and $\mathcal{M}_{23}$ follow immediately (by contracting on one or two points), and we indicate (without all the details) how their uniqueness is demonstrated.

The Mathieu designs (including the small ones in Chapter 4) are a fundamental feature of combinatorics and algebra. They have relationships to many other interesting combinatorial objects (e.g. as in Exercise 7.20, the Golay codes, the Leech lattice, etc.), their automorphism groups are central in group theory, and they are fascinating and rich in themselves. In Section 8.4 we look at some of their relationship with coding theory: after a very brief introduction to the concept of binary linear codes we use $\mathcal{M}_{24}$ to construct the Golay codes in a simple way.

Sections 8.5 and 8.6 provide an insight into the fruitful interplay between the combinatorial properties of the large Mathieu designs and group theory, with strongly regular graphs coming into the picture. These sections only touch upon a very large area, and we give an indication of the flavour through developing a few interesting results. Chapter 8 includes rather more group theory (especially permutation groups) than the earlier chapters, and we refer the reader to [5] or [4] for the (relatively elementary) theory of permutation groups used.

8.2 Properties of a 5-(24, 8, 1)

In this section we shall suppose that a 5-(24, 8, 1) $\mathscr{M}$ exists and examine some of its properties. First, some elementary results.

Lemma 8.1. The constants λ_i for $\mathscr{M}$ are: $b=\lambda_0=759$, $r=\lambda_1=253$, $\lambda_2=77$, $\lambda_3=21$, $\lambda_4=5$ and, if X_1, X_2, X_3 are distinct points of $\mathscr{M}$, then

(a) $\mathscr{M}_{X_1}$ is a 4-(23, 7, 1), with 253 blocks;

(b) $\mathscr{M}_{X_1,X_2}$ is a 3-(22, 6, 1), with 77 blocks;

(c) $\mathscr{M}_{X_1,X_2,X_3}$ is a 2-(21, 5, 1) and hence is a symmetric design isomorphic to the (unique) projective plane of order 4.

Proof. Use Theorem 1.2 for the λ_i. The uniqueness of a projective plane of order 4 is in Exercise 3.17. □

Lemma 8.2. In the terminology of Lemma 8.1:

(a) Two distinct blocks of $\mathscr{M}_{X_1,X_2}$ meet in 0 or two points;

(b) Two distinct blocks of $\mathscr{M}_{X_1}$ meet in one or three points;

(c) Two distinct blocks of $\mathscr{M}$ meet in 0, two or four points.

Proof. Since $\mathscr{M}_{X_1,X_2}$ is an extension of the symmetric design $\mathscr{P}\cong\mathscr{P}_2(4)$, we know that if two of its blocks meet in a point Y, then they are blocks, i.e. lines, of $\mathscr{P}$. Hence they meet in one more point. This gives (a). (See also the proof of Theorem 4.2.)

To prove (b) we note that $(\mathscr{M}_{X_1})_Y$ is a 3-(22, 6, 1). So if two distinct blocks of $\mathscr{M}_{X_1}$ meet in a point Y, they meet 0 or two times more in $(\mathscr{M}_{X_1})_Y$, and so they meet in one or three points of $\mathscr{M}_{X_1}$. We must now show that no pair of blocks of $\mathscr{M}_{X_1}$ fail to meet.

Let y be a fixed block of $\mathscr{M}_{X_1}$. If A is a point on y, then A is on $77-1=76$ blocks w different from y. If A, B, C are three distinct points on y, then they are on five blocks in $\mathscr{M}_{X_1}$, so on four blocks w, $w\neq y$. If A is fixed, then each of the $\binom{6}{2}$ choices of B, C on y gives us four blocks w which meet y in A and in the two additional points B, C. Hence $\binom{6}{2}=60$ blocks on A meet y in two more points, and $76-60=16$ blocks on A do not meet y again.

Thus $7\cdot 16=112$ blocks of $\mathscr{M}_{X_1}$ meet y in just one point, and $\binom{7}{3}\cdot 4=140$ blocks of $\mathscr{M}_{X_1}$ meet y in three points. This gives us $140+112=252$ blocks which meet y, and thus since $\mathscr{M}_{X_1}$ has 253 blocks, no block of $\mathscr{M}_{X_1}$ fails to meet y. This gives (b).

For (c), we note that if two blocks of $\mathscr{M}$ meet, in X_1 say, then they are blocks of $\mathscr{M}_{X_1}$, and so meet one or three more times. So two distinct blocks of $\mathscr{M}$ meet in 0, two or four points. □

Choosing three distinct points X_1, X_2, X_3 of $\mathscr{M}$ we write $\mathscr{P} = \mathscr{M}_{X_1,X_2,X_3}$; then $\mathscr{P}$ is a projective plane of order 4. The blocks of $\mathscr{M}$ are of four types:

(I) blocks on all three of X_1, X_2, X_3, with five points in $\mathscr{P}$;
(II) blocks on exactly two of X_1, X_2, X_3, with six points in $\mathscr{P}$;
(III) blocks on exactly one of X_1, X_2, X_3, with seven points in $\mathscr{P}$;
(IV) blocks on none of X_1, X_2, X_3, with eight points in $\mathscr{P}$.

Clearly, the blocks of type (I), being blocks of $\mathscr{P}$, are simply the 21 lines of $\mathscr{P}$, each with the three points X_1, X_2, X_3 adjoined. Our next problem is to find out which 6-sets, 7-sets and 8-sets in $\mathscr{P}$ give rise to the blocks of types (II), (III) and (IV), respectively. The answers are neat and simple, but, in order to exhibit them, we must make some definitions.

An *oval* in $\mathscr{P}$ is a set of six points, no three on a line (i.e. no three collinear). (In Chapter 7 this was called a *hyperoval*, the term oval being reserved for a set of five points, no three collinear; but for convenience we use the simpler term.) A *Baer subplane* is a set of seven points and seven lines in $\mathscr{P}$ which form a projective plane of order 2, i.e. a 2-(7, 3, 1). Then we use the term *BSP* to refer to the set of seven points of a Baer subplane. Finally, if y and z are distinct lines of $\mathscr{P}$, we define $y * z$ to be the set of eight points of $\mathscr{P}$ which are on y or on z, but not both (that is, the symmetric difference of the two point sets); $y * z$ is called a *double line.*

Exercise 8.1. Show that if $\mathscr{A}$ is a set of seven points in $\mathscr{P}$, then some three points of $\mathscr{A}$ are collinear.

Exercise 8.2. Show that if $\mathscr{O}$ is an oval in $\mathscr{P}$, then any line of $\mathscr{P}$ meets $\mathscr{O}$ in 0 or two points.

Exercise 8.3. Show that if $\mathscr{B}$ is a BSP in $\mathscr{P}$, then any line of $\mathscr{P}$ meets $\mathscr{B}$ in one or three points.

Exercise 8.4. Show that if $y * z$ is a double line in $\mathscr{P}$, then any line of $\mathscr{P}$ meets $y * z$ in 0, two or four points.

Theorem 8.3. The point sets of the blocks of $\mathscr{M}$ are made up as follows:

(a) a block on all of X_1, X_2, X_3 is a line in $\mathscr{P}$, and all lines in $\mathscr{P}$ arise in this way;

(b) a block on exactly two of X_1, X_2, X_3 is an oval in $\mathscr{P}$;

(c) a block on exactly one of X_1, X_2, X_3 is a BSP in $\mathscr{P}$;

(d) a block on none of X_1, X_2, X_3 is a double line in $\mathscr{P}$.

(The theorem is worded slightly abusively, but should be clear.)

Proof. (a) has been pointed out already.

For (b), if y is a block of $\mathscr{M}$ on X_1, X_2, say, but not on X_3, then let $\mathscr{Y}$ be the set of six points of y in $\mathscr{P}$, $\mathscr{Y}$ can meet no line z of $\mathscr{P}$ in three or more points, for then y would meet $z \cup X_1 \cup X_2 \cup X_3$ (which is a block of $\mathscr{M}$) in five or more points. Hence $\mathscr{Y}$ is an oval.

Now for (c), suppose y is a block of $\mathscr{M}$ on X_1 but not on X_2 or X_3, and let $\mathscr{Y}$ be the set of seven points of y in $\mathscr{P}$. For every line w of $\mathscr{P}$, $w \cup X_1 \cup X_2 \cup X_3$ is a block of $\mathscr{M}$ which meets y in X_1, but not in X_2 or X_3. Thus, by Lemma 8.2, w must meet $\mathscr{Y}$ in one or three more points. Consider the set $\mathscr{Y}'$ consisting of the seven points of $\mathscr{Y}$ and of all the lines of $\mathscr{P}$ which meet $\mathscr{Y}$ in at least two (and hence exactly three) points of $\mathscr{Y}$. Clearly, two points of $\mathscr{Y}'$ are on exactly one line of $\mathscr{Y}'$, so $\mathscr{Y}'$ is a 2-(7, 3, 1), i.e. a Baer subplane. So $\mathscr{Y}$ is a BSP.

Now we come to (d). Let y be a block of $\mathscr{M}$ on no X_i, and $\mathscr{Y}$ the set of eight points of y (all of which are in $\mathscr{P}$). From Lemma 8.2, every line of $\mathscr{P}$ meets $\mathscr{Y}$ in 0, two or four points. If P is a point of $\mathscr{Y}$, then every line of $\mathscr{P}$ on P must contain one or three more points of $\mathscr{Y}$. If each of the five lines on P contained only one more point of $\mathscr{Y}$, then $\mathscr{Y}$ would contain only six points; thus one line (exactly) on P contains three more points of $\mathscr{Y}$, and the other four lines on P contain just one. Suppose z is a line of $\mathscr{P}$ containing four points of $\mathscr{Y}$, and let Q be a point of $\mathscr{Y}$, Q not on z. Then there is exactly one line w on Q containing four points of $\mathscr{Y}$, and w meets z in a point R. If R is in $\mathscr{Y}$, then R is in two lines which contain four points of $\mathscr{Y}$, which is impossible. So y must be the double line $w * z$. □

We shall postpone until the next section the question of the existence of ovals and BSPs in $\mathscr{P}$, etc., and continue with our analysis of what must happen if $\mathscr{M}$ is to exist. We can partition both the ovals and BSPs which 'occur' in $\mathscr{M}$ into classes, as follows:

> $\mathbf{O}_i$ is the set of ovals $\mathscr{O}$ in $\mathscr{P}$ such that $\mathscr{O} \cup X_j \cup X_k$ is a block, where $j \neq i$, $k \neq i$.
>
> $\mathbf{B}_i$ is the set of BSPs $\mathscr{B}$ in $\mathscr{P}$ such that $\mathscr{B} \cup X_1$ is a block.

We call an oval $\mathscr{O}$, or a BSP $\mathscr{B}$, *admissible* if $\mathscr{O} \cup X_i \cup X_j$, or $\mathscr{B} \cup X_l$, for some i, j, or for some l, is a block of $\mathscr{M}$. Then $\mathbf{O}_i$, for $i = 1, 2, 3$, partitions the admissible ovals, and $\mathbf{B}_i$, for $i=1, 2, 3$, partitions the admissible BSPs. In addition we have:

Lemma 8.4

(a) Two admissible ovals are in the same class $\mathbf{O}_i$ if and only if they meet in an even number of points;

(b) two admissible BSPs are in the same class $\mathbf{B}_i$ if and only if they meet in an odd number of points;

(c) if $\mathcal{O}$ is an admissible oval in $\mathbf{O}_i$ and $\mathcal{B}$ is an admissible BSP in $\mathbf{B}_j$, then $i=j$ if and only if $\mathcal{O}$ meets $\mathcal{B}$ in an even number of points.

Proof. The lemma is an immediate corollary of Lemma 8.2. For example, suppose that $\mathcal{O}_1$ and $\mathcal{O}_2$ are admissible ovals, both in $\mathbf{O}_1$. Then $\mathcal{O}_1 \cup X_2 \cup X_3$ and $\mathcal{O}_2 \cup X_2 \cup X_3$ are blocks of $\mathcal{M}$ and meet in the two points X_2, X_3. So $\mathcal{O}_1$ and $\mathcal{O}_2$ must meet in 0 or two (or six) more points in $\mathcal{P}$. But if $\mathcal{O}_1 \in \mathbf{O}_1$, $\mathcal{O}_2 \in \mathbf{O}_2$, then $\mathcal{O}_1 \cup X_2 \cup X_3$ and $\mathcal{O}_2 \cup X_1 \cup X_3$ are blocks of $\mathcal{M}$, meeting only in X_1 outside of $\mathcal{P}$, hence meeting one or three more times in $\mathcal{P}$. That is, $\mathcal{O}_1$ meets $\mathcal{O}_2$ in an odd number of points.

The other two parts of the lemma are similar. □

Exercise 8.5. Show that each oval class $\mathbf{O}_i$ contains 56 admissible ovals, and that each BSP class $\mathbf{B}_i$ contains 120 admissible BSPs. (Hint: this involves counting the number of blocks of $\mathcal{M}$ on two specified points, but not on a third specified point, or on one specified point and on neither of two others.)

The results of Exercise 8.5 are relevant because we shall see that $\mathcal{P}$ contains 168 ovals and 360 BSPs. This will then imply that if $\mathcal{M}$ exists, then all ovals and all BSPs are admissible. But we do not need this yet.

Exercise 8.6. Let $\mathcal{D}$ be a 3-(22, 6, 1). Show that $\mathcal{D}$ must be constructed from $\mathcal{P}$ (a projective plane of order 4) by adjoining one point X to the 21 points of $\mathcal{P}$, and that the 77 blocks of $\mathcal{D}$ are:

(a) the 21 point sets $y \cup X$, where y is a line of $\mathcal{P}$, and

(b) 56 ovals of $\mathcal{P}$ with the property that any two of them meet in 0, two or six points.

Exercise 8.7. Let $\mathcal{S}$ be a 4-(23, 7, 1). As in the preceding exercise, show that $\mathcal{S}$ is constructed from $\mathcal{P}$ by adding two points X_1, X_2 to the 21 points of $\mathcal{P}$, and then choosing two classes $\mathbf{O}'_1$ and $\mathbf{O}'_2$ of 56 ovals each in $\mathscr{P}$ and one class $\mathbf{B}$ of 120 BSPs of $\mathscr{P}$, where:

(1) two of the chosen ovals are in the same class $\mathbf{O}'_i$ if and only if they meet in an even number of points;

(2) any chosen oval meets a BSP of $\mathbf{B}$ in an odd number of points;

(3) two BSPs of $\mathbf{B}$ meet in an odd number of points.

Then the blocks of $\mathscr{S}$ are:

(a) $y \cup X_1 \cup X_2$, where y is a line of $\mathscr{P}$, and

(b) $\mathcal{O} \cup X_i$, where $\mathcal{O} \in \mathbf{O}'_i$, and

(c) $\mathscr{B}$, for all $\mathscr{B} \in \mathbf{B}$.

The point of the last two exercises is that $\mathscr{D}$ and $\mathscr{S}$ are constructed in somewhat the same way that $\mathcal{M}$ must be, but only using one or two of the three oval and BSP classes that are required in the construction of $\mathcal{M}$. We shall see in the next section that the oval and BSP classes required for the construction of $\mathcal{M}$ not only exist, but are uniquely defined by their combinatorial properties (that is, their intersection numbers) inside of $\mathscr{P}$; then $\mathscr{D}$ and $\mathscr{S}$ can be seen to be constructible by using only certain of these classes, and that these constructions are similarly unique.

8.3 Existence of $\mathcal{M}_{24}$

In this section we shall show that the ovals of $\mathscr{P}$ can in fact be partitioned into three classes of 56 each such that two ovals are in the same class if and only if they meet in an even number of points; similarly, the BSPs of $\mathscr{P}$ can be partitioned into three classes of 120 each such that two BSPs are in the same class if and only if they meet in an odd number of points. In addition, these classes can be so indexed that they satisfy the conclusions of Lemma 8.4. This will imply that if a 5-(24, 8, 1) exists, then it is unique. However, more is required to show that it exists: we will still have to show that any five points of $\mathscr{P} \cup X_1 \cup X_2 \cup X_3$ are in a unique common block. To accomplish this, we shall prove a number of technical lemmas: then when we set out to show that $\mathcal{M}$ is a 5-(24, 8, 1), we will find that we have all the necessary information.

Part of our work is combinatorial, or geometric. We look more closely at $\mathscr{P}$ to understand how its ovals and BSPs behave. But a crucial aspect of the problem will involve some group theory, and we review here the results and definitions that we need. First, some definitions and notation:

D(1). $GL = GL(3, 4)$ is the group of all non-singular linear transformations of the 3-dimensional vector space V over $GF(4)$; given a basis of V, we can represent GL by the group of all 3-by-3 non-singular matrices over $GF(4)$, and we will feel free to represent GL in either way;

D(2). $SL = SL(3, 4)$ is the subgroup of GL of elements of determinant equal to one (and is, of course, independent of the basis of V);

D(3). Z is the centre of GL, and is hence the group of all matrices kI, where $k \neq 0$, in GL;

D(4). $PGL = PGL(3, 4)$ is the automorphism group of $\mathscr{P}$ induced by GL; $PSL = PSL(3, 4)$ is the automorphism group of $\mathscr{P}$ induced by SL;

D(5). a *quadrangle* of $\mathscr{P}$ is an unordered set of four points of $\mathscr{P}$, no three collinear; a *frame* is an ordered quadrangle;

D(6). a *triangle* of $\mathscr{P}$ is an unordered set of three non-collinear points of $\mathscr{P}$; if ordered, we speak of an *ordered* triangle;

D(7). a *double triangle* of $\mathscr{P}$ is a set $\{A, B, C, D\}$ of four distinct points of $\mathscr{P}$, such that A, B, C are collinear, but D is not on the line containing A, B and C; if A, B, C are ordered, then we speak of an *ordered* double triangle.

Next, some results:

Result 8.1. PGL is transitive on the frames (and hence on the quadrangles) of $\mathscr{P}$.

Result 8.2. $|PGL| = 21 \cdot 20 \cdot 16 \cdot 9$, and the number of frames in $\mathscr{P}$ is equal to $|PGL|$.

Result 8.3. $|PSL| = 21 \cdot 20 \cdot 16 \cdot 3$, and PSL is normal in PGL.

Result 8.4. Z is contained in SL, and $PGL \cong GL/Z$, $PSL \cong SL/Z$.

Result 8.5. PSL is simple; also, the alternating group A_6 of degree 6, and the group $PGL(3, 2)$ of all 3-by-3 non-singular matrices over $GF(2)$ are simple.

Of these results, 8.1–8.4 are easy to prove:

Exercise 8.8. Prove Results 8.1, 8.2, 8.3, 8.4.

Now we choose a quadrangle $\mathscr{F}$ consisting of the four points $A = \langle(1, 0, 0)\rangle$, $B = \langle(0, 1, 0)\rangle$, $C = \langle(0, 0, 1)\rangle$, $D = \langle(1, 1, 1)\rangle$, as in Figure 8.1.

Lemma 8.5. The seven points, A, B, C, D, X, Y, Z in Figure 8.1 are the seven points of a BSP $\mathcal{B}$, and $\mathcal{B}$ is the only BSP containing $\mathcal{F}$.

Proof. All the incidences are trivial to verify, including in particular those on the 'exceptional' line w which contains X, Y, Z. □

Lemma 8.6. There are exactly two points E, F in $\mathcal{P}$ such that $\mathcal{F} \cup E \cup F$ is an oval, and these are the two points on the line w in Figure 8.1 (other than X, Y, Z).

Proof. The BSP $\mathcal{B}$ is the point set of a Baer subplane $\mathcal{B}'$, and the seven lines of $\mathcal{B}'$ each contain two points of $\mathcal{P}$ not in $\mathcal{B}$ and, since any pair of these seven lines meet in $\mathcal{B}'$, they cannot meet again. So there are $7 \cdot 2 = 14$ points of $\mathcal{P} \backslash \mathcal{B}$ on these seven lines, hence every point of $\mathcal{P}$ is on (at least) one line of $\mathcal{B}'$ (note, by the way, that we have solved Exercise 8.3). All these 14 points, excepting precisely the two on the line w, are thus collinear with two points of $\mathcal{F}$, and can lie in no oval that contains $\mathcal{F}$. But the two points not in $\mathcal{B}$ but on w form, with $\mathcal{F}$, an oval. □

Theorem 8.7

(a) Any quadrangle of $\mathcal{P}$ is in exactly one BSP.

(b) Any quadrangle of $\mathcal{P}$ is in exactly one oval.

(c) A set of five points of $\mathcal{P}$, no three collinear, is in exactly one oval.

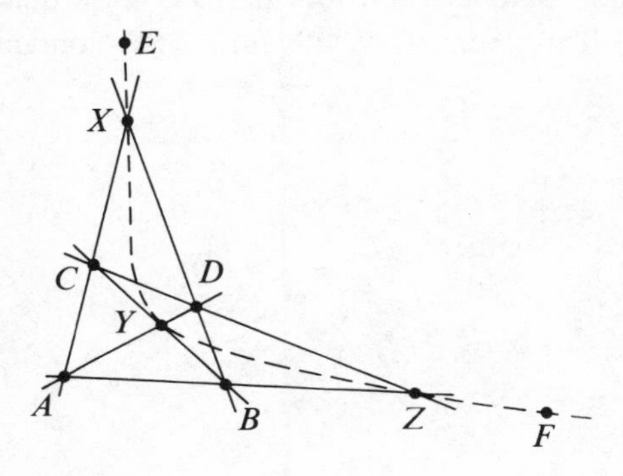

Fig. 8.1

Proof. Since PGL is transitive on quadrangles (from Result 8.1), (a) and (b) follow from Lemma 8.6. To see (c), note that in the proof of Lemma 8.6, if A, B, C, D are four of the points (in a set of five, no three collinear), then E or F must be the fifth. □

Lemma 8.8 A triangle of $\mathscr{P}$ is in exactly nine quadrangles and three ovals.

Proof. First we show that if $\mathscr{T}$ is a triangle, it is in nine quadrangles. Let $\mathscr{T} = \{A, B, C\}$. Then there are three points $\neq A, B, C$ on each 'side' of $\mathscr{T}$, so nine points of $\mathscr{P}$, not in $\mathscr{T}$, but collinear with two points of $\mathscr{T}$. Thus $21-9-3=9$ points of $\mathscr{P}$ are not in $\mathscr{T}$ and on no side of $\mathscr{T}$, so any one of these nine points, adjoined to $\mathscr{T}$, gives a quadrangle containing $\mathscr{T}$.

With $\mathscr{T}$ fixed, let us count pairs $\{\mathscr{F}, \mathscr{O}\}$, where $\mathscr{O}$ is an oval, $\mathscr{F}$ is a quadrangle, and $\mathscr{T} \subset \mathscr{F} \subset \mathscr{O}$. If there are m ovals on $\mathscr{T}$, then there are $3m$ such pairs: for choosing an oval $\mathscr{O}$ on $\mathscr{T}$, the choice of any one of the three points of $\mathscr{O}$ not in $\mathscr{T}$ gives a quadrangle in $\mathscr{O}$ and on $\mathscr{T}$. But there are nine choices of $\mathscr{F}$ on $\mathscr{T}$, from the first part of the proof, and each $\mathscr{F}$ is in exactly one oval $\mathscr{O}$. So $3m=9$, hence $m=3$. □

Lemma 8.9. Any double triangle of $\mathscr{P}$ is in exactly three BSPs.

Proof. Consider the configuration in Figure 8.2, where $\mathscr{D}=\{A, B, C, D\}$ is a double triangle. Each BSP $\mathscr{B}$ which contains $\mathscr{D}$ also contains a third point X on the line AD, as shown, and hence contains Y. Conversely, each of the three choices of X on AD determines a unique Y and, for each such Y, the quadrangle A, B, X, Y is in a unique BSP. Thus there are exactly three BSPs containing the double triangle $\mathscr{D}$. □

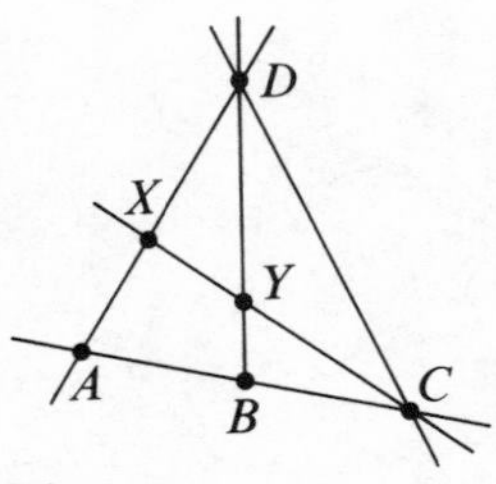

Fig. 8.2

Lemma 8.10. The number of ordered triangles in $\mathscr{P}$ is $21 \cdot 20 \cdot 16$, the number of ordered double triangles is $21 \cdot 20 \cdot 16 \cdot 3$, the number of ovals is 168, and the number of BSPs is 360.

Proof. The first two counts are easy: e.g. an ordered triangle can have its first point chosen 21 ways, its second 20, and then the third point can be any of the 16 points not on the line through the first two. (Note that the proof of Result 8.2 is essentially the same.)

Let N be the number of ovals. Each oval contains $6 \cdot 5 \cdot 4 \cdot 3$ frames, and each frame of $\mathscr{P}$ is in one oval. So counting pairs $\{\mathscr{F}, \mathscr{O}\}$ where $\mathscr{F}$ is a frame in the oval $\mathscr{O}$, it is easy to see that $N = 168$. A similar proof shows that $\mathscr{P}$ contains 360 BSPs. □

Theorem 8.11. *PSL* is transitive on ordered triangles and on ordered double triangles.

Proof. Let $\mathscr{T}$ be the ordered triangle $\{\langle(1,0,0)\rangle, \langle(0,1,0)\rangle, \langle(0,0,1)\rangle\}$, and let H be the subgroup of PSL fixing $\mathscr{T}$. If $A \in SL$ induces an element of H, then A has the form:

$$A = \begin{vmatrix} a & 0 & 0 \\ 0 & b & 0 \\ 0 & 0 & c \end{vmatrix}$$

(why?), and $abc = 1$ because $A \in SL$. Since $c^3 = 1$ for all non-zero c in $GF(4)$, we have $c^{-1}I \in SL$, so $c^{-1}A \in SL$. Hence we may assume that $c = 1$ in the representation of A above. Then $b = a^{-1}$, so there are three choices of a, and thus $|H| = 3$.

The number of images of $\mathscr{T}$ under PSL is $|PSL|/|H| = 21 \cdot 20 \cdot 16$, which is the number of ordered triangles in $\mathscr{P}$, so PSL is transitive on ordered triangles.

By adjoining $\langle(1,1,0)\rangle$ to $\mathscr{T}$ we have an ordered double triangle $\mathscr{D}$. It is easy to see that the subgroup of PSL fixing $\mathscr{D}$ is the identity, and now the same argument shows that PSL is transitive on ordered double triangles. □

Theorem 8.12. *PSL* has three orbits, of equal size, on quadrangles.

Proof. Let $\mathscr{F}$ be the quadrangle $\{\langle(1,0,0)\rangle, \langle(0,1,0)\rangle, \langle(0,0,1)\rangle, \langle(1,1,1)\rangle\}$, and let H be the subgroup of PGL fixing $\mathscr{F}$. Now PGL is transitive on the frames of $\mathscr{P}$, and in particular the subgroup H fixing $\mathscr{F}$ is transitive on the $4! = 24$ frames in $\mathscr{F}$. Only the identity in H (even in PGL) can fix a frame, so $H \cong S_4$, the symmetric group of degree 4. But in $GL(3, 2)$ there exists a subgroup which induces a 4-transitive group on the four points of $\mathscr{F}$, since $PGL(3, 2)$ is equally well transitive on the frames of $\mathscr{P}_2(2)$, and hence this subgroup of $GL(3, 2)$ is a subgroup of GL $(= GL(3, 4))$, and all its elements have determinant one, because they are all in $GF(2)$ (and not zero). So H is in PSL, and thus the number of images of $\mathscr{F}$, under PSL, is $|PSL|/|H| = 840$, which is one-third the total number of quadrangles in $\mathscr{P}$. The choice of basis is irrelevant, so the result is true for any quadrangle $\mathscr{F}$. □

Now we know that PSL has three orbits of quadrangles and, since each quadrangle is in one oval and in one BSP, it is natural to ask whether PSL has three orbits of ovals and BSPs as well: these then might be reasonable candidates for the three classes of ovals and BSPs that we need to build $\mathscr{M}$. Since a triangle is in three ovals, it should then follow that these are in three different oval orbits, and a similar conclusion should be expected from double triangles and BSPs. Everything, in fact, works out this way, as we now proceed to show.

Lemma 8.13. If $\mathscr{O}$ is an oval, then all the quadrangles in $\mathscr{O}$ are in the same orbit under PSL.

Proof. PGL is transitive on the $6 \cdot 5 \cdot 4 \cdot 3$ frames contained in $\mathscr{O}$, and the subgroup of PGL fixing one of these frames is the identity. Since each frame in $\mathscr{O}$ is in a unique oval, it follows that K, the subgroup of PGL fixing $\mathscr{O}$, is 4-transitive on the six points of $\mathscr{O}$, and has order $6 \cdot 5 \cdot 4 \cdot 3$. So K, acting on $\mathscr{O}$, must be isomorphic to A_6, the alternating group of degree 6. Consider $PSL \cap K$; since PSL is normal in PGL, it follows that $PSL \cap K$ is normal in K. But K is simple, so $PSL \cap K = 1$ or K.

If $PSL \cap K = 1$, then only the identity in PSL fixes $\mathscr{O}$, so $\mathscr{O}$ has $|PSL|$ images under PSL. Since $|PSL| > 168$, this is impossible (for $\mathscr{P}$ only contains 168 ovals). Hence $PSL \cap K = K$, so PSL is transitive on the quadrangles (even on the frames) contained in $\mathscr{O}$. □

Lemma 8.14. If $\mathscr{B}$ is a BSP, then all the quadrangles in $\mathscr{B}$ are in the same orbit under PSL.

Proof. This is similar to the proof of Lemma 8.13, excepting the subgroup K fixing a BSP is isomorphic to the simple group $PGL(3, 2)$ this time. □

Corollary 8.15

(a) PSL has three orbits of ovals, and each triangle of $\mathcal{P}$ is in one oval of each orbit.

(b) PSL has three orbits of BSPs, and each double triangle of $\mathcal{P}$ is in one BSP of each orbit.

Proof. Consider (a). Clearly Lemma 8.13 implies that PSL has three oval orbits. Since any oval contains a triangle, and PSL is transitive on triangles, every triangle is in at least one oval of each orbit. But a triangle is in only three ovals by Lemma 8.8, so it is in exactly one oval of each orbit.

(b) is exactly the same. □

We now have all the components necessary to construct $\mathcal{M}_{24}$. We call the three PSL-orbits of quadrangles $\mathbf{Q}_1$, $\mathbf{Q}_2$, $\mathbf{Q}_3$, and then let $\mathbf{O}_i$ be the PSL-orbit of ovals whose quadrangles are in $\mathbf{Q}_i$, and similarly let $\mathbf{B}_i$ be the PSL-orbit of BSPs whose quadrangles are in $\mathbf{Q}_i$. Next we define a structure $\mathcal{M}_{24}$ as follows:

(a) the points of $\mathcal{M}_{24}$ are the points of $\mathcal{P}$, plus three new points X_1, X_2, X_3;

(b) the blocks of $\mathcal{M}_{24}$ are:

 (i) $y \cup X_1 \cup X_2 \cup X_3$, for each line y of $\mathcal{P}$;

 (ii) $\mathcal{O} \cup X_i \cup X_j$, where $\mathcal{O}$ is an oval in $\mathbf{O}_k$, and $\{i, j, k\} = \{1, 2, 3\}$;

 (iii) $\mathcal{B} \cup X_i$, where $\mathcal{B}$ is a BSP in $\mathbf{B}_i$;

 (iv) the double lines $y * z$, for all unordered pairs of distinct lines y, z in $\mathcal{P}$.

Theorem 8.16. $\mathcal{M}_{24}$ is a 5-(24, 8, 1).

Proof. $\mathcal{M}_{24}$ certainly has 24 points, and every block has eight points. We now consider various sorts of 5-sets of points.

(a) P, Q, X_1, X_2, X_3, where $P, Q \in \mathcal{P}$. These are on the block $y \cup X_1 \cup X_2 \cup X_3$, where y is the line of $\mathcal{P}$ on P and Q.

(b) P, Q, R, X_1, X_2, say, where $P, Q, R \in \mathcal{P}$. If P, Q, R are collinear in $\mathcal{P}$, then these points are on $y \cup X_1 \cup X_2 \cup X_3$, where y is the line of $\mathcal{P}$ on P, Q, R. If, on the other hand, P, Q, R is a triangle, then there is a unique oval $\mathcal{O} \in \mathbf{O}_3$ on P, Q, R, so the five points are on $\mathcal{O} \cup X_1 \cup X_2$.

(c) P, Q, R, S, X_1, where $P, Q, R, S \in \mathscr{P}$.

(i) If P, Q, R, S, are on a line y in $\mathscr{P}$, then the five points are on $y \cup X_1 \cup X_2 \cup X_3$.

(ii) If P, Q, R are on a line y in $\mathscr{P}$, S not on y, then $\{P, Q, R, S\}$ is a double triangle and is in a unique BSP $\mathscr{B} \in \mathbf{B}_1$, so the five points are in $\mathscr{B} \cup X_1$.

(iii) If no three of P, Q, R, S are collinear, then $\mathscr{F} = \{P, Q, R, S\}$ is a quadrangle; if $\mathscr{F} \in \mathbf{Q}_1$, then $\mathscr{F}$ is in a unique BSP $\mathscr{B} \in \mathbf{B}_1$ and so the five points are in $\mathscr{B} \cup X_1$; if $\mathscr{F}$ is not in $\mathbf{Q}_1$, suppose it is in $\mathbf{Q}_2$, so that $\mathscr{F}$ is in a unique oval $\mathscr{O} \in \mathbf{Q}_2$, and hence the five points are in $\mathscr{O} \cup X_1 \cup X_3$.

(d) $P, Q, R, S, T \in \mathscr{P}$.

(i) If the five given points are on a line y in $\mathscr{P}$, then they are in the block $y \cup X_1 \cup X_2 \cup X_3$.

(ii) If P, Q, R, S are on a line y, but T is not on y, then there is a unique line z on T which contains none of P, Q, R, S, and so the five points are on $y * z$.

(iii) If P, Q, R are on a line y, but S, T are not on y, then let z be the line on S and T. If z meets y in a point other than P, Q or R, then the five points are on $y * z$; if z meets y in R, say, then $\{P, Q, S, T\}$ is a quadrangle, contained in a unique BSP $\mathscr{B}$, and $\mathscr{B}$ is in $\mathbf{B}_i$, so the five points are in $\mathscr{B} \cup X_i$.

(iv) If no three of P, Q, R, S, T are on a line of P, then there is a unique oval $\mathscr{O}$ of $\mathscr{P}$ on the five points; since $\mathscr{O} \in \mathbf{O}_i$, the five points are in $\mathscr{O} \cup X_j \cup X_k$, where $j \neq i$, $k \neq i$.

In each case above we have seen that there is a block on the five given points. The reader can easily verify that no other block also contains the five points, so $\mathscr{M}_{24}$ is a 5-(24, 8, 1). □

Corollary 8.17

(a) Two ovals in $\mathscr{P}$ meet in an even number of points if and only if they are in the same PSL-orbit.

(b) Two BSPs of $\mathscr{P}$ meet in an odd number of points if and only if they are in the same PSL-orbit.

Proof. This is a consequence of Theorem 8.16 and Lemma 8.4. □

We saw in Section 8.2 that the existence of $\mathscr{M}_{24}$ depended, among other things, upon the ovals and BSPs of $\mathscr{P}$ falling into classes having the properties of

Corollary 8.17. We have constructed $\mathcal{M}_{24}$ without directly proving that these intersection properties are valid, but, since it follows that they are, it also follows that our method of constructing $\mathcal{M}_{24}$ is the only way it could have been done. In other words, $\mathcal{M}_{24}$ is unique up to isomorphism. But inspection of Exercises 8.6 and 8.7 (which are not difficult, under any circumstance) shows that a 4-(23, 7, 1) and a 3-(22, 6, 1) can be constructed from $\mathscr{P}$ in only one way as well. Thus $\mathcal{M}_{22}$, a 3-(22, 6, 1), is constructed from $\mathscr{P}$ by adjoining the 56 ovals of one of the oval classes $\mathbf{O}_i$; since *PGL* is transitive on the three *PSL*-orbits of ovals, it does not matter which oval class is used.

Theorem 8.18. There is, up to isomorphism, a unique 5-(24, 8, 1), a unique 4-(23, 7, 1), and a unique 3-(22, 6, 1). □

These three designs, denoted $\mathcal{M}_{24}$, $\mathcal{M}_{23}$ and $\mathcal{M}_{22}$, respectively, are the *large Mathieu designs.*

8.4 The Golay codes

Given a vector space $V = V_n(K)$ of dimension $n < \infty$ over the field K, with a fixed basis specified for V, a *code* is a subset of V (it is important to notice that a code depends on the basis of V). A code is *linear* if it is a subspace of V, and it is *binary* if $K = GF(2)$. The vectors in the code are sometimes called *code-words.* We represent the vectors of V as n-tuples, using the given basis, and define the *weight* $wt(\mathbf{v})$ of a vector $\mathbf{v}$ in V to be the number of non-zero coordinates of $\mathbf{v}$. The *distance*, or *Hamming-distance*, between two vectors $\mathbf{v}, \mathbf{w}$ in V is $d(\mathbf{v}, \mathbf{w}) = wt(\mathbf{v} - \mathbf{w})$.

Exercise 8.9. Show that the function d satisfies:
(a) $d(\mathbf{v}, \mathbf{w}) = d(\mathbf{w}, \mathbf{v})$;
(b) $d(\mathbf{v}, \mathbf{w}) \geqslant 0$, and $d(\mathbf{v}, \mathbf{w}) = 0$ if and only if $\mathbf{v} = \mathbf{w}$;
(c) $d(\mathbf{v}, \mathbf{u}) + d(\mathbf{u}, \mathbf{w}) \geqslant d(\mathbf{v}, \mathbf{w})$.
(In other words, d obeys the rules of a true distance function.)

The *minimum weight* of a code $\mathscr{C}$ is the least non-zero weight of any vector in $\mathscr{C}$, and the *minimum distance* of a code $\mathscr{C}$ is the least of all $d(\mathbf{v}, \mathbf{w})$ where $\mathbf{v}, \mathbf{w} \in \mathscr{C}$, $\mathbf{v} \neq \mathbf{w}$.

Exercise 8.10. If $\mathscr{C}$ is a linear code, show that its minimum weight equals its minimum distance.

The theory of linear codes, especially binary linear codes, is very highly developed, and we cannot even begin to go into it here. But one of the desirable features of a 'good' code is that it should be as large as possible and yet also have a large minimum distance, and we shall construct two codes in this section which do this extremely well. (The reader interested in more material about coding theory might like to consult [2, 3].) But we will also see that the use of codes has implications in other parts of mathematics by proving additional facts about $\mathscr{M}_{24}$ and its automorphism group through some relatively simple coding theory. Our codes will be both linear and binary, so we use the term 'code' in the rest of this section to mean 'binary linear code'.

Let $V = V_{24}(2)$ be a 24-dimensional vector space over $GF(2)$, represented as 24-tuples. If A is an incidence matrix for $\mathscr{M}_{24}$, then we will let our first code $\mathscr{C}$ be generated by the 759 rows of A^{T}. Those vectors in $\mathscr{C}$ which are rows of A^{T} will be called *block-vectors*; they all have weight 8. If $\mathbf{c}$ is any vector of $\mathscr{C}$ (or of V, for that matter), then the *support* of $\mathbf{c}$ is the set of $wt(\mathbf{c})$ points in $\mathscr{M}_{24}$ such that $\mathbf{c}$ has 1s in the positions corresponding to those points. Hence the support of a block-vector is precisely the block that 'defined' it.

If we define the *dot-product* (or *inner product*) of two vectors in V in the natural way (i.e. $(x_1, x_2, \ldots, x_{24})\cdot(y_1, y_2, \ldots, y_{24}) = x_1y_1 + x_2y_2 + \cdots + x_{24}y_{24}$, the arithmetic being modulo 2 of course), then this turns V into its own dual space. From this we want one fact: if W is a subspace of V and $\mathscr{A}(W) = \{\mathbf{v} \in V \mid \mathbf{w}\cdot\mathbf{v} = 0 \text{ for all } \mathbf{w} \in W\}$, then $\dim(W) + \dim(\mathscr{A}(W)) = 24$. Then:

Lemma 8.19. $\mathscr{A}(\mathscr{C}) \geqslant \mathscr{C}$, and hence (a) $\dim(\mathscr{C}) \leqslant 12$ and (b) if $\mathbf{c}_1, \mathbf{c}_2 \in \mathscr{C}$, then the supports of $\mathbf{c}_1$ and $\mathbf{c}_2$ meet in an even number of points.

Proof. Since two blocks y, z of $\mathscr{M}_{24}$ meet in an even number of points, the block-vectors $\mathbf{y}$ and $\mathbf{z}$ have the property $\mathbf{y}\cdot\mathbf{z} = 0$. Thus any two code-words of $\mathscr{C}$ have the same property. So $\mathscr{C} \leqslant \mathscr{A}(\mathscr{C})$. Hence $\dim(\mathscr{C}) \leqslant \dim(\mathscr{A}(\mathscr{C}))$, and so $24 = \dim(\mathscr{C}) + \dim(\mathscr{A}(\mathscr{C})) \geqslant 2(\dim(\mathscr{C}))$; thus $\dim(\mathscr{C}) \leqslant 12$, which is (a). And (b) is merely a restatement of $\mathscr{C} \leqslant \mathscr{A}(\mathscr{C})$. □

Lemma 8.20. All vectors in $\mathscr{C}$ have weight divisible by 4.

Proof. Certainly the generators of $\mathscr{C}$ have weights divisible by 4. If $\mathbf{c}_1, \mathbf{c}_2 \in \mathscr{C}$ have weights $4s_1, 4s_2$, respectively, then their supports meet in an even number of points, $2t$ say, from Lemma 8.19. Suppose, without loss of generality:

$$\mathbf{c}_1 = (1\ 1\cdots1 \quad 1\ 1\cdots1 \quad 0\ 0\cdots0 \quad 0\ 0\cdots0)$$

$$\mathbf{c}_2 = (\underbrace{1\ 1\cdots1}_{2t} \quad \underbrace{0\ 0\cdots0}_{4s_1-2t} \quad \underbrace{1\ 1\cdots1}_{4s_2-2t} \quad 0\ 0\cdots0)$$

then

$$\mathbf{c}_1 + \mathbf{c}_2 = (0\ 0\cdots0 \quad 1\ 1\cdots1 \quad 1\ 1\cdots1 \quad 0\ 0\cdots0)$$

clearly has weight $(4s_1 - 2t) + (4s_2 - 2t) = 4s_1 + 4s_2 - 4t$, which is divisible by 4. So the lemma is true for all vectors in $\mathscr{C}$. □

Now recall that $\mathbf{j}$ is the vector of all $+1$s, 24 long in our case. Define N_i to be the number of vectors in $\mathscr{C}$ of weight i.

Lemma 8.21. $\mathscr{C}$ contains $\mathbf{j}$, and if $\mathbf{c} \in \mathscr{C}$ then $\mathbf{j} + \mathbf{c}$ (the 'complement' of $\mathbf{c}$) is in $\mathscr{C}$. Thus it follows that $N_i = N_{24-i}$.

Proof. Every point of $\mathscr{M}_{24}$ is in 253 blocks and $253 \equiv 1 \pmod 2$, so the sum of all 759 block-vectors of $\mathscr{C}$ must be $\mathbf{j}$. Then $\mathbf{c} \in \mathscr{C}$ implies $\mathbf{j} + \mathbf{c} \in \mathscr{C}$, and obviously the mapping $\mathbf{c} \to \mathbf{j} + \mathbf{c}$ is a bijection from the vectors of weight i to those of weight $24 - i$. □

Lemma 8.22. $\mathscr{C}$ has no vectors of weight 4 or 20, and every vector in $\mathscr{C}$ of weight 8 is a block-vector (thus $N_8 = N_{16} = 759$).

Proof. If we show that there are no vectors of weight 4, then from Lemma 8.21 there can be none of weight 20. Suppose $\mathbf{c} \in \mathscr{C}$ has weight 4. Then the support of $\mathbf{c}$ is a set of four points P_1, P_2, P_3, P_4 in $\mathscr{M}_{24}$. Let y be a block of $\mathscr{M}_{24}$ on P_1, P_2, P_3, but not on P_4 (such as y must exist: why?), and let $\mathbf{y}$ be its block-vector. Then $\mathbf{y} + \mathbf{c}$ has for its support exactly P_4 and the five points of y other than P_1, P_2, P_3. Thus $wt(\mathbf{y} + \mathbf{c}) = 6$, which contradicts Lemma 8.20. Thus $\mathbf{c}$ does not exist.

Now let $\mathbf{c} \in \mathscr{C}$, where $wt(\mathbf{c}) = 8$. Choose five points X_i, $1 \leqslant i \leqslant 5$, in the support of $\mathbf{c}$, and let $\mathbf{y}$ be the block-vector of the block y of $\mathscr{M}_{24}$ on the five X_i. Suppose $5 + t$ points of y are in the support of $\mathbf{c}$, so there are $3 - t$ points in the support of $\mathbf{c}$ not in y, and $3 - t$ points of y not in the support of $\mathbf{c}$. Then $wt(\mathbf{y} + \mathbf{c}) = 2(3 - t) = 6 - 2t$. But $6 - 2t \neq 4$, since no vector of $\mathscr{C}$ has weight 4, so $6 - 2t = 0$, thus $t = 3$, and $\mathbf{c} = \mathbf{y}$, and $\mathbf{c}$ is a block-vector. Hence $N_8 = N_{16} = 759$. □

Now we know almost all the 'weight distribution' of $\mathscr{C}$: $N_i=0$ except for $N_0=N_{24}=1$, $N_8=N_{16}=759$, and N_{12} is yet to be determined. If $\mathbf{c}\in\mathscr{C}$ has weight 12, then we call the support of $\mathbf{c}$ in $\mathscr{M}_{24}$ a *dodecad* of $\mathscr{M}_{24}$.

Lemma 8.23. If $\mathscr{A}$ is a dodecad and y a block, in $\mathscr{M}_{24}$, then if y meets $\mathscr{A}$ in more than four points, it follows that y meets $\mathscr{A}$ in exactly six points and that there is a unique block z in $\mathscr{M}_{24}$ such that $\mathscr{A}$ is the symmetric difference $y*z$.

Proof. Suppose $\mathscr{A}$ is the support of $\mathbf{a}$ in $\mathscr{C}$, and let $\mathbf{y}$ be the block-vector for y. Suppose y meets $\mathscr{A}$ in t points. Then $wt(\mathbf{y}+\mathbf{a})=(8-t)+(12-t)=20-2t$. But if $t>4$, then $20-2t<12$, hence $20-2t=8$ and $t=6$. So $\mathbf{z}=\mathbf{y}+\mathbf{a}$ is a block-vector, $\mathbf{a}=\mathbf{y}+\mathbf{z}$, and $\mathscr{A}=y*z$. Clearly, z is unique. □

Now we set a straightforward counting exercise:

Exercise 8.11. Show that if y is a block of $\mathscr{M}_{24}$, then exactly 448 blocks of $\mathscr{M}_{24}$ meet y in precisely two points. (Hint: count the number of blocks that meet y in a *fixed* pair of two points by utilising the construction of $\mathscr{M}_{24}$ from $\mathscr{P}$.)

Lemma 8.24. If y is a block of $\mathscr{M}_{24}$, then there are 448 dodecads $\mathscr{A}$ such that y meets $\mathscr{A}=y*z$ for some block z, and hence 448 dodecads $\mathscr{A}$ such that y meets $\mathscr{A}$ in exactly six points.

Proof. This is an immediate corollary of Lemma 8.23 and Exercise 8.11. □

Theorem 8.25. There are 2576 dodecads in $\mathscr{M}_{24}$, hence $N_{12}=2576$, $\mathscr{C}$ has dimension 12 and contains 2^{12} vectors.

Proof. We count pairs $(y,\mathscr{A})$, where y is a block and $\mathscr{A}$ is a dodecad meeting y in exactly six points. There are 759 blocks y and 448 dodecads $\mathscr{A}$ for each y. On the other hand, if $\mathscr{A}$ is a dodecad, then, for each of the $\binom{12}{5}$ sets of five points in $\mathscr{A}$, there is one block y on the five points, and thus y meets $\mathscr{A}$ in six points. Each such block y is counted six times in this way, and so

$$759 \cdot 448 = N_{12}\binom{12}{5}\Big/6.$$

Then $N_{12} = 2576$.

Then the number of vectors in $\mathscr{C}$ is $N_0 + N_8 + N_{12} + N_{16} + N_{24} = 4096 = 2^{12}$, which finishes the proof. □

From a coding theory point of view, $\mathscr{C}$ has dimension 12, minimum distance 8, and length 24; curiously, it might seem, $\mathscr{C}$ can be 'improved by choosing one coordinate place and suppressing it in all vectors (of $\mathscr{C}$, and of V). Then $\mathscr{C}^*$, the resulting code, has length 23, dimension 12 and minimum weight 7; $\mathscr{C}^*$ is called the *Golay code*, while $\mathscr{C}$ is the *extended Golay code.*

Exercise 8.12. If N_i^* is the number of vectors of weight i in $\mathscr{C}^*$, then show that $N_i^* = 0$ except for: $N_0^* = N_{23}^* = 1$, $N_7^* = N_{16}^* = 253$, $N_8^* = N_{15}^* = 506$, $N_{11}^* = N_{12}^* = 1288$.

Exercise 8.13. If $\mathbf{v} \in V_{23}(2)$, show that there is a unique vector $\mathbf{c} \in \mathscr{C}^*$ such that $d(\mathbf{v}, \mathbf{c}) \leqslant 3$. (Hint: use Exercise 8.9(c) to show that there could not be as many as two such vectors $\mathbf{c} \in \mathscr{C}^*$; then count the number of vectors in $V_{23}(2)$ which are at distance 3 or less from some fixed vector.)

In view of Exercise 8.13, $\mathscr{C}^*$ is a *perfect* 3-*error correcting code.* This means that any vector in $V_{23}(2)$ has distance at most 3 from a unique code-word $\mathbf{c} \in \mathscr{C}^*$, and that, if the vectors of $\mathscr{C}^*$ are sent as messages and no more than three errors occurs in the transmission, then the received vector can be 'returned' to its rightful place with no error.

8.5 The Mathieu groups

The automorphism groups of the large Mathieu designs have many interesting properties, and by playing them off against the designs, the coding theory of Section 8.4, and strongly regular graphs, many important properties can be deduced. In this and the next section we give a small sample of the kind of things that can be done. First a fairly elementary group theory result (followed by an exercise for the reader with the group theory background required).

Result 8.6. Suppose G is a permutation group on a set $\mathcal{S}$, with $n=|\mathcal{S}|$, and suppose s, t are positive integers satisfying $n-s\geqslant 2$. If, for every set of s distinct elements of $\mathcal{S}$, the subgroup of G fixing all s of the elements is t-transitive on the remaining $n-s$ elements, then G is $(s+t)$-transitive on $\mathcal{S}$.

Exercise 8.14. Prove Result 8.6. (Hint: show first that G is necessarily transitive, then proceed by induction on s.)

Theorem 8.26. Aut $\mathcal{M}_{24}$ is 5-transitive on the points of $\mathcal{M}_{24}$.

Proof. Let us write $\mathcal{M}$ for $\mathcal{M}_{24}$. If X_1, X_2, X_3 are any three points of $\mathcal{M}$, then $\mathcal{M}$ is constructed from $\mathcal{P}=\mathcal{M}_{X_1,X_2,X_3}$ as in Section 8.3. Now, clearly, PSL is an automorphism group of $\mathcal{M}$, fixing X_1, X_2, X_3, since PSL preserves the oval classes and BSP classes (and, of course, the double lines) used to build $\mathcal{M}$ from $\mathcal{P}$. Hence, since PSL is 2-transitive on the 21 points of $\mathcal{P}$, the subgroup of Aut $\mathcal{M}$ fixing any three points is 2-transitive on the rest, so by Result 8.6 Aut $\mathcal{M}$ is 5-transitive on the points of $\mathcal{M}$. □

Corollary 8.27. Aut $\mathcal{M}_{23}$ and $\mathcal{M}_{22}$ are, respectively, 4-transitive and 3-transitive on the points of $\mathcal{M}_{23}$ and of $\mathcal{M}_{22}$. □

These results are extremely important: apart from the alternating and symmetric groups, the automorphism groups of $\mathcal{M}_{11}$, $\mathcal{M}_{12}$, $\mathcal{M}_{23}$ and $\mathcal{M}_{24}$ are the only 4-transitive groups known (and are presumably the only finite ones which can exist). We write $M_{24}=\text{Aut } \mathcal{M}_{24}$, $M_{23}=(M_{24})_X$, $M_{22}=(M_{23})_Y$, for a choice (and hence any choice) of a pair of distinct points X, Y in $\mathcal{M}_{24}$.

Exercise 8.15. Show:

(a) Aut $\mathcal{M}_{23}=M_{23}$;

(b) M_{22} is a normal subgroup of index 2 in Aut $\mathcal{M}_{22}$;

(c) PSL is a normal subgroup of index 6 in Aut $\mathcal{P}$.

(Hint: for (b), for example, it is required to show that M_{24} contains elements fixing $\mathcal{M}_{22}=(\mathcal{M}_{24})_{X,Y}$ which do not fix X and Y, etc.)

The additional most significant fact about the three large Mathieu groups, that they are simple, depends upon the simplicity of PSL, and is purely group theoretic (i.e. it uses facts already deduced but, from here on in, does not use any design theory). We sketch a proof for completeness.

Theorem 8.28. M_{22}, M_{23} and M_{24} are simple.

Proof. We use the following basic results from group theory:

(a) a non-identity normal subgroup of a 2-transitive group is, itself, transitive;

(b) if N is a transitive normal subgroup of the transitive group G acting on $\mathscr{S}$, and if a stabiliser $N_X=1$ (so N is *regular*), then G_X induces an automorphism group of N which is permutation isomorphic to G_X acting on $\mathscr{S}$;

and hence

(c) if N is a regular normal subgroup of a 2-transitive group G acting on the set $\mathscr{S}$, then $|N|=|\mathscr{S}|=$ a power of a prime; if in addition, G is 3-transitive, then $|N|=|\mathscr{S}|=3$ or a power of 2.

Now suppose $N\neq 1$ is a normal subgroup of M_{22}. Since $PSL\cong(M_{22})_X$, it follows that $N_X=N\cap PSL$ is normal in the simple group PSL. If $N>PSL$, then $N=M_{22}$ is immediate. If $N\cap PSL=1$, then N is regular, so from (c) above, $|N|=22$ is a prime power. This is not so, and hence M_{22} is simple.

Now repeat the argument on M_{23}, then on M_{24}, to prove their simplicity. □

The groups M_{24}, M_{23} and M_{22} also act as permutation groups on the blocks of their respective Mathieu designs, and hence:

Lemma 8.29. M_{24}, M_{23} and M_{22} are, respectively, transitive on the 759, 253 and 77 blocks of $\mathscr{M}_{24}$, $\mathscr{M}_{23}$ and $\mathscr{M}_{22}$.

Proof. Since M_{24} is 5-transitive on the points of $\mathscr{M}_{24}$, and since any five points of $\mathscr{M}_{24}$ are in exactly one common block, M_{24} must be transitive on the blocks. The same proof works for M_{23} and M_{22}. □

Let us define B_8, B_7 and B_6 to be the subgroup of M_{24}, M_{23} and M_{22}, respectively, fixing a block. Then:

Lemma 8.30. The orders of B_8, B_7 and B_6 are, respectively, $16(8!/2)$, $16(7!/2)$ and $16(6!/2)$.

Proof. $|B_8| = |M_{24}|/759$, since M_{24} is transitive on 759 blocks, and B_8 is the stabiliser of this representation. Since $|M_{24}| = 24 \cdot 23 \cdot 22|PSL| = 24 \cdot 23 \cdot 22 \cdot 21 \cdot 20 \cdot 48$, the order of B_8 follows immediately. The proofs for B_7 and B_6 are the same. □

Theorem 8.31. In their representations on the i points of a fixed block, the groups B_i are isomorphic to the alternating groups A_i, for $i = 6, 7, 8$.

Proof. In the proof of Lemma 8.13, we saw that the subgroup of *PSL* fixing an oval is transitive on frames, hence 4-transitive on the six points. Since the subgroup of *PSL* fixing a frame must be the identity (since it is even true in the larger group *PSL*), it follows that the subgroup of *PSL* fixing an oval is exactly 4-transitive, and hence it must be permutation isomorphic to A_6, on the six points. This subgroup is contained in M_{22}, so B_6, on the block, is either A_6 or S_6, the symmetric group on the six points. But if B_6, on the block, were (isomorphic to) S_6, then the subgroup fixing a point would be S_5, and would be contained in *PSL*. Then it is easy to see that *PSL* would contain a non-identity element fixing a frame. So B_6 is isomorphic to A_6 in its representation on a block, and from this it immediately follows that B_7 and B_8 are permutation isomorphic to A_7 and A_8, in their actions on a fixed block. □

Exercise 8.16.* Let T be the kernel of the representation of B_8 on its fixed block (i.e. T is the subgroup of B_8 fixing all eight points of the block). Show that T is elementary abelian of order 16 and is transitive on the 16 points off the fixed block. Then show that the same statement is true of B_7 and B_6.

In fact, Exercise 8.16 indicates yet more interesting group theory inside of the large Mathieu groups. For it implies that the B_i have regular normal subgroups in their actions on the 16 points off the fixed block, and then using result (b) of the proof of Theorem 8.28 it can be seen that the A_i (for $i = 6, 7, 8$) act as automorphism groups of the elementary abelian group T. These are in some technical sense 'exceptional' places to find the alternating groups acting, and the combinatorics and the group theory can be exploited to examine these representations and deduce important and interesting theorems of a more or less purely group theoretic kind. We do not follow these up here, since they would take us into more group theory than we are concerned with.

8.6. Some more combinatorics

Among the wealth of combinatorial information that is contained in the large Mathieu designs, we pick out a few topics in this section. First we recall that the block intersection numbers of $\mathcal{M}_{24}$ are 0, 2 and 4, and look into this in a little more detail.

Theorem 8.32. If y is a block of $\mathcal{M}_{24}$, then there are 30 blocks that do not meet y. If y and z are a pair of non-intersecting blocks, then there is a unique block w which meets neither y nor z.

Proof. Given a block y, we choose three points X_1, X_2, X_3 on y, let $\mathscr{P}=(\mathcal{M}_{24})_{X_1,X_2,X_3}$, and then note that if a block z of $\mathcal{M}_{24}$ fails to meet y it can only be a double line in $\mathscr{P}$, and a double line of the form $s * t$, where s and t are lines of $\mathscr{P}$ which meet in a point Q lying on y. There are five points on y (in $\mathscr{P}$ as well), and for each of these we have $\binom{4}{2}$ choices of a pair of lines (neither equal to y, of course) which will lead to a double line not intersecting y. So there are $5\binom{4}{2}=30$ blocks not meeting y.

Suppose y and z do not meet in $\mathcal{M}_{24}$. We could show the existence of a unique block w in $\mathcal{M}_{24}$ which meets neither, by the same sort of argument that was used in the first half of the proof of this theorem. But let us consider instead the code $\mathscr{C}$. The block-vectors $\mathbf{y}$ and $\mathbf{z}$ give us a vector $\mathbf{y}+\mathbf{z}$ in $\mathscr{C}$, and $wt(\mathbf{y}+\mathbf{z})=16$. So $wt(\mathbf{y}+\mathbf{z}+\mathbf{j})=8$, thus $\mathbf{y}+\mathbf{z}+\mathbf{j}=\mathbf{w}$ is a block-vector, and clearly w is the unique block not meeting y or z. □

Exercise 8.17

(a) Show that each block of $\mathcal{M}_{24}$ is in 15 block-triples (a block-triple is a set of three mutually non-intersecting blocks).

(b) Show that $\mathcal{M}_{24}$ contains 3795 block-triples.

(c)* Show that M_{24} is transitive on the 3795 block-triples in $\mathcal{M}_{24}$.

The block intersection numbers of $\mathcal{M}_{23}$ are 1 and 3, and hence by Theorem 3.46 this enables us to construct an SR graph; indeed, the same is true of $\mathcal{M}_{22}$, whose blocks meet in 0 or 2 points.

Theorem 8.33. The block-adjacency graph of $\mathscr{M}_{23}$, $\mathscr{M}_{22}$, respectively, leads to an SR graph with parameters (253, 112, 36, 60), (77, 16, 0, 4) respectively.

Proof. We leave this as an exercise:

Exercise 8.18. Prove Theorem 8.33. (Hint: it is, in fact, possible to show, using merely the SR graph theory of Chapter 3, that an SR graph with parameters $(253, 112, \lambda, \mu)$ must have $\lambda=36$, $\mu=60$.) □

Another similar example of these ideas, and of considerable interest in design theory, is given by:

Theorem 8.34. Let X be a point of $\mathscr{M}_{22}$. Then the block-adjacency graph of $\mathscr{M}_{22}^X$ leads to the existence of an SR graph with parameters (56, 10, 0, 2), and hence to the existence of a biplane with parameters (56, 11, 2).

Proof. First $\mathscr{M}_{22}^X$ is a 2-(21, 6, 4) (see Theorem 1.15), with 56 blocks. Since any two blocks of $\mathscr{M}_{22}$ meet in 0 or two points, and the blocks of $\mathscr{M}_{22}^X$ are the blocks that do not contain X, it follows that any two blocks of $\mathscr{M}_{22}^X$ also meet in 0 or two points. Given a block y of $\mathscr{M}_{22}^X$, on each pair of its six points there are three blocks not equal to y, hence $3\binom{6}{2}=45$ blocks meet y in two points. So 10 blocks fail to meet y. Thus the SR graph can be assumed to have parameters $(56, 10, \lambda, \mu)$, and we can compute λ and μ easily: since $10(10-\lambda-1)=\mu(56-10-1)$, we have $2(9-\lambda)=9\mu$. So $\lambda=0$ and $\mu=2$. Now see Theorem 3.45 for the existence of the biplane. □

Exercise 8.19. Show that *PSL* is a transitive automorphism group of the biplane with parameters (56, 11, 2) constructed in Theorem 8.34.

Now let us consider the dodecads that the coding theory of Section 8.4 has given us. Let $\mathscr{A}$ be a fixed dodecad. For each block y of $\mathscr{M}_{24}$ that meets $\mathscr{A}$ in exactly six points, let y' be the set of six points of y in $\mathscr{A}$.

Lemma 8.35. For each block y meeting $\mathscr{A}$ in six points, there is a unique block z meeting $\mathscr{A}$ in six points such that $y' \cup z' = \mathscr{A}$.

Proof. This is merely a restatement of Lemma 8.23. □

Theorem 8.36. The structure whose points are the 12 points of $\mathscr{A}$ and whose blocks are the 6-sets y', for blocks y meeting $\mathscr{A}$ in six points, is a 5-design for (12, 6, 1).

Proof. Any set of five points of $\mathscr{A}$ is in a unique block y of $\mathscr{M}_{24}$ and, by Lemma 8.23, y must meet $\mathscr{A}$ in six points, and hence y' is the unique block on the five points. Hence we have a 5-structure but, since $\lambda = 1$, it must be a 5-design. □

The 5-design of the last theorem has the same parameters as $\mathscr{M}_{12}$, and is in fact isomorphic to it. We shall not attempt to prove this, since it depends on the uniqueness of $\mathscr{M}_{12}$. But an interesting exercise is:

Exercise 8.20. Let H be the subgroup of M_{24} fixing the dodecad $\mathscr{A}$. Then show that H has order $12 \cdot 11 \cdot 10 \cdot 9 \cdot 8$ and is 5-transitive on the points of $\mathscr{A}$.

References

[4] and [5] are standard references for permutation groups, while [2] and [3] are both excellent texts on coding theory.

[1] Hughes, D. R. & Piper, F. C. *Projective Planes.* Berlin–Heidelberg–New York, Springer, 1973.
[2] van Lint, J. H. *Introduction to Coding Theory.* Graduate Texts in Mathematics 86, Springer-Verlag, 1982.
[3] MacWilliams, F. J. & Sloane, N. J. A. *The Theory of Error-correcting Codes.* North Holland, 1977.
[4] Passman, D. S. *Permutation Groups.* W. A. Benjamin Inc., 1968.
[5] Wielandt, H. *Finite Permutation Groups.* Academic Paperbacks, 1964.

Common abbreviations

BRC theorem: Bruck–Ryser–Chowla theorem, 55
BSP: set of 7 points of a Baer subplane in a projective plane of order 4, 214
GQ: generalised quadrangle, 193
$\mathscr{H}$-matrix: Hadamard matrix, 101
MOLS: mutually orthogonal latin squares, 94
SR graph: strongly regular graph, 119
SSD: semisymmetric design, 196
STS: Steiner triple system, 173

Index

Page numbers in italics represent definitions.